VICE PRESIDENT AND PUBLISHER	Jay O'Callaghan
EXECUTIVE EDITOR	Ryan Flahive
ASSISTANT EDITOR	Courtney Nelson
SENIOR PRODUCTION EDITOR	Valerie A. Vargas
MARKETING MANAGER	Danielle Torio
CREATIVE DIRECTOR	Harry Nolan
DESIGNER	Michael St. Martine
PRODUCTION MANAGEMENT SERVICES	Elm Street Publishing Services
SENIOR ILLUSTRATION EDITOR	Sandra Rigby
SENIOR PHOTO EDITOR	Lisa Gee
MEDIA EDITOR	Lynn Pearlman
COVER DESIGNER	Jim O'Shea
COVER PHOTO	Ron Niebrugge/Mira.com/drr.net

This book was set in 10/12 Janson by Thomson Digital and printed and bound by R.R.D.-Jefferson City. The cover was printed by R.R.D.- Jefferson City.

This book is printed on acid free paper. ♾

The opinions expressed by the authors are theirs and do not reflect the official position of the United States Military Academy, the University of North Carolina, the Department of the Army, or the Department of Defense.

To order books or for customer service please, call 1-800-CALL WILEY (225-5945).

ISBN-13: 978-0-470-09826-4

Printed in the United States of America

10 9 8 7 6 5 4 3 2

7 TH EDITION

REGIONAL LANDSCAPES OF THE UNITED STATES AND CANADA

STEPHEN S. BIRDSALL
University of North Carolina at Chapel Hill

EUGENE J. PALKA
United States Military Academy at West Point

JON C. MALINOWSKI
United States Military Academy at West Point

MARGO L. PRICE
University of North Carolina at Chapel Hill

JOHN WILEY
WILEY

To our family, friends, and colleagues
Who continue to provide love and support

PREFACE

ew would argue that the past five years have been a time of great change for North America. Economic flux, demographic shifts, and catastrophic natural disasters have dramatically affected the landscapes of the United States and Canada. These changes present great challenges to us as authors as we seek to keep this text relevant for students. Fundamentally, and as we noted in the previous edition, we feel that the need for geographic understanding, whether at home or abroad, has perhaps never been greater. The core of this seventh edition is still the regional organization that has made this book successful, but we have tried to ensure that this edition reflects the changes that a new generation of college-level geographers witness as they grow as scholars.

When we published the previous, sixth edition of *Regional Landscapes*, we made dramatic changes to the look and content of the text, including introducing the first full-color edition and two new authors. With this edition, we sought to build on those changes to ensure that the book would remain relevant and up-to-date, yet we maintained the organization and regional framework that have been so well received over the years.

First, we have updated nearly all the relevant statistics and tables using the latest available economic and demographic information. We feel this is crucial to understanding the dramatic changes affecting some regions of the realm. Population growth in the Southwest and the implosion of the pineapple industry in Hawaii are just two examples.

Second, we have updated many of the maps and photos to ensure that they are both relevant and engaging. Additionally, we have updated the entire Agricultural Map Index. Compared to other texts and previous editions of *Regional Landscapes*, we feel that this is the most visually enhanced North America text available.

Third, we have introduced two features to assist student comprehension of the material. First, we have introduced a short section at the beginning of each chapter highlighting the key points of the narrative. This summary of the region's distinguishing characteristics will help focus student reading of each chapter. Second, we have introduced review questions at the end of each chapter to ensure that each student has grasped the key concepts relative to each North American region.

Finally, we have addressed several important events and issues that have emerged since the previous edition. For instance, we have addressed the impact of Hurricane Katrina, the immigration issue along the Southwest border, and the potential effects of global warming on the Northlands.

We hope that students and instructors will find the seventh edition of *Regional Landscapes* as timely and engaging as any past edition and encourage feedback to make future editions even better. We know that the great pace of change in both countries will challenge us to keep pace for many years, and many editions, to come.

ACKNOWLEDGEMENTS

We are continually pleased that friends and colleagues take time to help us make this a better book. With each edition the list of those to whom we are in debt grows. Even when we did not adopt their suggestions, people who sent us their thoughts about the past edition helped us improve this new version. We appreciate deeply the encouragement and the assistance we have received over the years.

In addition to the many reviewers who contributed to improvements in earlier editions, we want to thank those who made helpful comments on our efforts for this seventh edition. Any weaknesses in this edition are a result of our own shortcomings, not those of the many people who have recommended ways to improve it. These people include:

Dawn S. Bowen, University of Mary Washington
Laurie Gasahl, Grand Valley State University
John Jakubs, University of South Carolina
William G. Laatsch, University of Wisconsin-Green Bay
Taylor E. Mack, Mississippi State University

Kenneth C. Martis, West Virginia University
Paul Phillips, Fort Hays State University
Steven Silvern, Salem State College
Forrest Wilkerson, Minnesota State University-Mankato
Donald C. Williams, Western New England College

The staff members at John Wiley & Sons have been helpful as always. We are especially indebted to our editor, Courtney Nelson, who helped us build on the success of the sixth edition. We also thank the production team at John Wiley & Sons for their efficient and effective assistance. A deeply felt thank you to Jay O'Callaghan, Ryan Flahive, Valerie Vargas, Sandra Rigby, Lisa Gee, Michael St. Martine, Lynn Pearlman, and Danielle Torio.

S.S.B.
E.J.P.
J.C.M.
M.L.P.

CONTENTS

CHAPTER 1

REGIONS AND THEMES

Preview

1. Although the United States and Canada can be subdivided into regions, many of them are shared by both countries.

2. Canada and the United States are largely urban, industrialized societies with highly mobile populations.

3. Both countries produce an abundance of food and raw materials, but both also have high rates of consumption.

4. The United States and Canada have complex political systems with multiple jurisdictions from the local to national level.

5. The presence of indigenous populations and centuries of immigration have created culturally complex societies.

Two of the world's largest and wealthiest countries span North America from the frigid arctic to the Mexican border. Both economically and militarily, the United States has had a dominant influence on the world for decades. Canada, larger than the United States in land area but with only one-tenth as many people, possesses great resource potential and is growing in worldwide economic influence.

Across their vast spaces, the United States and Canada have great landscape diversity. The diversity is partly a result of physical factors, such as landforms and climate. But a place's landscape is also made by the people who live there. Because each country lured people of many cultural heritages, and these heritages blended with those of indigenous Native Americans, the landscapes they created drew on disparate histories and were subject to many distinct environmental conditions.

Each country became wealthy and, although the wealth is unevenly distributed, their general affluence has had two broad geographic consequences. First, the goods and services most people benefit from are widely available. Second, the paired goals of producing and consuming wealth play an enormous role in determining where most settlements are located and strongly influence how most landscapes evolve. Indeed, the mix of cultural heritages and environments in the United States and Canada, together with the ways these countries produce and use wealth, has created a mosaic of landscapes across almost 20 million square kilometers (7.5 million square miles).

In geographers' parlance, the term landscape has a particular meaning. They use it to refer to the look of a place and to the many messages inherent in what is visible. Evidence of residents' past and present goals and accomplishments is often apparent in a place's landscape. The place's location and surrounding territory, and the opportunities inherent in each, often affect landscape evolution. Sometimes the broader patterns and tensions of a nation's culture come to the fore. This is why landscapes in various parts of the United States and Canada are so dissimilar even when they seem, overtly, much alike. In each place, the local environment and local history are different.

The closer geographers look at a place, the more different parts they see. Details about the environment become apparent, both the naturally occurring ones and the place's human constructions. Geographers take note of topographic characteristics, the presence of rivers and streams, climate, vegetation, and soils, as well as the nature of human constructions such as buildings, roads, fields, and fences. Then there are the people themselves: Who are they? When did they come and why? What did they bring with them, and what lasting marks have they made?

To use this flood of information and make sense of a landscape, geographers typically seek to identify broad patterns while remaining aware of particulars that contribute to a place's personality. For example, although Vancouver, Miami, and Boston look very different to visitors, all share certain characteristics. There are many reasons for the cities' contrasting appearances. For example, each grew to city size during a different period and attracted the bulk of its immigrants from different places; each built downtowns, residential housing developments, streets, and highways, and established industry and commerce at unique times and in unique ways; each has a different climate; and each city's government has responded to local opportunities and problems in its own manner.

As varied as they are, however, Vancouver, Miami, and Boston also have much in common. All three anchor metropolitan areas exceeding 1 million people; all are active ocean ports; all wrestle with traffic flow and waste disposal, along with crime, racism, and poverty; all compete for tourist dollars, federal program support, and new business employment.

Geographers studying a place and its landscape, therefore, use two conceptual scales simultaneously. One scale looks only at broad patterns, such as a possible relationship between city size and proximity to water. The other scale is attuned to the myriad of details that give a place its special character. For example, New Orleans' French Quarter derives its

From back roads to urban centers, the landscapes of the United States and Canada are rich with diversity and provide insights into regional differences.

Scale

Scale is a central concept for geographers, especially when places are compared. Unfortunately, the word's meaning as used by geographers in a technical manner is often opposite to its meaning in everyday parlance.

When a geographer refers to a large-scale map, he or she usually means the scale ratio (or how many inches on the ground are represented by an inch on the map) is large. A scale ratio, such as 1:2400, where 1 map inch represents 200 feet on the ground is called large because the mathematical fraction is larger than that for a scale of 1:240,000, where 1 map inch represents 20,000 feet, or about 3.8 miles. Thus, a large-scale map at the 1:2400 scale will show a small geographic area, while the small-scale map at the 1:240,000 scale will cover a broad area.

Geographers' use of map scale ratio terminology when speaking about areas has confounded students for decades. The problem arises because "large scale" is casually synonymous with "big." For clarity's sake, we adopt general usage in our text and take scale to mean "a range of sizes" or "a different level of viewing." Thus, small scale will mean small in size or extent, and large scale will indicate large in size or extent.

singular cachet from its unique history, cultural mix, river location, tourist attractions, and international reputation. This is why geographers emphasize the importance of scale when thinking about places and their landscapes.

For decades, geographers have subdivided Canada and the United States into broad areas of distinctive character, terming these areas "regions." Often, the subdividing process begins when a single classification category is selected, such as physiography, climate, ethnicity, or economic activities. Areas containing similar qualities or quantities of the chosen category are identified as (physiographic, climatic, etc.) regions.

More rarely, each region is identified as the area where one or two facets underlying much of the region's character can serve as descriptive regional themes. One region may be predominantly urban, another basically agricultural, a third affected by its long isolation, and a fourth by its cultural and economic history. The thematic regionalization we use, then, is an approach that tries to capture what is fundamentally significant in a region rather than creating regions according to predetermined categories.

Geographers have learned that as a region's identity develops, the themes underlying its identity help define the region's landscape characteristics. Thus, a region's landscape results from its location, its environmental features, and the cultural origins of its residents, along with the imprint of innumerable individual aspirations and decisions accumulating over time.

REGIONS

Students in a geography class at Ohio State University were asked to identify parts of the United States that meant something to them as regions and to locate them on a map (Figure 1.1). From their answers, it became apparent that certain sections of the country were clearly defined in the students' minds. They consistently named and located the Southwest, New England, and the South, and they generally but consistently identified the regions' borders. However, other parts of the country, such as the Middle Atlantic states, were usually not part of student thinking, and so they excluded them from the regionalization entirely.

With minor variations, the students' pattern would be repeated by people from most parts of the country. New England and the South are regions with firmly established images that most Americans recognize and

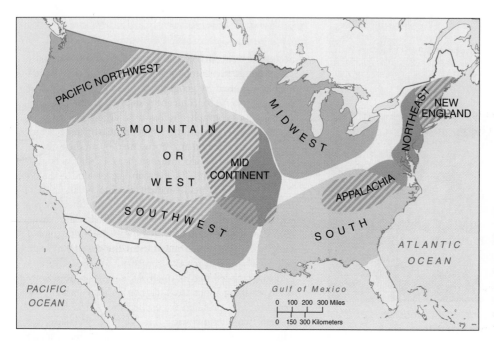

Figure 1.1 Perceived regions of the United States. This set of perceptions is taken from students in Ohio. How might they differ from regions perceived by students in other parts of the country?

can elaborate in general terms. Similarly, to most Canadians, British Columbia and the Maritimes conjure up definite impressions, although residents of Alberta and Ontario may differ in the specifics of these impressions.

Creating Regions

A notable geography text, written decades ago by J. Russell Smith, began with the observation that "Hell is hot."[1] This was not a veiled threat to students who might treat the book's material too lightly. Rather, Smith was making a significant comment about one of his culture's more widely accepted regions—Hell.

Smith observed that all available information about the region came from written materials, the words of others, folktales, and beliefs, not from individual experience. He maintained that even though no one had experienced Hell personally, most people in his culture were nonetheless confident of the general nature of the place and the quality of the experience to be had there. In short, Hell was a "hot" region.

From a geographer's perspective, the point is this: Hell, like any region, is a mental construct. One person may divide an area for study one way; another may envision the same area differently. In part, different regionalizations follow from the researcher's goals or the special information he or she wants to get. A district sales manager wishing to determine how his or her sales staff should work market territories would create a regionalization pattern different from that of a botanist studying variations in tree type. Similarly, an economic geographer's regionalization scheme based on agricultural patterns will bear little resemblance to one based on manufacturing activities. Thus, geographers, when subdividing the United States and Canada, may create maps noticeably different from each other because each might select different elements to study.

Because maps begin with the mapmakers' goals, no one regional scheme is necessarily better than another. A regionalization is acceptable if the scheme meets its creator's requirements, if those requirements have been thought out reasonably, and if the map is executed correctly and factually.

Geographers use regions as a system of categorization, a way of organizing large, complex sets of facts about places into a compact, meaningful set of areas. As with any categorization, regions are satisfactory if they identify patterns, make combinations of discrete information understandable, and help clarify spatial complexity.

[1] J. Russell Smith, *North America* (New York: Harcourt Brace, 1925).

Complex Regions and Their Boundaries

Geographers define regions as either nodal or uniform, and as either single-featured or multifeatured. A nodal region, sometimes called a functional region, is characterized by a set of places connected by lines of communication or movement to another place that serves as a focus (or node) for the entire set. Places in a nodal region, therefore, are associated with each other because they share a common focus and have the same functional purpose. It is the focusing and shared function that puts such places in the same nodal region. A newspaper's circulation area is an example of such a region, as is a milkshed, the set of places surrounding a city that send raw milk to its markets.

In comparison, a uniform region, sometimes called a formal region, is a territory where one or more features are present throughout the area and absent or unimportant beyond it. For instance, the entire area where the cottonwood constitutes a significant part of the total tree cover is a single-featured, uniform region. Other examples of single-featured, uniform regions are areas administered by a given level of government, where Spanish is the first language of at least 40 percent of the population, and the drainage basin of the Mississippi River.

A uniform region may also be multifeatured, representing an area's combined physical or cultural elements. Because multifeatured uniform regions summarize different types of information for a given area, they are general and frequently subjective. We use multifeatured uniform regions to organize this book's general structure. Furthermore, because the elements significant in a given region contribute to its distinctive landscape, we refer to our structure as one of regional landscapes.

A multifeatured uniform region encompasses several distinct zones. Deep within the region is a core area where the major identifying regional features are most obvious. Away from the core and toward the region's margins, the identifying features are less prominent and distinct. For example, although nearly everyone accepts central Alabama and Mississippi as a part of the South, there is much less agreement about whether southern Illinois, the Louisville area of Kentucky, or northern Virginia belong in the South.

Boundaries between adjacent multifeatured regions are seldom sharp and well defined on the

vernacular

Recent urban expansion in both countries, such as this development in Orange County, California, often ignores the rectangular grid system that dominated the American cities from the early nineteenth century.

landscape. Regions blend into each other, and travelers are often well into a new region before realizing they left the previous one behind. Boundaries between such regions are transition zones, not sharp lines, which makes them difficult to map. If regions were simply categories of places, they might occupy mutually exclusive portions of the earth's surface.

Rigid boundaries and territorially exclusive regions do not fit the landscape of the United States and Canada. A given portion of the continent's territory may be occupied by parts of two or more multifeatured thematic regions. And the boundaries of many regions are fairly broad transitional zones (Table 1.1 and Figure 1.2). These zones, containing a region's characteristics though at diminished strength, indicate where the mix of characteristics is so subtle or complex that it is difficult to assign the area to any one region. The margin between the Midwestern agricultural core and the Great Plains is an example of such a subtle change, as are parts of the transition between the agricultural core and the South. The Midwest also has both urban-industrial and rural-agricultural characteristics. Both themes are complex and important enough to address separately (see Chapters 5 and 11).

Regions and regional boundaries are not static. Settlement patterns shift, society develops new technological abilities, and political patterns are altered. Regions reflecting these changes may expand, contract, appear, or disappear. Regional complexity is also reflected on the landscape. Every landscape is a composite of historically accumulated themes and consequences set within a distinctive physical environment. This is why a regionalization of North America in 1492 would be quite different from one for 1776, 1865, or 2009. There is no reason to believe that the continent's regional pattern in 2109 will be at all similar to that in 2009.

A FEW BASIC THEMES

Regions exist at different scales. Often, themes significant in defining a larger region are present, but perhaps not pertinent, in defining smaller regions within it. Thus, any change of scale in thinking about regions also involves a change of theme. New York City's dominant theme, for example, is its urbanness. The theme for New York's Chinatown is its Chinese

TABLE 1.1

Regional Themes		
Chapter	**Chapter Title**	**Themes**
4	Megalopolis	Urban growth and sprawl, urban problems
5	North America's Manufacturing Core	Industrialization, how regions change
6	Canada's National Core	Federalism, cultural pluralism
7	Bypassed East	Relative isolation, regional economic problems
8	Appalachia and the Ozarks	Resource dependence, governmental intervention
9	Changing South	Regional culture, regional dynamics
10	Southern Coastlands	Amenities and economic resources, development conflict
11	Agricultural Core	Rural culture and agriculture's economic changes
12	Great Plains and Prairies	Environmental perception, human/land interaction
13	Empty Interior	Settlement in a difficult environment
14	Southwest Border Area	Ethnic diversity
15	California	Perception of amenities, population redistribution, environmental dependence
16	North Pacific Coast	Regional independence, environmental impact
17	Northlands	Settlement in a difficult environment, resource extraction, governmental intervention
18	Hawaii	Ethnic diversity, environmental impact

cultural imprint, even though Chinatown is also very urban.

Several general cultural patterns in Canada and the United States pervade all of the thematic regions into which the two countries are divided. These general patterns are continental themes that help distinguish Canada and the United States from much of the rest of the world. They cut across regional and political boundaries and, in many cases, ignore major environmental differences. They also indicate the characteristic ways Canadians and Americans organize their national territories and, therefore, underlie the formation of North American landscapes.

Urbanization

Citizens of the United States and Canada hold dearly to their nations' rural roots. In both countries, people

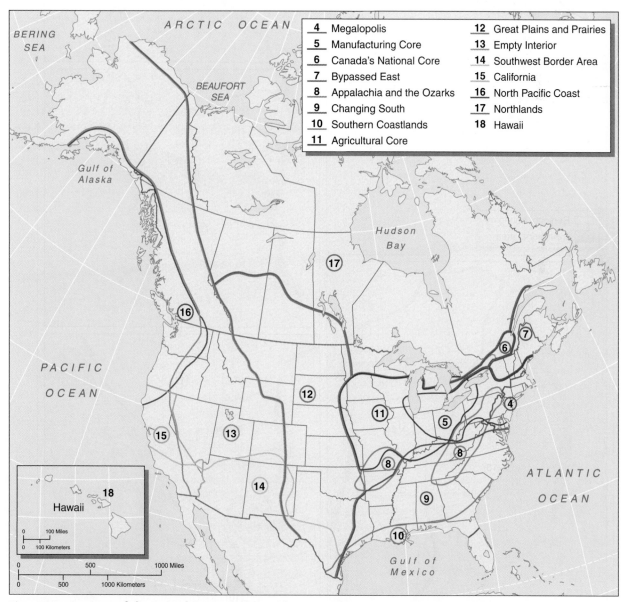

Figure 1.2 **Regions of the United States and Canada.** Separated by theme but with overlapping areas, the regions used in this book reflect the complex variations that characterize the geography of these two countries.

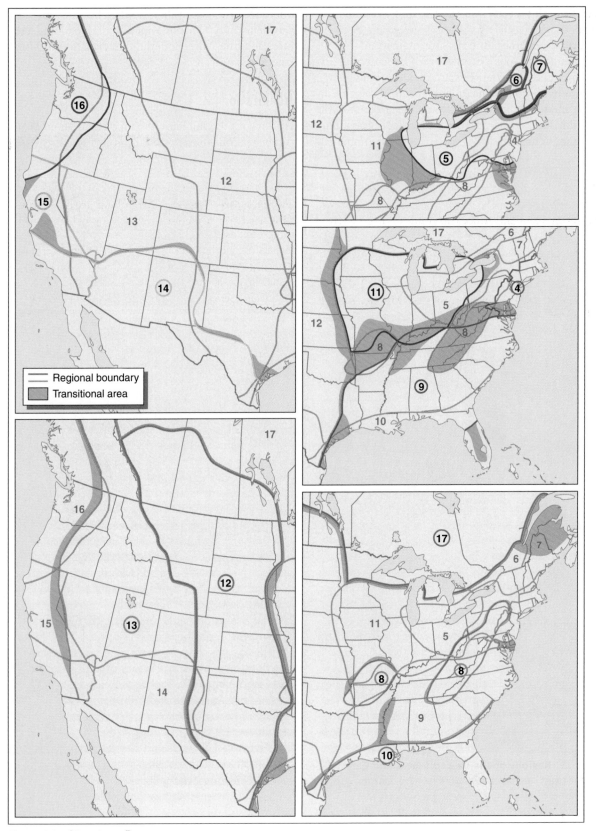

Figure 1.2   (Continued)

widely believe that life in a rural setting is wholesome and good, whereas life in an urban environment is less good, if not downright unhealthy and bad. This attitude is so prevalent that millions of Americans and Canadians, most of them urbanites, still consider the United States and Canada as basically rural places. They hold tight to the notion that this rurality sets them apart from Europeans, providing the two North American countries with a national vigor Europe has lost.

This view of rural dominance is no longer justified in either country. Although most of the land is rural, most people do not live on most of the land. By the countries' own census definitions, more than 75 percent of Canadian and American residents live in urban areas. But the urbanized population is not evenly distributed. In the year 2006, for example, more than 45 percent of the U.S. population lived in just 25 of the largest urban regions (Figure 1.3).

Clearly, rural life with rural occupations, though held as an ideal, is fading. Only about 3 million Americans operated farms in 2002 according to the U.S. Department of Agriculture. Now representing less than 1 percent of the country's people, that number has declined steadily since the first national census in 1790 when more than 90 percent of Americans were farmers. Today, roughly 40 percent of the United States' population lives in Metropolitan Statistical Areas (MSAs) with 1 million or more people.

Canada exhibits a similar trend. Its farm population of 727,125 in 2001 accounted for 2.5 percent of the country's people, and the total number of farms decreased by over 17,000 between 2001 and 2006. More than 79 percent of Canadians live in urban areas. One-third of Canadians now live in one of the country's three largest metropolitan areas—Toronto, Montreal, or Vancouver. Canada's urbanization is even more obvious when the list of top metropolitan areas lengthens to include Ottawa-Hull, Edmonton, Calgary, Quebec, Winnipeg, and Hamilton. Half of all Canadians live in one or another of these nine metropolitan areas.

Urbanization, as a theme, has several elements geographers emphasize. One is the rectangular-grid street layout typical in most American cities. The grid

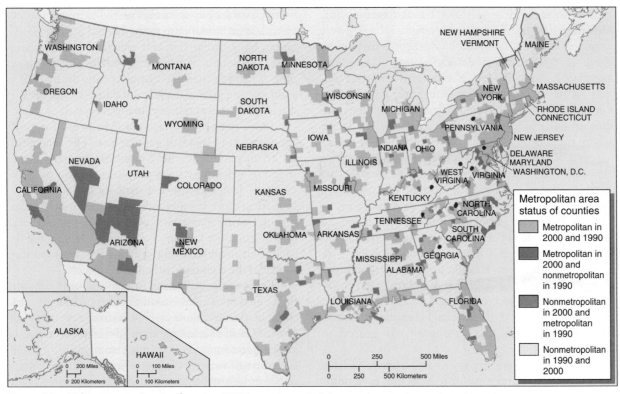

Figure 1.3   Urban expansion in the year 2000. By the end of the past century, large urban clusters in the United States occupied extensive areas between the Great Lakes and the Atlantic Coast, and in California and Florida.

Urban housing built before the advent of the family automobile usually consisted of closely packed houses and walk-up apartments like these units on Boston's Beacon Hill. Afterward, housing spread outward and cities exploded as people commuted great distances by car.

system stems partly from cultural ideals about order and symmetry, partly from the fact that the grid pattern is easy to survey and record land for sale.

Types of urban activities also tend to cluster within cities into zones, each with its distinctive landscape. Within cities, industry clusters in some areas, while commercial centers and warehouses aggregate and dot the landscape in others; residential areas fill intervening spaces; and so on. Typically, modern shopping centers locate in fringe suburbs near major highways. Geographers are interested in why certain activities cluster and why any particular city's activities locate where they do.

Most large cities grew for specific reasons, reasons that still sustain them and give them their special character. A city may have an important transportation role, such as New Orleans through its ocean port, or Omaha, Nebraska, through its railroad connections. A city may provide key administrative functions, such as Ottawa, the Canadian federal capital, or New York City, the headquarters location for many corporations. A city can also trace its vitality to recreation; prominent examples of such cities are Las Vegas, Orlando, and Atlantic City. Others, like Hamilton, Ontario, and Detroit, Michigan, were primarily geared to manufacturing.

Thousands of smaller towns and cities in Canada and the United States exist for reasons that have nothing to do with national acclaim. Yet they thrive because they provide goods and services to their own residents and to those of a surrounding trade territory.

New problems have arisen as a result of the cities' sprawl. For example, the total urbanized area of Canada and the United States is currently increasing at a rate of about 890,312 hectares (2.2 million acres) each year. So much rural acreage is being gobbled up that some observers are concerned that adequate amounts of good agricultural land may not be available in the future. Other problems arise when sprawl leads urban areas to coalesce, where different, independent cities meet and merge physically. As they meld, and borders blur, the role and meaning of urban political organization change.

During the past 100 years, the pattern of continuing city growth has been a key element in American and Canadian geography. Coupled with increasing personal mobility, it stimulated the spread of urban landscapes in both countries.

Industrialization

No other country in the world comes close to matching the volume of the United States' manufacturing output. Nor can any country match its product variety. Canada, much smaller than the United States in population, has a proportionately smaller industrial enterprise. Although relatively powerful in terms of output, Canada lacks the manufacturing diversity of its southern neighbor, largely because of its smaller domestic market and lower financial capital.

Industrialization has consequences for both American and Canadian geographies. The fundamental urban pattern in the United States and Canada, which reflects where most people in the two countries currently live, was affected strongly by industrialization. Most cities were founded and grew quickly when manufacturing industries became the primary economic stimulus. Change has been underway for decades, but manufacturing remains the overwhelming source of the countries' urban patterns.

Regional specialization in manufacturing exists in the United States and Canada, in part because of variations in the availability of industrial raw materials. North Carolina, for example, is a major tobacco-growing state and has been a leader in cigarette manufacturing. Iowa has long been a major corn-producing state, and corn-based ethanol manufacturing plants have been sprouting across the state's landscape in recent years. Quebec, with its vast interior forests, leads both countries in pulp and paper production.

Economic Engine

The highly industrialized character of the United States and Canada has far-reaching economic implications. The most important economic implications are those related to the distinction geographers make between basic and nonbasic economic activities.

Economic activities that bring income to a region from outside are called basic activities. Because most manufactured items are sold in places other than where they are made, money flows into the area where the manufacturing takes place. Manufacturing, therefore, is a basic activity. In contrast, nonbasic activities recirculate money that is already in the region. Local retail sales and services, such as a barber shop or grocery store, are among the many nonbasic services in a community.

Basic activities are not limited to manufacturing. For example, a university that attracts most of its students from out of state is also a basic activity. Whenever knowledge and accumulated experience lead to activities that people outside a region are willing to pay for, the activities are basic.

Even so, Canadian and American manufacturing is significant because many people are employed in manufacturing, and this basic employment activity is important to the countries' long-term growth. Only by bringing in wealth from outside can more activities—basic and nonbasic—be supported.

Besides access to raw materials, North America's pattern of manufacturing industries is also affected by industrial linkages. In essence, manufacturers want to minimize total movement costs while remaining

The complexity of the U.S. and Canadian economies is represented by this image. Agriculture and manufacturing continue to be vital parts of both national economies.

responsive to demands for their products. Therefore, those industries that make component parts for an item locate near each other as well as near the final assembly site. The many automobile parts manufacturers in northern Ohio and southern Michigan are a classic example.

Other important sources of regional variation in manufacturing include differences in labor availability or labor skills, quality of transportation facilities, and local political attitudes. Regions also tend to specialize in producing whatever can be most competitively made locally. Within the United States and Canada, regional specialization has spawned regional interdependence. Few, if any, places in either country are truly self-sufficient in manufacturing.

Because manufacturing is so widespread and so fundamental to the U.S. and Canadian economies, especially in and around urban areas, few city landscapes lack the visible consequences of manufacturing industries. Large, low buildings of frequently indeterminate purpose squat near sites where raw materials are stored. Heavy truck and rail traffic spout exhaust as smokestacks exhale plumes.

High Mobility

U.S. and Canadian transportation networks are extensive and vital in maintaining each country's high levels of economic activity. They allow goods and people to move easily and relatively inexpensively from place to place. They also enhance regional interdependence and reduce isolation. When a region is isolated, this isolation is important in understanding the region's economy (see Chapter 7).

Residential mobility is also high in both countries. Nearly all Americans move more than once during their lives. On average, 20 percent change their residence in any given year. Most moves are local, with individuals or families moving just a few blocks or a few miles. But many are interregional. Usually job-related, they reflect another way distant parts of each country are linked.

Since many factors influence a person's decision to move, general migration patterns suggest which influences currently lead to a call for a moving van. Population geographers and others who study migration have dubbed the influences push factors and pull factors. That is, the decision to migrate is a mix of the place of origin's apparent negative characteristics (push factors) and the potential destination's positive characteristics (pull factors).

Traditionally, people move to areas they think offer greater economic opportunity. Until the end of the nineteenth century in the United States and the 1910s in Canada, people were pulled toward frontier agricultural lands and so migrated westward. But as manufacturing's share of the national economies grew, the location of opportunity shifted. By the early twentieth century, migration within the United States and Canada was oriented overwhelmingly toward urban areas where most industry was located.

Recently, the two countries' national economies have entered a postindustrial phase, with employment growth occurring primarily in professions and services rather than in manufacturing. Because service and professional industries have more flexibility as to where they locate, employment opportunities (and migration destinations) are growing more rapidly in areas that contain greater amenities (Figure 1.4).

Resource Abundance and Resource Dependence

The United States and Canada export and make domestic use of many types of raw materials. For both countries, foodstuffs lead raw material exports. In 2005, U.S. agricultural exports exceeded $238 billion. The United States is the world's leading exporter of most major grains, including wheat, corn, and rice. It also exports significant amounts of soybeans, tobacco, and cotton. Measured in terms of calories, the United States and Canada grow enough food in a normal year to feed more than twice their combined populations. About 25 percent of the United States' total land in row crops produces export crops; Canada's percentage is even higher.

Canada is also a substantial exporter of other raw or semimanufactured materials, such as petroleum, coal, pulp, lumber, iron ore, natural gas, copper, and nickel. The United States is easily Canada's top buyer for these raw materials, although other countries purchase large amounts. Japan, for example, buys

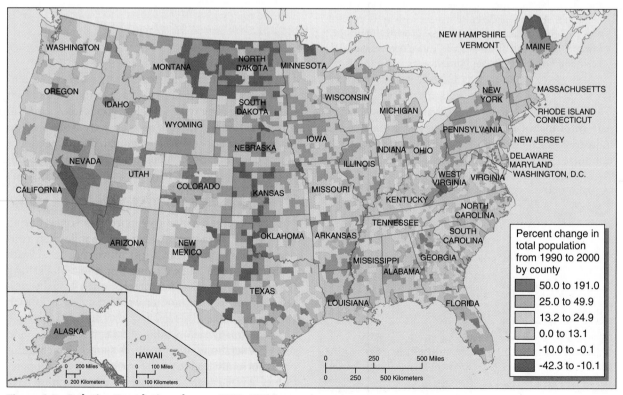

Figure 1.4   Relative Population change, 1990–2000. Populations in most states in the Southeast and West are expected to grow well above the national average into the next century. How do you explain this pattern of growth and decline?

Americans value what they perceive as their rural or small town background. The essence of that perception is captured by small New England villages such as this one in Vermont.

coal and forest products from Canada's West Coast sites.

The United States supplies the international markets with a few nonagricultural raw materials; it is, for instance, the world's leading exporter of coal. But while the United States has a wide array of natural resources, it also has an enormous demand for industrial raw materials. Much of that demand, but not all, can be satisfied from domestic sources. Within the context of resource independence, demand in the United States and Canada is clearly as relevant as natural supplies.

High Income and High Consumption

Resource development has a direct impact on the land. As a result, the interplay between features of a place's physical environment and how humans live there are clearly visible and are a principal source of landscape variation.

Canada and the United States are rich in terms of inherited natural resource wealth and the disposable income of their inhabitants. Americans and Canadians have annual per capita incomes that rank among the highest in the world. They earn good money, and they spend it. Their demands generate massive consumption and propel the countries' economies.

The economies of both countries are increasingly dependent on energy, education, and innovation. These facets feed each other in an economic cause-and-effect chain. High national incomes are linked to worker productivity, and this productivity, in turn, requires machines, a skilled labor force, and computer technology. Inanimate energy sources fuel modern machines and, at a personal level, literally fuel individual mobility. Education drives labor force skills and technology innovation. All this opens up opportunity, but it also increases national energy consumption. For example, in 2007 the United States and Canada accounted for nearly 28 percent of the world's daily oil consumption, with most of that used in the United States.

In spite of tremendous resources, high consumption makes both countries increasingly dependent on imported resources. The United States imports half the petroleum it uses, as well as an increasing share of iron ore and natural gas consumed. In addition, all of its tin and aluminum come from imports, along with large quantities of many other mineral ores.

Canada is less dependent on foreign sources, but it, too, purchases from other countries what is not

Large shopping malls are climate-controlled settings that concentrate shoppers and typify the high consumption of U.S. and Canadian societies.

available domestically, as well as some of the petroleum it needs. In 2004, for example, Canada produced 3.1 million barrels of oil per day and imported 1.2 million barrels per day.

A high-consumption society manifests itself on the landscape in many ways. For example, travelers are bombarded by a constant series of ads and signs urging them to "buy me." Retail establishments are carefully designed to make purchasing pleasant and easy. Streets and parking are designed for individual shopping convenience and spontaneity rather than landscape aesthetics or low social costs.

The implications of high consumption in a finite environment and high consumption's natural partnership with high income will be among the most important, and most difficult, issues facing the United States and Canada during the next few decades. American and Canadian resources are consumed rapidly in an attempt to satisfy their populations' appetites for manufactured goods and specialized foods. The United States' proven oil reserves stayed relatively constant during the 1990's while natural gas reserves increased slightly by increasing imports from foreign sources. Canada's search for untapped resources is moving into more fragile environments where disruption is less easily absorbed without long-term damage.

In spite of high overall income, poverty remains a nagging, serious problem in both countries (Figure 1.5). Many rural areas in French Canada and rural sections of

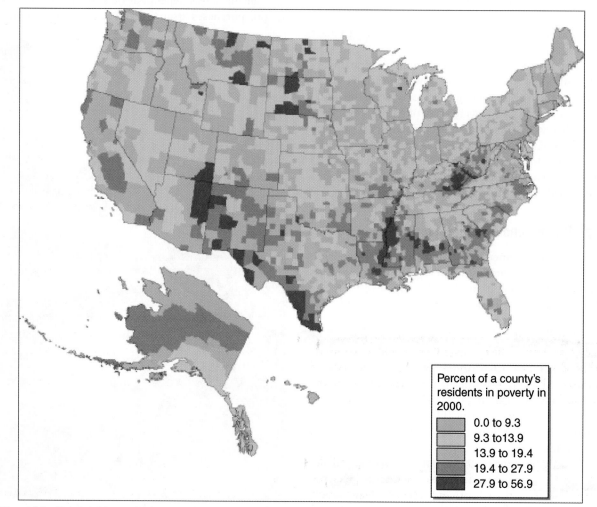

Figure 1.5   Poverty rates, 2000. Household income is not evenly distributed across the United States. Lower incomes and poverty have persisted for decades in some parts of the country. How might this be explained?

Landscapes often form according to the cultural lens of the people who live in a place. Toronto's Chinatown clearly reflects the culture of the Chinese immigrants who live and work here.

the Atlantic Provinces have income averages that are little more than half those found in Toronto or Vancouver. Much of the rural U.S. South, Mexican-American and American Indian areas of the Southwest, sections of Appalachia, and nonwhite ghettos in most American cities have populations whose average income is far below the national average. The persistence of these poverty areas is a major problem; it means that a large number of people are not sharing in their country's wealth, and substantial personal poverty represents a real economic loss to the two countries' national economies.

Political Complexity

Both the United States and Canada have complex political systems, with jurisdiction over an activity or territory divided among many different decision-making bodies. To compound the complexity, some decision-making bodies are elected and some are appointed.

Below the state and provincial levels, Canada and the United States have similar political structures. The structural complexity in both countries presents major problems in effective and efficient distribution of governmental services. Each county, city, town, and sometimes township is governed by its own elected officials. Many special administrative units, some appointed and some elected, oversee the provision of

specific services, such as education (school boards), planning (planning boards), fire protection (fire districts), public transportation (transit authorities), and water supply (water and sewer authorities). Often the resulting administrative pattern is nearly impossible to comprehend. At any location, overlapping jurisdictions provide different administrative boards with control over one service or another.

Adding to the governmental muddle is the fact that once a jurisdictional unit is created, it almost never disappears and is seldom willing to surrender its powers. It is also rare for political units to join efforts and form administrative organizations under a new structure. In both reflections of bureaucratic resistance to change, problems arise as populations grow and merge spatially. Thus, although the underlying spatial pattern of administration remains the same, service inefficiencies increase over time (see Figure 4.6).

Variety of Cultural Origins

The people of Canada and the United States spring from diverse cultural backgrounds. In Canada, the distinction between British and French cultural areas may be the country's most widely recognized feature of cultural difference, but Canada is much more diverse than this. Numerous Inuit and Indian peoples occupy northern territories and live among other Canadians in southern settlements. In addition, Canada receives immigrants from European countries other than Great Britain and France, along with those from numerous Asian countries.

In the United States, obvious long-term cultural contributions include the important ones made by African Americans to the national culture, particularly in the South, along with those made by Native Americans and substantial numbers of people from every European country.

Distinctive culture areas exist in certain parts of the United States. In the Southwest, for example, many American Indian tribes admix with Hispanic Americans and Anglos to create a unique region. In cities such as San Francisco and New York, Chinese, Japanese, and people from other Asian countries settled in distinctive parts of each urban area early in U.S. history. Most recently, Latino immigrants from various countries in Latin America are moving into communities across the United States.

Cultural diversity adds to the distinctive character of many places in Canada and the United States. When cultural diversity remains evident through an ethnic group's customs and language, and yet does not threaten the social fabric, diversity is enriching. Diversity can coexist with unity in a society because at the larger scale, people assimilate enough to function in the society and keep it functioning. The rich and dynamic cultural fabrics of the United States and Canada help make landscapes of each country distinctive.

Environmental Impact

Most themes identified thus far—especially urbanization, industrialization, high mobility, resource dependence amid resource abundance, and high consumption accompanying high incomes—have an inevitable consequence: They disrupt the physical environment.

Resources can rarely be removed from the natural landscape without leaving an impression, most of the time a starkly visible one. When people live in dense urban settlement concentrations, they transform the landscape. Manufacturing industries consume material resources, and where resources are used heavily with exclusive attention to immediate returns and short-term costs, the air and water are harmed. High personal mobility and personal consumption create patterns of land use designed largely for individual convenience rather than aesthetics or long-term social benefit.

As urbanization, industrialization, and consumption levels have risen and become more widespread in the United States and Canada, the environmental impacts of these activities have become more severe and more obvious. This problem, in turn, has enlivened age-old arguments about the merits of economic development versus the merits of resource and environmental conservation. Arguments during the past quarter-century took a political turn, stimulating greater governmental involvement in both development and conservation. Federal and state/provincial governments are attempting to find a middle ground between the divergent viewpoints.

Because each position in the argument has its proponents and adherents, middle-ground government policy shifts as one group or another gains temporary ascendancy. Government's problem is frequently described as that of establishing regulations providing appropriate and meaningful environmental protection without raising extraction or land development costs so much that a necessary resource cannot be developed.

In the past decade, the issues argued in both countries have shifted to how much and what kind of development, what resources are necessary to meet society's goals, and what those goals are. With domestic resources becoming more and more scarce and costs of extraction and production increasing, the importance of these arguments will grow. The entire field of environmental impact promises to be a major regulatory battleground for the next several decades.

Review Questions

1. Why do geographers employ regional schemes?
2. What are the general themes used to divide North America into 15 regions?
3. How are the various themes manifested on the landscape?
4. What is the impact of scale on our geographic analysis?
5. What are the challenges inherent in any regional scheme that is superimposed on the United States and Canada?
6. How are the many regions of the United States and Canada similar? What makes the landscapes of any two regions different?
7. How does the supply and demand for energy in the United States and Canada reflect the general themes identified in this chapter? How are energy supply and demand shown on North America's landscapes?

CHAPTER 2

GEOGRAPHIC PATTERNS OF THE PHYSICAL ENVIRONMENT

Much of the western United States and Canada is characterized by extreme differences in the physical environment across very small distances. This Utah scene from Arches National Park is typical.

The physical environment, modified by human effort, impinges on our lives regardless of where we live. Where you live affects how much and what part of the physical environment you notice, but the amount of attention is governed by how much the physical environment is modified. For example, climate may be the environmental condition most noticed by people who live in North America's large cities because the original topography and vegetation have been modified beyond recognition in many areas. Residents of smaller towns and rural areas, on the other hand, attend to their local topography, water conditions, and perhaps soils and vegetation as well.

As described in this text, regions are expressions of landscape patterns formed primarily through human activity and human organization. Thus, the physical environment is only one of several factors that help explain the character of geographic distributions in the United States and Canada. Even so, the physical environment is integral to the makeup of all regions.

THE PHYSICAL ENVIRONMENT

Topography

Compared to a continent such as Europe, where very different landforms are close to each other, the topographic regions of Canada and the United States

are extensive (Figure 2.1). It is possible to drive an automobile in North America for days and encounter little change in landform appearance. Even so, because most long trips are east-west and the major topographic features extend north-south across the two countries, travelers often cut across the continental grain through several topographic regions each day.

In its most general form, the topography of North America consists of broad lowland plains—some flat and others gently rolling—with long mountainous zones paralleling the East and West coasts. The lowlands differ in the ways they were formed, however similar they may appear to our eyes. In contrast, highland and mountainous terrain is extremely varied because local conditions causing the variation differ from place to place.

Lowlands

The continent's interior is a vast, sprawling lowland stretching from the Gulf of Mexico to the Arctic. This broad lowland is up to 5000 kilometers (3000 miles) wide and encompasses more than half the continent's total land area. Geomorphologists, as scientists with an interest in landform development, assign this expanse

Figure 2.1   Physiographic regions (adapted from Fennemen).

of flat land and gently rolling hills to three or four different physiographic regions—the Atlantic and Gulf coastal plain, the interior lowland, and the Canadian (or Laurentian) Shield. Some geomorphologists divide the interior lowland further, into the Great Plains and the interior plains. The processes that led to each region's appearance and their subsurface bedrock materials are the basis for regional differentiation. To the untrained eye, however, the landforms in each lowland region are very similar. Features other than the topography, such as climate and cultural landscape, give the regions their distinctive appearances.

The Atlantic and Gulf coastal plain extends from northern Mexico along the East Coast of the United

States to the southern margins of New England. Underlying this area are beds of young, soft, easily eroded rock formed by deposition in recent geologic time when sea level rose and covered the land. The continental shelf, extending seaward as much as 400 kilometers (250 miles) beyond the shore, is physiographically part of the Atlantic and Gulf coastal plain but lies beneath the current sea level.

The coastal plain's flat terrain places unusual demands on the region's farmers. In many locations, for example, large drainage ditches, some more than 5 feet deep, must be cut around farm fields so that fields are not flooded by the region's heavy rains.

The interior lowland, bounded by the Appalachian Mountains on the east and the Rocky Mountains on the west, occupies the continent's interior. Noticeably hillier than the coastal plains, the lowland has almost no rough terrain. This region is shaped like a large saucer, turned up at the edges. The land is underlain to great depth by hard, durable sedimentary rock in nearly horizontal beds. Local stream erosion or, in the north, glacial action during the recent Ice Age caused the small amount of topographic variation found across the interior lowland. Visitors accustomed to mountainous landscapes are struck by the lowland's overwhelming topographic blandness.

Within the lowlands, the Great Plains' geology differs little from that of the interior plains. Sedimentary beds dominate throughout, although in the north they are broken by eroded domes, most notably the Black Hills of western South Dakota. Nearly horizontal, the sedimentary beds dip gently westward to a trough at the foot of the Rocky Mountains. Both Denver and Colorado Springs, for example, are located over the trough. The tilt of the sedimentary beds is not apparent at the land's surface because the dip is masked by a mantle of loose sediments eroded from the Rockies. The boundary between the Great Plains and the interior plains is marked by a series of low escarpments that indicate the eastern edge of the mantle (see Figure 2.1). The sediment cover creates a gentle eastward slope from an elevation of 1750 meters (5500 feet) at the foot of the Rockies to 650 meters (2000 feet) at the section's eastern edge. One result of the gradient's regularity is a parallel west-to-east stream drainage pattern across the U.S. Great Plains.

The extent and character of the interior lowland have influenced the economic and settlement history of the United States. First, because the climate over much of this large region supports farming, the lowland has vast agricultural potential. Second, half of the country can be crossed west to east without encountering significant topographic barriers, helping to integrate both the lowland and the distant West with East Coast growth centers. Third, almost all of the interior lowland is drained by the Mississippi River and its tributaries. This, too, assisted regional integration by giving the land west of the Appalachians a transport focus. To many of its residents, the interior lowland is America's heartland—vast, economically powerful, productive, buffered from outside influences that enter the country from the coasts. Their case may be overstated, but many popular images of America are based on life in this region.

The Shield

The Canadian Shield lies north and northeast of the interior lowland. It is the two countries' second largest physiographic region occupying more than 2.8 million square kilometers (1 million square miles). Old, hard crystalline rocks lie at the Shield's surface, unlike the rocks in the interior lowland which are covered by sedimentary beds deposited under a sea that once filled the continent's midsection.

More than any other North American physiographic region, the Shield's landforms were remolded and shaped by massive continental glaciers. During the last 1 million years, vast sheets of ice originating on either side of Hudson Bay crept outward and then melted back repeatedly. At their greatest extent the massive ice sheets covered nearly 13 million square kilometers (5 million square miles). This is two-thirds of the continent's total land area. These continental glaciers covered most of Canada and Alaska east of the Rocky Mountains and the Coast Ranges. They extended southward to approximately the present Missouri and Ohio river valleys. The ice was up to 1.6 kilometers (1 mile) thick in places. Its massive weight gradually depressed the land on which it rested. Much of the Shield is still rebounding from the last glaciation, the Shield's elevation slowly rising in response to the ice's meltback about 10,000 years ago.

As the glacial ice moved slowly outward from its source areas on the Shield, it changed the landscape over which it flowed. The ice plucked rocks weighing many tons off the surface and carried them great distances. Massive boulders called *erratics*, often many feet in diameter, were strewn across the Shield when the glaciers melted. Preexisting drainage patterns meant little to the glaciers. The ice's passage greatly disrupted old patterns. And when the glaciers melted, major rivers were created and cut broad new pathways to the sea.

Glaciation scoured much of the Shield's surface, leaving behind a thin or, in places, nonexistent soil cover. Streams on the Shield wander across a labyrinth of lakes and swamps rather than accumulating into large rivers that flow directly to the sea. Thus, most of the Shield is a rocky, thinly vegetated landscape crisscrossed with small streams flowing between thousands of lakes. The portion of the Shield in central and northern Minnesota, for example, gives that state its nickname, "Land of 10,000 Lakes."

The continental glaciation's landscape impact extends far beyond the Shield. Toward the south, where the ice was not as thick and its erosive force correspondingly less, glaciers were diverted or channeled by higher elevations. For example, the ice's advance was largely blocked in central New York by highlands south of the Mohawk River. Narrow glacial probes pushed up tributary streams, gradually broadening and deepening the valleys. As the ice around the glaciers' peripheries melted faster than new ice flowed outward, the glacial margins retreated. The debris collected by the glaciers and then deposited at the ice's margins created a series of natural dams across the tributary streams. Today New York State's deep and narrow Finger Lakes fill such glacially enlarged and dammed valleys and form a truly beautiful landscape.

Along and beyond the glaciers' southern edges, landscapes were transformed by deposition rather than erosion. Large areas across the continent's central lowland are covered by a mantle of *glacial till*, or drift, the mix of rocks and soil dropped by the glaciers. Till covers the lowland to depths ranging from a meter or less to more than 100 meters (330 feet). Each time the glaciers paused before retreating, higher hills, called *moraines*, were created. In the east, Staten Island, Long Island, Martha's Vineyard, Nantucket, and Cape Cod are end moraines marking the farthest major extension of the glaciers' advance. Landscape south of the Great Lakes, too, is laced with long, low, semicircular moraine ridges and other glacial deposits.

One section of the generally glaciated portion of the interior lowland puzzled geographers and geomorphologists for many years. Wisconsin's southwestern quarter and 400 kilometers (250 miles) of the adjacent Mississippi River apparently escaped glaciation, even though this area is surrounded by land that was subjected to the moving ice sheets. The Driftless Area, as this small region is called, has a local landscape with

The Canadian Shield was scoured by glaciers and lies in an area too cold for rapid soil formation. The Shield's forested and tundra-clothed landscape is characterized by meandering streams haphazardly connecting ponds and lakes.

more angular, fragile rock formations, including natural bridges and arches destroyed in glaciated areas. The Driftless Area was apparently spared glaciation because the Superior upland to the north acted as a partial barrier against the flowing ice, channeling the main glacial lobes into the deep valleys of Lakes Michigan and Superior. Because the Driftless Area has no till cover, its more rugged topography and poorer quality soil make the region less useful agriculturally than surrounding glaciated areas.

As the glacial front retreated, massive lakes were created where normal drainage patterns were blocked. The St. Lawrence River, the Great Lakes' current outlet, was dammed for a time as ice in the south melted more rapidly than ice in the north; the Lakes' flow was south into the Mississippi and Susquehanna rivers. On the northern Great Plains, Lakes Agassiz and Regina together covered an area larger than today's Great Lakes. As the glaciers disappeared, these lakes shrank to small remnants. Their earlier extent is now marked by the former lake beds, a flat area covering parts of Manitoba, Saskatchewan, North Dakota, and Minnesota.

Sea level was significantly lower during periods of widespread glaciation. This lowered the *base level* of many rivers, thus allowing deeper valleys to be eroded by the rivers' flows. Many of these valleys extend well out into today's ocean floor. Like other river valleys, the lower reaches of the Susquehanna and Hudson rivers' deeply cut valleys were gradually filled by the ocean as the sea level rose. Two of the world's finest harbor areas were formed in this way: New York Bay with the deep Hudson River and the protective barriers formed by Staten Island and Long Island, and Chesapeake Bay with the drowned valley of the Susquehanna River and some of its major former tributaries, such as the Potomac and the James rivers.

Mountains

Two extensive mountainous regions extend along the continent's eastern and western margins. The western region is complex, comprised of numerous different mountain chains of varying age and orographic origin with broad intermontane basins and plateaus. This region reaches southward into Mexico and Central America.

In the east, coastal plains lie between the southern Appalachian Highlands and the sea but are gradually squeezed against the coast as one moves north until the plains disappear entirely at Cape Cod. From there, northeast to Newfoundland, the Appalachians provide the coastal landscape. The Appalachian Highlands are old, eroded remnants of what were once much higher mountain ranges. Today they reach elevations less than half those of their interior western counterparts. Mount Mitchell, in North Carolina, for example, is the tallest peak in the east at 2036 meters (6684 feet), while many mountains in the Rockies exceed 14,000 feet elevation.

The Appalachian Highlands are important despite their relatively low elevations. The settlement and economic history of the two regions encompassing the Appalachians—the Bypassed East and Appalachia—were profoundly affected by the mountains. Most Appalachian soils are shallow, and, as Robert Frost once observed, rocks are often a farmer's most dependable crop. Steep slopes, difficult to farm under any circumstances, are not suited to modern agricultural practices emphasizing mechanization. Large-scale urban or industrial growth is cramped in small, local lowlands. Pittsburgh lays claim to the distinction of being the hilliest large city in North America. Between the Mohawk River in New York and northern Alabama,

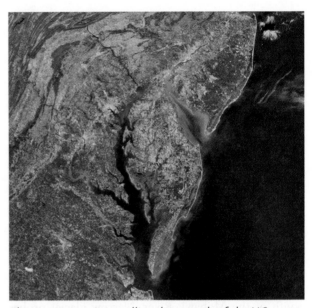

This is a composite satellite photograph of the U.S. coast from New York City to Norfolk, Virginia. The excellent harbor facilities, seen here in Chesapeake Bay, Delaware Bay, and New York Harbor, have played an important role in the economic growth and urbanization of this region.

the Appalachians proved to be a surprisingly effective barrier for early American settlers attempting to migrate west. There are few breaks in the mountains, and inadequate transportation remains a major regional problem.

Although generally rugged, the Appalachians contain several different kinds of landforms. South of New

Extensive areas in the Appalachian Mountains are underlain by sedimentary rock that has been contorted into folds by immense pressure.

England, the eastern Appalachians describe an intricate pattern of folded linear ridges and valleys. The western Appalachians, on the other hand, offer a jumbled landscape that was created as streams cut into the region's level-bedded sedimentary rocks. In Newfoundland and the Atlantic Provinces, massive igneous rock domes have been heavily eroded by water and ice. These domes are most obvious in the highlands of central New Brunswick and on Cape Breton Island in Nova Scotia. New England's landscape represents a little of everything, with perhaps a bit more of the southern Appalachians' combination.

If North America's physical landscape east of the Great Plains is characteristically a gently rolling terrain punctuated by a few mountains, the West is mountainous with sudden, great changes in local elevation. Few places in the West are beyond sight of mountains. Rather than giving relief to a generally flat landscape, the western mountains form and mold the region's landscape, dominating all of its other elements.

The West's physiography is arranged in a series of three north-south trending bands, with the Rocky Mountains on the east, a series of high, heavily dissected plateaus west of the Rockies, followed by the Pacific coastlands' mountains and valleys. These landscape regions can appear homogeneous on a map. But

in fact, each contains the great visual diversity characteristic of this section of the continent.

With few exceptions, the Rocky Mountains present a massive face to the Great Plains, such as along central Colorado's Front Range, with peaks rising a mile or more above the plains. In Idaho's northern Rockies, the mountain range's typical north-south linearity is masked by massive igneous domes irregularly eroded into rugged ranges containing the largest remaining wilderness area in the United States outside Alaska. Farther north still, parts of the Canadian Rockies were carved by alpine glaciation during the Pleistocene and contain some of the continent's most picturesque mountain scenery. Elsewhere, as in south-central Wyoming, the Rockies are not at all apparent. Travelers driving across southern Wyoming on Interstate 80 are likely to be unaware of the high elevations through which they are passing. Complex geologically, the Wyoming Basin serves as a western peninsula of the Great Plains and has allowed millions of east-west travelers to circumvent the Rockies' more rugged stretches.

The interior West's high plateaus are also varied in their origin and appearance. In the southernmost subsection, the Colorado Plateau, a series of thick sedimentary beds rise more than 1000 meters (3300 feet) above the adjacent lowlands' elevation, tilting upward toward

Western landscapes are dramatic in both countries with few places beyond the sight of mountains. Many thousands of visitors each year, not content to look only from the roadside, hike into mountains such as these in Banff National Park.

the northeast. The Grand Canyon of the Colorado Plateau is most dramatically viewed from the south rim because the north rim is higher and the relatively flat, less exciting land beyond is not visible. The Colorado Plateau may be the most geologically colorful part of the continent. It is a land of spectacular canyonlands, volcanic peaks, sandy deserts, and the Painted Desert. More than a dozen national parks and monuments help preserve portions of this region's natural treasures.

Northwest of the Colorado Plateau, north-south trending fault-block mountains are scattered across barren lowlands in the basin and range region. This area is characterized by interior drainage, with many of the area's small and intermittent streams ending their courses in lakes and dry sinks. Much of the basin and range region is covered with a thick mantle of alluvium washed down from the mountains.

The Columbia-Snake basin lies north of the basin and range region. The Basin was filled over time to a depth of more than 1000 meters (3300 feet) by repeated lava flows. The Basin surface was then eroded deeply by rivers flowing west from the Rockies to the Pacific Ocean. Volcanic cones also dot portions of the region, especially across south-central Oregon and in the Snake River valley in Idaho.

The plateaus between the Rockies and the Pacific ranges effectively disappear near the international boundary of Canada and the United States, but they reappear farther north. Gradually widening toward the north through a complex of lava plateaus, low mountains, and river plains in British Columbia and the Yukon River valley in Alaska, is a broad, flat, poorly drained lowland.

The Pacific Coast's physiographic region is complex but consists for the most part of two north-south

trending mountain chains separated by a discontinuous lowland. The Coast Range is fairly massive in southern California with peaks reaching 3000 meters (10,000 feet), but north from about the Oregon border the mountains are lower, seldom rising above 1000 meters (3300 feet) in elevations. This is also California's major fault zone and a region of frequent earthquake activity. Along the California-Oregon border, the Klamath Mountains are higher than California's Coast Range, extensive, and very rugged—much like the Idaho Rockies. The Coast Ranges in the rest of Oregon and Washington are low and hilly rather than mountainous, with the notable exception of the Olympic Mountains on the Olympic Peninsula. In Canada, along the coast of British Columbia, the coastal mountains are higher once again and called the Insular Mountains.

The interior lowlands in this physiographic region—the Central Valley of California, the Willamette Valley in Oregon, and the Puget Sound lowland in Washington and British Columbia—are the only extensive lowlands near the Pacific Coast. Possessing relatively good soils, these lowlands support much Pacific Coast agriculture. California's Central Valley, in particular, has some of the flattest land to be found anywhere in the United States. Though not as extensive as the Central Valley, the Willamette Valley and Puget Lowland contain productive farmland and are also home for most of the North Pacific Coast's population.

East of the lowlands are the Sierra Nevada and the Cascade mountain ranges. The Sierra Nevadas were born as the earth's crust cracked and a massive section of the crust was tilted upward in what is called a *fault block* (Figure 2.2). The highest and sharpest side of the

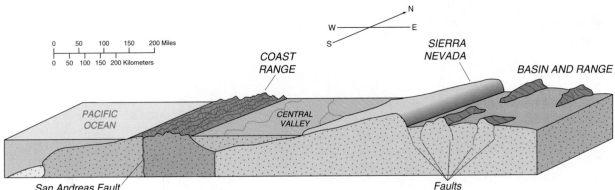

Figure 2.2 California cross-section schematic. North America's west is geologically complex. In California, the fault-block structure of the Sierra Nevadas and the sedimentary Central Valley are two of the larger features that contribute to the region's complicated physical landscape.

fault block faces east, while the western slopes are reasonably gentle. The eastern face of the Sierra Nevadas, for example, can rise more than 3000 meters (10,000 feet) within a few linear miles. In contrast to the origins of the Sierra Nevada, volcanic activity was more important in the Cascade range's formation. Some of the continent's best known volcanic peaks, such as Washington's Mount Rainier and Mount St. Helens and Oregon's Mount Hood are found here. Northern California's Mount Shasta anchors the southern Cascades. As this range continues into Canada, the Cascades are called the Coast Mountains.

Climate

Climate is the aggregate of day-to-day weather conditions, a characterization of weather conditions prevailing across many years. The most important components of climate are temperature and precipitation, and the climatic patterns that interest geographers result from the interaction of three geographic influences.

The first geographic influence is *latitude*, a measure of the distance north or south of the equator. As the sun's energy reaches the earth, that energy is most concentrated where the surface is perpendicular to the incoming energy. From our perspective on earth, the energy is most concentrated where the sun can be directly overhead at midday. People who live at high latitudes—those far from the equator—know that the sun is never directly overhead at midday, even in midsummer when it rises highest above the horizon. Thus, locations farther from the equator receive less solar energy on the average than locations closer to the equator.

People living at high latitudes also observe that the midday sun rises higher in the sky during the summer than the winter. This occurs because the earth's axis of daily rotation is tilted relative to our planet's annual orbit about the sun. As the earth moves around the sun each year, the axis's tilt alternately presents the Northern Hemisphere and then the Southern Hemisphere to the sun's more direct rays. From our perspective in the Northern Hemisphere's mid- to higher latitudes, the sun rises high in the sky at midday during our summer, and we experience longer days and daily receive considerable solar radiation. During our winter when the sun's rays are more directly striking the Southern Hemisphere's latitudes, our days are less bright, shorter, and colder. The earth's axis tilt thus establishes seasonality.

Seasonal differences between summer and winter would be greater than they are except for another consequence of latitude. Because the earth is spherical, the transition from winter to summer adds more hours of daylight at high latitudes than at mid-latitudes. Someone living north of the Arctic Circle (66.5 degrees North latitude), for example, experiences daylight for 24 hours at midsummer but little or no daylight at all at midwinter. As a result, the much longer summer days in northern latitudes allow more heating than we could expect solely from latitude. For example, Canadians and residents of the northeastern United States who journey to Florida in January may find temperatures averaging 15°C or 20°C (roughly 30°F or 40°F) warmer than those they left behind. Floridians living in the subtropics and making the reverse trip in

A place's weather is part of what gives that place its character. To the residents of the northeastern United States, extreme weather events, such as this snow storm, may be more important than long term averages.

summer are likely to find temperatures only slightly (perhaps 6°C, or 10°F) lower than those in Florida. Thus, temperature seasonality is greatest in the middle and high latitudes where winters can be cold but warm to hot summers are widespread.

The second geographic influence on climate is based on the different ways land and water absorb solar energy. Water is more reflective than most land surfaces, mixes its warmer and colder areas through ocean currents and wave motion, and simply absorbs and loses heat less quickly than many land materials. As a result, land heats more rapidly and cools more rapidly than water. In a tendency called *continentality*, places far from large bodies of water experience greater seasonal temperature extremes than do coastal communities. People living in the northern Great Plains or the Prairie Provinces, for example, experience annual temperature ranges close to 65°C (150°F). Annual temperature differences of as much as 100°C (180°F)—from +50°C (+120°F) to −50°C (−60°F)—between summer and winter have been recorded. The larger the midlatitude landmass, the greater is this annual temperature range at interior locations. North America is a very large landmass in the middle latitudes.

The converse of continentality occurs at maritime locations, especially on west coasts of midlatitude continents. Coastal locations have much smaller temperature ranges than continental interiors because of what is called a *maritime influence*. Horizontal and vertical ocean currents minimize seasonal variations in the surface temperature of the water. In turn, the moderated water temperature reduces temperature extremes in the air mass above the water surface. The maritime influence occurs wherever air moves onshore, so west coasts that regularly receive onshore wind systems experience the greatest moderation of summer and winter extremes. Vancouver, British Columbia, for example, has an average annual temperature range of only 15°C (27°F).

Proximity to large water bodies also tends to produce more precipitation, with coastal locations receiving generally higher amounts. Moisture is drawn into the atmosphere as water evaporates, and large water bodies provide more water for evaporation. With more moisture in the atmosphere, the possibility of precipitation increases as the air moves onshore. There are notable exceptions to this general rule, however. The dry coast of southern California provides one example. Moisture-producing weather systems developing over the North Pacific and moving southward are blocked during the long summer months by a large, stable air mass that moves northward from Mexico each spring. Southern California's coast, therefore, experiences hot, dry summers, with wetter winters after the stable air mass moves southward again in late fall. The Arctic coastline of Canada and Alaska is also very dry. The long winters minimize atmospheric humidity in this region. Very cold, stable air dominates the Arctic for long periods and reduces the likelihood of snowfall.

Topography is the third geographic influence on climate. The most obvious connection between climate and topography has to do with elevation and

A desert landscape rich in a variety of cacti, such as those pictured here at Organ Pipe Cactus National Monument in Arizona—so much a part of the image of the interior West after countless western movies and television shows—is, in fact, uncommon in the United States.

temperature. Higher elevations are cooler than lower elevations by about 6.5°C per kilometer (or 3.5°F per 1000 feet) of greater elevation, although this effect may be overwhelmed locally by other climatic conditions. Topography's influence on climate is not simply a matter of elevation, however. Because the amount of moisture air can carry is governed by the air's temperature, when moist air is forced to cool the water condenses and may fall as precipitation. Precipitation results if the cooling causes the relative humidity to reach 100 percent.

If a major mountain chain lies astride the prevailing wind direction, the air rises and cools. Moisture falls on the windward side of the mountains. Conversely, as the air descends on the mountains' lee side, the air warms and little precipitation occurs. North America's wettest region lies along the Pacific Coast between Oregon and southern Alaska. Moisture-laden winds off the northern Pacific are forced over the Coastal Ranges and the Insular Mountains. Average annual precipitation throughout the area exceeds 200 centimeters (80 inches) and in some places is greater than 300 centimeters (120 inches) (Figure 2.3). The central and northern interior West is arid primarily because the region's north-south trending mountain ranges form a barrier to precipitation.

In addition to the mountains' effect on precipitation, they also reduce the maritime influence on temperature. Within the United States and Canada, the Western Cordillera (mountain mass) confines West Coast maritime climatic conditions to that coast. Some of the continent's greatest differences in both precipitation and temperature found across small distances exist along the west and east sides of the Coast Ranges.

Topography's effect on precipitation almost disappears east of the Rockies. The eastern mountains are much lower and are less of a barrier to moving air. More importantly, much of the interior's weather results from the conflict between two huge air masses, one flowing north and northeast from the Gulf of Mexico and the other flowing south and southeast out of Canada. The eastern half of the United States and southeastern Canada lie in the contact zone for these two air masses. Without major mountain barriers, these air masses confront each other and frequently produce violent weather across the region.

Beyond the three geographic influences of latitude, water-land contrasts, and topography, the impact of different air masses on each other represents a fourth complex influence on climate. Much of the weather in the two countries is affected markedly by the confrontation between polar continental air masses (usually cold, dry, and stable) and tropical maritime air masses (warm, moist, and unstable). Polar air pushes farther south in winter, whereas tropical air reaches farther north in summer. Weather systems tend to move west-to-east across North America, carried by prevailing air currents at midlatitudes. As a result, the two countries' East coasts experience some of the effects of the interior's continental climate. Coastal New England and the Atlantic Provinces, for example, have a climate with much greater average seasonal variations in temperature than do similar latitudinal locations along North America's West Coast.

Together these climatic controls create a regional pattern that is decidedly different between the eastern and western halves of the United States and Canada. Temperature is the principal element in the East's climatic variation; precipitation differences dominate the West. In the East, climate regions are defined as different largely because of growing season length—the period from the average date of the last frost in spring to the first frost in fall—and the average summer maximum or winter minimum temperature. In the West, average annual precipitation is the key, although moderated temperatures are important in defining the Marine West Coast climate. To be sure, it is drier in the Northeast than the Southeast, and it is colder in the Northwest than the Southwest. Still, the keys to understanding climatic patterns in the East are temperature and latitude, whereas in the West, the keys are precipitation and topography.

Climate can be classified in many different ways. Greek scholars in the first or second century B.C., for example, divided their known world into three climate zones: torrid, temperate, and frigid. These zones were simply latitudinal bands. The classification system most widely used today was developed by Wladimir Köppen (1846–1940), a German climatologist and amateur botanist (Figure 2.4). Köppen observed that major vegetation associations were sensitive to general climatic factors, and he suggested that the climatic pattern was reflected in the distribution of major plant associations. Thus, Köppen's regional climatic boundaries represent the limiting effects of precipitation and temperature on vegetation complexes. Köppen used only mean annual and monthly temperature and precipitation data in constructing his system. Compare Köppen's map for North America with the map of

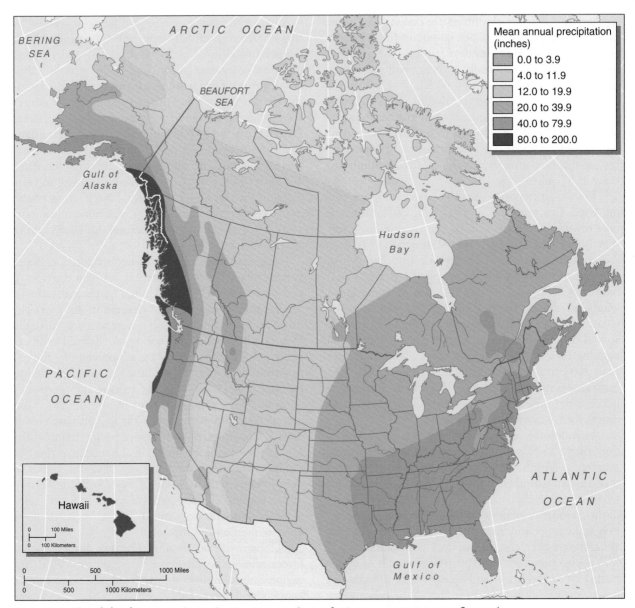

Figure 2.3 **Precipitation.** From the Rocky Mountains to the Pacific Ocean, precipitation is influenced by topography; east of the Rockies most precipitation is generated by storm systems. One result is a pattern of great local variation in the West and broad areas with similar average moisture conditions in the East.

vegetation (Figure 2.5). The fit is not perfect, but the two patterns are clearly similar.

Vegetation

When botanists speak of *climax vegetation*, they mean the assemblage that will grow and reproduce indefinitely at a place, given a stable climate and average conditions of soil and drainage. It is basically a "what if" concept intended to define natural vegetation—the vegetation that would exist somewhere without human and other short-term influences. For most of North America's inhabited portions today, the concept has little meaning. Climax vegetation probably never existed, but any approximation of it is now substantially removed, rearranged, and replaced. Pine trees presently

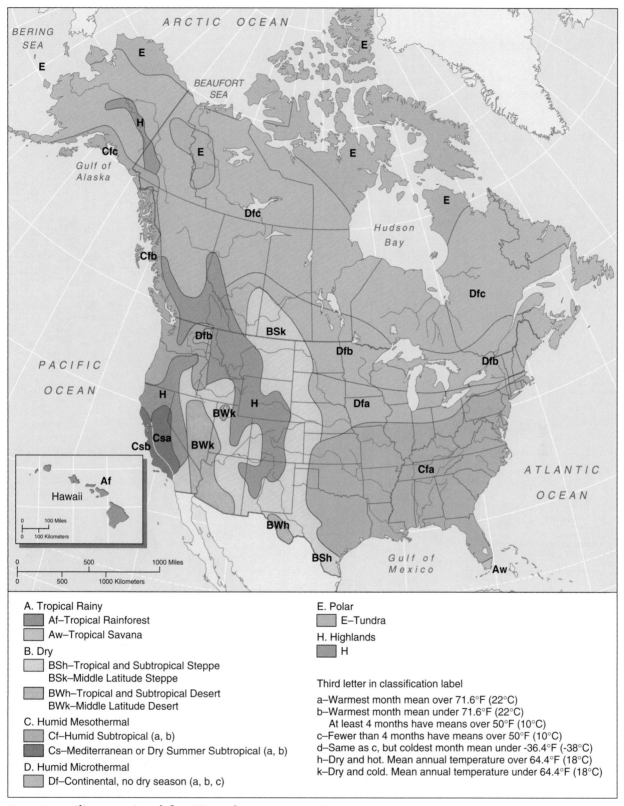

Figure 2.4 **Climate regions (after Köppen).** Climate results from the combination of long-term moisture and temperature conditions.

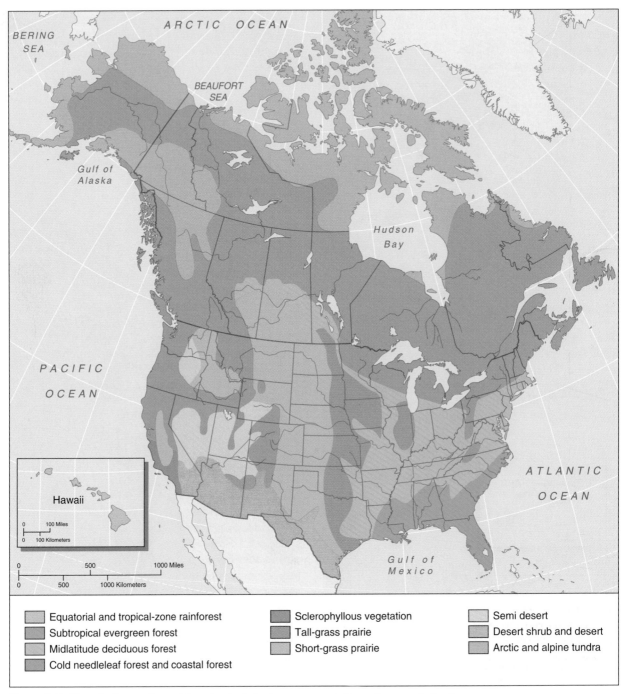

Figure 2.5 **Vegetation regions.** Note the similarity between vegetation patterns and climate.

	Equatorial and tropical-zone rainforest		Sclerophyllous vegetation		Semi desert
	Subtropical evergreen forest		Tall-grass prairie		Desert shrub and desert
	Midlatitude deciduous forest		Short-grass prairie		Arctic and alpine tundra
	Cold needleleaf forest and coastal forest				

grow across much of the Southeast, for example, because they give greater economic returns than the original mixed broadleaf and needleleaf forests. The grasses of the interior's plains and prairies are now mostly European imports. Native grasses are gone either because they offered inferior grazing for farm animals or because they could not withstand the weeds imported from Europe and Asia. Most of the two countries' remaining climax vegetation is located in the mountain west and the north. Logging and settlement expansion continue to reduce the small remnant that remains.

Even though most natural vegetation is now gone, vegetation regions are worth noting for two reasons. First, the central role of climate in vegetation growth is clearly indicated by the pattern similarity of climatic and vegetative regions. Knowledge of one pattern tells us about the other. Second, even with all the changes, the pattern of vegetation is similar to that of the pre-European era. Areas that were forested continue to have substantial tree cover today, and areas that were grasslands still have few trees. In both cases, however, domestic varieties have supplanted what was once a natural vegetation mix.

Perhaps the simplest way of creating vegetation regions is to use three broad categories: forest, grasslands, and scrublands (see Figure 2.5). Forests once covered most of the East, the central and northern Pacific Coast, higher elevations of the West, and a broad band across the interior North. Pacific Coast forests, those of the interior West, the North, and a narrow belt in the Deep South were all needleleaf and composed of many different trees. Much of the Ohio and lower Mississippi river valleys and the middle Great Lakes region was covered by a deciduous broad-leaf forest. Between the broadleaf and needle-leaf forests were mixed forest belts with needleleaf and broadleaf varieties.

Much of the interior lowlands was grass covered, including nearly all of the Great Plains from Texas and New Mexico to central Alberta and Saskatchewan. Because precipitation amounts provided in this area's generally subhumid climate are not adequate to support tree growth, grasses dominate the landscape. An eastward extension of the grasslands, referred to as the Prairie Wedge, extends across Illinois to Indiana. Precipitation in the Prairie Wedge is adequate to support tree growth; thus, the fact that it exists has generated considerable speculation. Current thinking suggests that fires set repeatedly by American Indians in order to improve buffalo grazing probably had a major impact on this eastern extension of grasslands.

Scrublands usually develop under dry conditions. They are concentrated in the lowlands of the interior West of the United States, with some minor extensions into Canada. Scrubland vegetation may include cacti in the Southwest, dense, brushy chaparral in southern California, and sage and mesquite in Texas. Scrubland has also expanded into some of the drier grassland sections of the Southwest as overgrazing destroyed the fragile grass cover (see Chapter 14).

In the Northlands, tundra grows in a climatic regime that is too cold and too dry for other vegetation. Tundra vegetation does not fit well into our three-part classification. Grasses, lichens, and mosses dominate the Northlands' vegetation. Tundra exists in small areas southward into the United States as well, where climatic conditions at high elevations are inhospitable to tree growth. The altitude above which trees will not grow, given sufficient moisture, is called the *treeline*. As one moves north, the altitudinal treeline is found at lower elevations. Eventually, the treeline is determined by latitude rather than altitude.

The close relationship between climate and vegetation is well illustrated by the treeline, clearly visible in this photograph of Mount Reynolds in Glacier National Park. Higher elevations experience climatic conditions too cold for tree growth.

Soils

Soil is basically a mixture of weathered rock material and organic matter. A place's soil owes its characteristics to such factors as the parent rock material, climate, topographic characteristics such as slope and orientation, and decaying plants and animals. Hundreds of different soil types result from the interaction of these elements. Although each soil is unique because of its mix of properties (such as color and texture) and composition (including such conditions as organic content and the action of soil colloids), general soil zones can be specified.

Color is perhaps the most obvious soil property: the red clays of Georgia, the black soils of the Red River valley in Manitoba—such colors may be the first characteristic to catch an observer's attention. A dark color usually indicates abundant organic materials, whereas red shows the presence of iron compounds. Generally, color results from the soil-forming process. For example, the northern needle-leaf forest's pale-gray soil indicates that organic matter and minerals were leached from the soil's surface layer.

Soil texture, a second soil property, refers to the ratio of different-sized particles in the soil. Soil texture depends largely on the parent material's composition. Sand has the coarsest soil texture, silt is intermediate, and clay has the finest texture. Texture is important because it determines a soil's ability to retain and transmit water. Soils called loams contain a substantial mix of each of the three particle grades and are considered best for agriculture. Loams possess a texture fine enough to hold moisture, yet not so fine that water cannot be taken up easily.

Colloids make an important contribution to a soil's composition. Colloids are small soil particles with complex properties that often exert significant influence on soil productivity. Soil acidity, for example, results from the alteration and integration of soil colloids. Acidity is measured along a scale from acid to neutral to alkaline. Acid soils are characteristic of moist climates; alkaline soils are typical in dry areas. Most soils in the agricultural zones of the eastern United States and Canada are moderately to strongly acidic. Lime must be added periodically to neutralize soil acidity; otherwise most row crops will not be productive.

Where soils have developed in place for a long period, they take on a layered appearance. These layers are called soil horizons. The uppermost layer, the A-horizon, is the zone from which certain colloids,

The soil of a place is not an unlimited resource. The quality of these different soil horizons differs substantially, with the greatest amount of nutrients commonly concentrated in the upper horizons.

chemical compounds, and other matter have been removed through leaching. The red clay found across much of the South's piedmont in the United States, for example, has its tint because the coloring materials other than iron have been removed from the A-horizon by the region's substantial precipitation. The B-horizon is the level below the A-horizon and is where the matter removed from the A-horizon accumulates. The C-horizon is below the B-horizon and is the parent material. In more humid regions, soils may also possess a thin O-horizon at the surface where organic matter is freshly decaying and actively adding organic matter to the soil.

Many attempts have been made to classify soils. Soil scientists usually prefer a genetic classification that focuses on how the soil was created. The most widely accepted classification system in the United States was formulated by the U.S. Department of Agriculture (USDA). Properly titled the United States Comprehensive Soil Classification System, it is best known as the Seventh Approximation because soil scientists rejected six earlier attempts by the USDA.

The Seventh Approximation is based on precise soil horizons that develop under particular environmental conditions. A general soil region based on the Seventh Approximation (Figure 2.6) or any regionalization of soils necessarily masks the tremendous soil complexity that can exist in small areas. For example, the system is hierarchical, with about 50 recognized suborders, 225 great groups, and about 12,000 soil series. For our purposes, the general regions remain

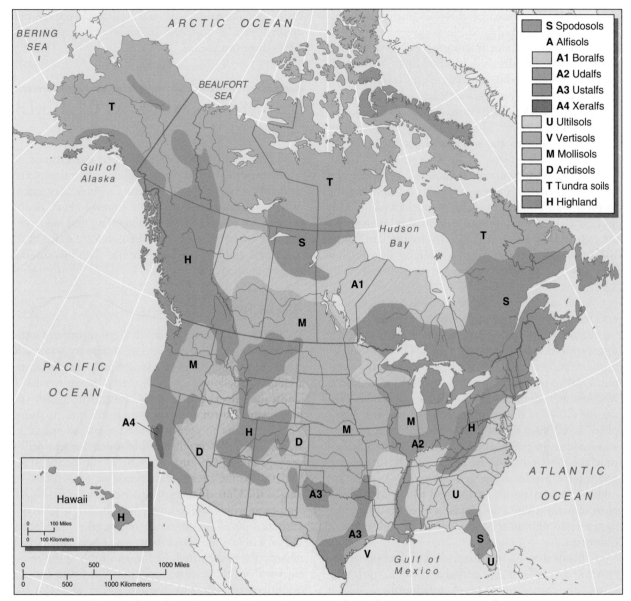

Figure 2.6 Soil types (based on the Seventh Approximation from the U.S. Department of Agriculture). This map portrays broad soil categories. The elements that together make up soil type are complex and can be localized, so that a single small field may contain several different soils.

satisfactory by indicating the most important distribution pattern of the soils.

Four extensive but generally unproductive soils are aridisols, spodosols, tundra soils, and highland soils. Aridisols gain their name from the word arid. These dry climate soils are low in organic content and have little agricultural value. Spodosols develop in cool, moist climates. In Canada and the United States,

spodosols are largely northern needleleaf forest soils. Quite acidic and low in nutrients, these soils are agriculturally valuable only for acid-loving crops such as potatoes or blueberries. Tundra soils are associated with a cold, moist climate. The soil is shallow, frequently water saturated, and has a subsurface that is perennially frozen (see Chapter 17). Tundra soils also have little agricultural value. Highland soils are found throughout high and rugged terrain, are little developed, and agriculturally worthless.

Three more productive and also extensive soils are the mollisols, alfisols, and ultisols. Mollisols are midlatitude grassland soils of semiarid and subhumid climates. These soils are characterized by thick dark brown or black A- and B-horizons, loose texture, and high-nutrient content. They are among the most naturally fertile soils in the world and produce most of North America's cereals.

Alfisols are second only to mollisols in agricultural value. They are soils of the midlatitude forest and the forest-grassland boundaries. Very much "middle" soils in a climatic sense, alfisols are located in areas moist enough to allow clay particles to accumulate in the B-horizon (their characteristic identifying feature), but they are not found in areas moist enough to heavily leach or weather the soil.

Based on their distribution, alfisols can be divided into four subcategories, each with its own characteristic climatic association. Boralfs are boreal needleleaf forest soils located in central Canada. They are usually thin, acidic, and have very modest agricultural potential. Udalfs are deciduous forest soils of the American Midwest and southern Ontario. Somewhat acidic, they are nevertheless highly productive when lime is applied. Ustalfs are found in warmer areas with strong seasonal precipitation variation. In the United States, they are most common in Texas and Oklahoma and are highly productive when irrigated. Xeralfs are highly productive soils. Associated with a Mediterranean climate where cool, moist winters and hot, dry summers occur, they are found in central and southern California.

Ultisols develop in areas with abundant precipitation and a long frost-free period and are taken to represent the ultimate stage of weathering and soil formation in the United States. The particle size of ultisols is small, and much of the soluble material and clay has been carried downward, away from the A-horizon. These soils can be productive, but high acidity, leaching, and erosion are often problems.

Finally, although entisols are nowhere sufficiently predominant to show up on the soils map used here, they are important locally. These are recent soils, too young to have been strongly modified by their surroundings. Entisols are widely scattered and of many types, ranging from the Sand Hills of Nebraska to the alluvial floodplains of the St. Lawrence and Mississippi river valleys. The agricultural potential of entisols

varies, but alluvial floodplain soils drawn from the rich upper layers of upstream soils are among the continent's most productive.

MINERAL RESOURCES OF NORTH AMERICA

A distinct association exists between the location of minerals needed by heavy manufacturing industry and the land's subsurface rock structure. Minerals economically useful to humans may be found in each of the three major forms of rock. Sedimentary and metamorphic rocks are the most prevalent and are more likely to contain minerals of broad utility than are igneous rocks, the third category.

Mineral Fuels

Sedimentary rock forms when small solid particles settle gradually in stationary or almost stationary water and are then compressed over a long time period. For example, if a shallow sea were located adjacent to an arid landscape subject to occasional rainstorms, sand particles would wash into the sea from time to time. Water currents and gravity would spread the particles across the sea's bottom. As this process continued—perhaps taking 500 times as long as the period since Imperial Rome's legions marched or 5000 times as long as the United States of America has existed—each sand layer would weigh heavily on the sandy mass deposited only a few thousand years earlier, squeezing and solidifying the layer beneath. Eventually, when the earth's crust shifted, lifting and folding the rock-hard seabed, the sandstone formerly beneath the sea would be exposed to weathering and erosion like any land surface. There are many sedimentary rocks formed from various parent materials and subject to local conditions at the time, but all were formed in this general fashion.

About 300 million years ago, during what is called the Paleozoic era's Carboniferous period, unusual conditions prevailed and unusual sedimentary sequences were created. Heavily vegetated regions that had remained swampy for eons, producing layer upon layer of decayed organic material, were drowned and covered with new sediment. In some cases, the decayed organic matter evolved into liquid form, trapped between folds of impermeable rock and eventually drawn

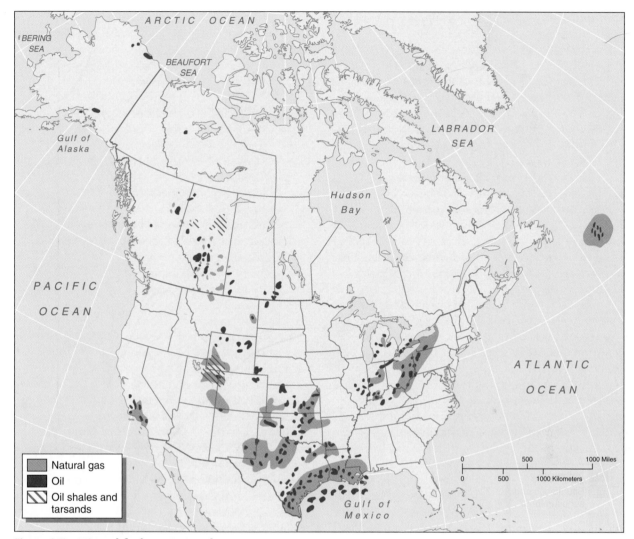

Figure 2.7   Mineral fuels except coal. North America is favored with abundant mineral fuel resources. The two countries, however, have a large demand for fuels, which generates a constant search for new sources.

off as petroleum. Most petroleum deposits are found in conjunction with another byproduct of the period— natural gas. In other cases, the organic matter became solid layers of coal, whether only inches thick or dozens of feet thick as occurs in some locations.

North America possesses vast regions underlain by sediments formed during the Carboniferous period. These areas where coal, oil, or natural gas might be found are located in the interior plains and Great Plains physiographic provinces (Figure 2.7), sections of the Gulf coastal plain, portions of the Pacific mountains and valleys, the arctic rimland, and in folded and

broken form along the western margins of the Appalachian Highlands and into the eastern Rockies.

Large mineral fuel deposits have been identified across extensive portions of these sedimentary lowlands. The continent's most important coal deposits have been mined in the rugged Appalachian field. Although the surface topography is irregular, the coal seams are easily worked. Most Appalachian seams are between 30 centimeters and 3 meters (1 and 9 feet) thick and are frequently located close to the surface. Both factors facilitate use of machinery in extraction. Mines throughout this nearly continuous field in

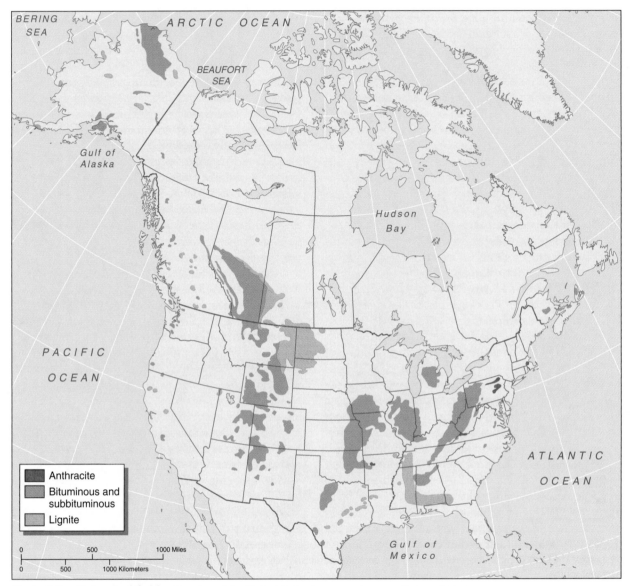

Figure 2.8   Major coal fields. Canada and the United States have a coal reserve that could last hundreds of years at present use levels. However, the increasing cost of alternative fuels and gradual decline in the use of nuclear power to generate electricity should result in increased coal production.

eastern Kentucky, West Virginia, and western Pennsylvania were the earliest brought into production, and they continue to supply more than one-third of the nation's coal needs. The reserves that remain in this field still amount to 20 percent of the United States' total reserves (Figure 2.8).

Until recently, most of the rest of the United States' mined coal was taken from the Eastern Interior Field. This large coalfield underlies most of Illinois and extends into western Indiana and western Kentucky. Although some of the Eastern Interior Field's coal was used in iron and steel production, it has a high sulfur content and is not optimal for steel production. The sulfur content restricted use of most coal from this field to heating and electric-power generation.

Eventually, the sulfur content of the Eastern Interior Field's coal became an air pollution concern. One byproduct of burning this coal is the noxious gas, sulfur dioxide. When dissolved in water, such as rain, fog, or snow, for example, sulfur dioxide forms sulfuric acid. Although not the only source of acid precipitation, the volume of coal burned to meet heavy demand from large populations produced the undesirable byproduct that has adversely affected the water quality of downwind lakes in southern Canada and the northeastern United States. Acid precipitation has also been associated with damage to masonry structures and various species of trees and vegetation.

Growing demand for coal energy and awareness of the consequences of burning high-sulfur coal turned the attention of mining interests westward. The Western Interior Field is almost as large as the Eastern Interior Field and is located under Iowa and Missouri, with a narrowing extension into eastern Oklahoma. The coal found in the Western Interior Field is slightly poorer quality than that found in eastern fields and has only recently begun to be mined. There are also many small and a few large bituminous deposits scattered through the Rocky Mountains and along their eastern margins. Extensive deposits in Wyoming and Montana came into production in the last three decades. Following two decades of rapid mining expansion, Wyoming now leads the United States in coal production (see Table 8.1). The northern Great Plains and the Canadian western prairies also contain several extensive fields of lignite (brown coal).

Petroleum and natural gas are not formed in the exact manner as coal, but because their genesis is similar (see Figure 2.7) the distributions of these mineral fuels often overlap. Petroleum and natural gas deposits are scattered throughout the Appalachian coalfield and include the United States' first commercial well at Titusville, Pennsylvania. Southern Illinois and south-central Michigan produce some petroleum, as do sites across the northern Great Plains and the northern Rockies.

Easily the most important petroleum fields, however, are those in the southern plains, along the Gulf Coast and in southern California. One great arc of producing wells is located along the full length of the Texas and Louisiana coasts. Another slightly broken arc extends from central Kansas south through Oklahoma and westward across central Texas to New Mexico. Between and beyond these two large areas lie two more fields of great importance, the East Texas field and the Panhandle field in northwest Texas. At some distance from this array of midcontinental fields, but also of major importance, are deposits located in

southern California. In the mid-1960s, exploitation of petroleum and natural gas deposits was begun along the north Alaska slope and in Canada's Arctic fringes; the development of significant deposits in Alberta increased that province's economic influence within Canada.

North America's broad pattern of mineral fuel deposits, then, is one of extensive coal south of the Great Lakes and west of the middle Appalachians accompanied by very large petroleum and natural gas fields southwest of the major coal reserves, stretching from the middle Great Plains south to the Gulf Coast. Scattered, occasionally significant deposits of each mineral fuel are located west and northwest of the major reserves.

Metallic Minerals

Metamorphic rock is formed differently than sedimentary rock. As deep currents gradually deform the earth's crust, tremendous pressure is exerted on rock in the crust. The internal structure of previously formed rocks can be metamorphosed in the process. So great is the pressure exerted during thousands of years and so great is the heat generated that the rock's molecular structure is altered. Such transformations indicate why economically extractable quantities of metallic minerals are located most often where metamorphic rock occurs. In North America, therefore, the primary deposits of metallic ores are located in the three major metamorphic rock zones: the Canadian Shield, portions of the Appalachian Highlands and their eastern Piedmont, and sections of the western mountains (Figure 2.9).

Many early metallic mineral mining sites on the Canadian Shield were located near the Shield's margins. The pattern of mineral production follows a long arc extending from the North Atlantic and St. Lawrence River estuary across the Great Lakes north to the Arctic Ocean. This arc defines the Shield's edge, a broad horsehoe shape around Hudson Bay. Mining sites include the cluster of iron deposits on the Quebec-Labrador border at Burnt Creek and Schefferville; a tremendous mineral concentration on the Quebec-Ontario border that included the nickel-rich region at Sudbury; and such self-explanatory places as Cobalt, Ontario, and Val d'Or (Valley of Gold), Quebec. The arc continues on both sides of Lake Superior; in northern Michigan, Wisconsin, and Minnesota where copper and iron are found and in Ontario with

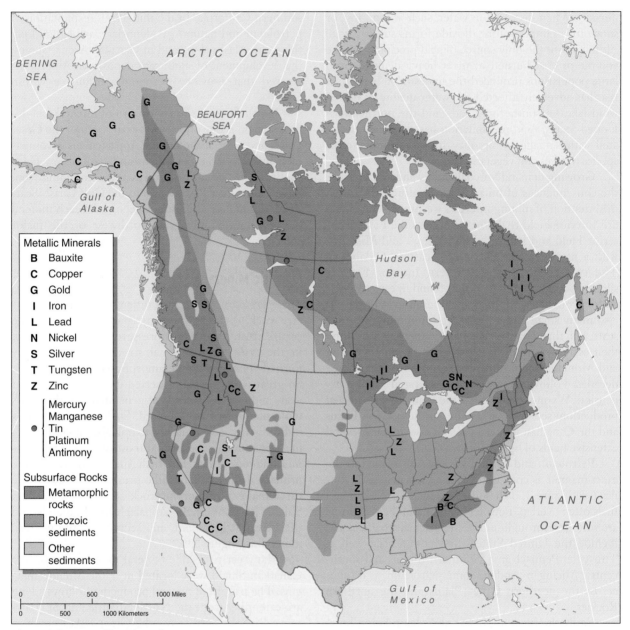

Figure 2.9 **Metallic Minerals.** The metamorphic rocks containing metallic minerals in Canada and the United States arc around North America's interior lowland and its fossil fuel deposits.

iron at Steep Rock. The arc swings north through Flin Flon (copper, lead, zinc) to such likely places as Uranium City, Yellowknife (gold), Port Radium, and Coppermine. Mineral deposits in the Shield's interior began to be developed in recent decades, blurring the arc's shape. But this zone remains important economically to both countries.

Metamorphic rock is located in a second zone along the eastern Appalachian Mountains. Iron mines at Wabana, Newfoundland, formerly supplied the ore for a small steelworks. Copper and iron were important locally to early New England colonists. The most active centers in this zone include Ducktown, Tennessee (still a source of copper), and Birmingham,

Alabama, where iron ore was mined at the Appalachian chain's southern end.

The Appalachians' narrow, linear band of metamorphic rock remains only moderately important as a metallic mineral source. Most small deposits were depleted as demand grew, and transportation eased shipment from more distant deposits that were more easily worked. But these eastern minerals were significant during the United States' early industrial growth.

A third and extensive metallic mineral region is formed by the western mountains. Gold and silver deposits scattered from south of the Mexican border to central Alaska, a few of them very rich, drew prospectors and mining companies to isolated locations, providing Hollywood writers the material for countless (even if mythical) adventures. Large deposits of copper, zinc, lead, and molybdenum are of great industrial importance. Uranium is also found in this western region. When this list of major deposits is expanded by adding minerals found in smaller deposits, such as tungsten, chromite, manganese, and many others, we can appreciate the great flexibility this diversity gave to growing industrial economies in the United States and Canada.

Even so, Canadian and U.S. industrial requirements were not met fully by the tremendous variety of minerals found in the continent's three major zones of metamorphic rock. A few minerals required by modern industry are not located in North America in sufficient quantities to satisfy domestic needs (e.g., tin, manganese, and high-grade bauxite for aluminum). In addition, as industrial capacity grew, it was matched by growth in demand for many minerals. Demand now often exceeds the diminishing domestic supply of these nonrenewable resources. Few other countries, however, ever equaled or even approached the original quantity and diversity of metallic minerals and mineral fuels located in the United States, especially when complemented by neighboring Canada's vast reserves. This mineral abundance and its distribution were critical in the development of an immense North American manufacturing-industrial complex (see Chapter 5).

ENVIRONMENTAL LIMITS CASE STUDY

Water Need: A Matter of Natural Supply and Artificial Demand

People need water and, in the United States and Canada, they expect it to be available where they live. The water must be clean and fresh—not just for drinking and washing, but for agriculture, construction, industry, recreation, electric-power generation, and many other activities. If people choose to live where there is not enough water, they demand that more be brought in. Or if the number of people living in an area grows, the demand for water grows.

Water shortages exist in the western third of the United States, where aridity and inadequate supplies have challenged human ingenuity for as long as people have lived there. Shortages also occur in the eastern halves of the United States and Canada, where enough rain falls each year to grow crops without irrigation. Although some people may blame these shortages on climate change, a more immediate explanation involves the huge amounts of water used by the average person in the United States and Canada.

The economically complex economies of Canada and the United States create very high levels of water consumption per capita, not because of personal consumption patterns alone, but because of numerous agricultural, industrial, and municipal uses. Many North Americans are accustomed to abundant natural water supplies, so they are profligate in their use even where natural supplies are limited.

Large and growing urban areas demand tremendous concentrations of water, and these demands strain supplies even in the east where moisture is naturally abundant. For example, the average single-family household in New York City uses 100,000 gallons of water per year. With a central city population of more than 8 million, the city must supply 1.09 billion gallons of water per day to its residents and businesses.

New York City's water, widely recognized as among the highest quality municipal water in the country, is drawn primarily from three large reservoir systems containing 18 individual reservoirs, the most distant being more than 100 miles north of the city in the Catskill Mountains (Figure 2.10). The city owns the land used to create the reservoirs as a result of a 1905 state law. But New York City's goal to provide enough water of good quality collides with the goals of the people living far from the city in the reservoirs' watersheds.

For decades, the Catskills were a thinly settled resort retreat for city dwellers. However, as more people moved into the area, and thus into the watersheds, questions arose on New York City's side about whether this development

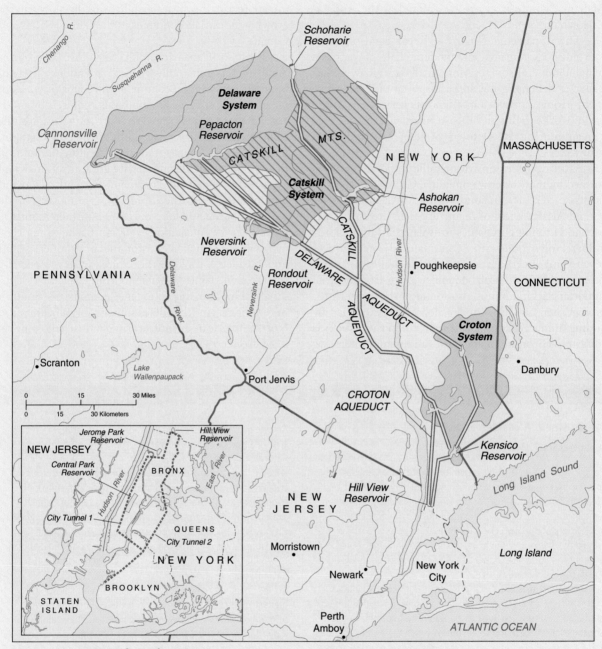

Figure 2.10 New York City's water system. New York City draws its water from distant sources using an interlocking network of reservoirs and aqueducts to supply the city's 8 million people. (Copyright © 1998 by *The New York Times.* Reprinted by permission.)

might compromise the quality of the city's water, and on the Catskills residents' side about whether New York City had the right to take, without compensation, what the residents viewed as local water resources.

More than simply a matter of water resource availability, this exchange raises serious and complex issues about equity, land pressure, political clout, and environmental limits. The case is one echoed in many forms across the United States and Canada.

Review Questions

1. What are the general physiographic regions of North America?

2. Identify and describe the four major climatic controls discussed in the text. What are their relationships to the modified Köppen climate classification system?

3. Where are the general locations of the major coal and metallic ore reserves? How might you explain this distribution?

4. How does the relationship between people and their natural surroundings vary from one region to the next?

5. What are the major water supply issues currently experienced by both the United States and Canada? Identify where the problems are most severe. Explain this pattern.

CHAPTER 3

FOUNDATIONS OF HUMAN ACTIVITY

Preview

1. Both the United States and Canada have been settled for thousands of years and still have sizable indigenous populations.

2. Immigration has been a constant part of the history of both countries, but the numbers and origins of the immigrants have varied widely in space and time.

3. For more than a century, European settlers hung close to the coasts, but expansionist policies, economics, and changing cultural attitudes prompted a steady stream of migrants to the interior.

4. Despite roughly equal land areas, the United States has 10 times Canada's population. The populations of both countries are not evenly distributed.

5. Both countries have highly urbanized and mobile societies marked by significant cultural diversity in terms of religious beliefs and language.

and constructed extensive irrigation systems to water corn, squash, and bean crops in the arid environment. Meanwhile, the Paiutes of the Great Basin pursued a seminomadic existence based on wild edible vegetation and small game. Their temporary homes were designed to be quickly assembled and disassembled. East of the Mississippi River, thousands upon thousands of farming people lived in farmsteads and palisaded hamlets located along streams, abandoned levee meanders, and fertile river valleys. In some places, like Moundsville, Alabama, between A.D. 1000 and 1500, people built bustling market and religious centers dominated by large earthworks crowned by temples, council houses, and homes for high-ranked families.

When Europeans reached North America, they brought with them diseases like smallpox and measles which were new to the people of North America. Although conflict and uprootedness took a stark toll on indigenous people, their bane was disease. With no accumulated immunities, populations in the East succumbed

Countless Native American tribes lived in Canada and the United States when Europeans began colonizing North America at the turn of the sixteenth century. Estimates of Native American Indian and Eskimo populations at that time range from 2 million to over 10 million, with perhaps four-fifths living in what came to be the United States.

Archeologists debate when and how the ancestors of these original inhabitants first peopled this new world. The conservative, increasingly challenged view puts the Native American ancestors' arrival at least 12,000 years ago when sea levels lowered enough to reveal the Bering Strait land bridge connecting Siberia and Alaska during the last Ice Age. But compelling evidence based on finds like the 9000-year-old Kennewick Man suggests the continent was populated well before this time. However archeologists ultimately piece together the puzzle of the peopling of the continent, what is clear is the earliest immigrants spread to all parts of North and South America to create rich and diverse cultures. Along the California coast alone, people spoke several hundred dialects by the time Europeans arrived. The diversity among the continent's indigenous peoples was more than linguistic. Different tribes categorized as Pueblo, for example, lived in what is now New Mexico and probably traded with the Aztecs far to the south in central Mexico. Pueblo peoples resided in multilevel adobe apartments

The human geography of North America shows the imprint of every culture that has come to the continent, from the earliest peoples to the most recent immigrants. Here, a Native American boy from the Cheyenne tribe dons traditional regalia for a community gathering.

in devastating numbers after contact. Many small tribes were wiped out. North America's native population declined to about 1 million by the start of the nineteenth century as large-scale European settlement moved west past the Appalachians.

American Indians sometimes slowed European settlement expansion, but for the most part their impact on the spread of European settlers was minimal. In some places, Indians contributed to the arriving Europeans' settlement transition, especially during the first decades. As the European settlement frontier moved west, so did many of the remaining Indians.

Canada's record of relations with the Indians was and is still better than that of the United States. For example, the Canadian government officially expressed profound regret in early 1998 for its past treatment of the country's indigenous peoples. In 1999, Canada created the northern province of Nunavut to settle decades of land claims issues with native Inuit in the region. But neither country deserves any awards for their treatment of their previous inhabitants. About two-thirds of Canada's American Indians reside on about 2200 small reservations scattered throughout the country. In the United States, about one-third of American Indians live on 281 reservations. The largest of these are located in the dry western portions of the Great Plains and the Empty Interior. In total, the United States Government recognizes 562 American Indian tribes.

Most nonreservation American Indians live in urban places, with the largest concentration in the New York and Los Angeles metropolitan areas. Many live as our next-door neighbors, integrated into our communities, yet keeping and reviving their cultural traditions.

EUROPEAN SETTLEMENT

Arguably the largest long-distance migration in human history brought different settlers to occupy the lands that the American Indians reached and inhabited thousands of years earlier. Although it is impossible to say how many people from Europe and, to a much lesser extent, from Africa entered what is now the United States and Canada, a reasonable estimate places the figure at close to 60 million.

The overall pattern of European immigration to North America is complex, but a few generalizations can be made. Most early immigrants came from northwestern Europe, primarily the British Isles (Table 3.1). When the first United States national census was taken in 1790, more than two-thirds of the country's white population was ethnically British. Germans and Dutch were next in numbers. Nearly all French migration to Canada occurred in the seventeenth century. The number of French making the trip was small, totaling fewer than 15,000. Even so, the number of French-heritage Canadians increased hundreds of times over in the ensuing 300 years.

Immigration to North America slowed significantly between 1760 and 1815 when intermittent warfare in Europe and North America reduced the flow.

Pre-European North America was home to thousands of tribes of American Indians. Groups in different parts of the continent developed distinctive lifeways that were expressed in such features as housing, apparel, food, and religion.

TABLE 3.1

	United States		Canada	
Decade	Number in Thousands	Principal Sources	Number in Thousands	Principal Sources
1820s	129	Ireland, Britain		
1830s	538	Ireland, Germany		
1840s	1427	Ireland, Germany		
1850s	2815	Ireland, Germany	253	
1860s	2081	Germany, Britain	156	
1870s	2742	Germany, Britain	329	
1880s	5249	Germany, Britain, Scandinavia	850	
1890s	3694	Eastern Europe, Italy, Germany	373	
1900s	8602	Eastern Europe, Italy, Russia	1401	Britain, United States
1910s	6347	Eastern Europe, Italy, Russia	1859	Britain, United States, Russia
1920s	4296	Canada, Latin America, Italy	1273	Britain, United States, Ireland
1930s	699	Canada, Germany, Italy	251	United States, Britain
1940s	857	Latin America, Canada, Britain	429	Britain, Eastern Europe
1950s	2300	Germany, Canada, Latin America	1540	Britain, Italy, United States
1960s	3212	Latin America, Canada, Asia	1375	Britain, Italy, United States
1970s	4493	Latin America, Asia	1589	Britain, United States, Hong Kong
1980s	8555	Latin America, Asia	1092	Asia, Europe
1990s	9095	Latin America, Asia	1830	Asia, Europe

Source: M. C. Urquart, ed., *Historical Statistics of Canada* (Toronto: Macmillan Co. of Canada, 1965); Statistics Canada, *Canada Year Book*, 1965, 1981, 1990; W. Zelinsky, *Cultural Geography of the United States* (Englewood Cliffs, N.J.: Prentice-Hall, 1976); *Statistical Abstract of the United States* (Washington, D.C.: U.S. Government Printing Office, 1970–1998).

Each decade of the century between 1815 and the start of World War I in 1914 witnessed a growing volume of migration to the United States and Canada, although there were times when war or widespread economic depression reduced movement. Between 1815 and 1820, annual migration totaled perhaps 10,000. By the last few years before World War I began, nearly 1 million immigrants entered the United States each year and another 400,000 streamed into Canada.

In the decade following the end of World War I, the United States passed its first major legislation to restrict immigration. Canada did the same in the 1930s. With enforced limitations on volume, coupled with the Depression in the 1930s and then World War II, immigration was cut drastically. In 1945, for example, only 45,603 people arrived in the United States; just 22,722 immigrants arrived in Canada. After 1945,

arrivals increased somewhat. Between 1991 and 2000 the United States averaged more than 900,000 new legal immigrants annually, and Canada, about 180,000 per year. Officials don't know how many illegal aliens enter the United States each year, and estimates vary greatly. Still unclear are the long-term effects on immigration of new controls enacted in the United States after the September 11, 2001, attacks, but in general, the number of persons obtaining legal permanent resident status in the United States was higher after 2001 than in most years of the late 1990s. In 2006, nearly 1.3 million people became permanent, legal residents of the United States.

The national origins of immigrants to the United States changed considerably over time. Each decade the major source areas of European immigrants gradually shifted south and east. For the first half of the 1815–1914 period, most migrants continued to come from northwestern Europe. The percentage from Britain declined, but more arrived from Ireland and Germany. By later in the nineteenth century, these regions became less important sources. In the 1880s, there was a major Scandinavian outflow, followed in subsequent decades by streams of people from southern and eastern Europe. By 1913, more than 80 percent of all immigrants were from southern and eastern Europe, especially Italy, Austria-Hungary, and Russia.

The southeast shift in the source of immigrants is based on the diffusion of the Industrial Revolution within Europe. Beginning in the eighteenth century in the British Isles and the adjacent Low Countries across the English Channel, the Industrial Revolution spread southeast across Europe during the next 150 years. Mortality decline and rapid population growth accompanied industrialization. Economies shifted from agriculture to manufacturing, and urbanization increased as the agricultural population decreased. The growth in demand for urban labor, as fast as it was, did not match the growth in the number of potential workers. As a result, there were many willing emigrants.

Within North America, these immigrants' destinations were largely the result of existing economic opportunity rather than a desire to duplicate their former environments. It might be easy to believe that migrants chose landscapes similar to those of their European homes. The substantial Scandinavian settlement in Minnesota and the Dakotas would be a case in point. Whatever small truth exists in this view, far more important was the fact that those states happened to represent the principal settlement frontier just when major Scandinavian immigration occurred. Did Germans move to the Texas Hill Country because it reminded them of home? It is unlikely. Did southern Italian immigrants find Boston or Newark much like their Mediterranean farms and villages? Certainly not. For the most part, North America's ethnic mosaic resulted from a movement toward opportunity—opportunity found first on the agricultural frontier and then in the cities.

Some immigrant groups were exceptions to the rule that settlement occurred where economic opportunity was greatest when the group arrived. French Canadians moving into the northeastern United States, Mexicans into the Southwest, and, more recently, Cubans into the Miami area are all examples of these exceptions. These groups all settled in areas close to their point of entry.

The immigrants who have come to North America during the past 150 years have added to the diversity of the culture of the two countries while at the same time been integrated into that which was already there. This family is arriving in New York City in 1905.

Not all North Americans enjoyed a choice of where they could live. The most obvious examples are American Indians who were restricted to reservations and African Americans who were enslaved (see Chapters 9 and 14). In a different manner, much early Irish immigration in the late 1840s and 1850s was also impelled. In a panic move by a destitute population, Irish immigrants fled starvation when the country's principal food crop, the potato, failed massively. The Irish lacked the funds to move beyond East Coast cities and became the first of many impoverished immigrant ethnic populations settling in America's rapidly growing cities.

NON-EUROPEAN SETTLEMENT

Until recently, most U.S. and Canadian immigrants came from Europe. Black settlement in the American South, resulting from slave labor for the region's plantations, was the major exception. Blacks were brought to the American South as part of a larger shackled transfer of Africans whose destinations included the Caribbean Basin and the northeast coast of South America.

Next to the European exodus, the African diaspora was probably the second largest long-distance movement in recorded human history. Perhaps 20 million left Africa during the slave trade, with approximately half perishing before their arrival in the Western Hemisphere. Fewer than 500,000 blacks may have entered the United States, although accurate statistics are not available. Most probably arrived from the Caribbean islands rather than directly from Africa, and the plantation South was their primary destination.

When the slave trade ended and European immigration increased, the balance of the U.S. population changed. One in every five Americans was of African origin when the country's first census was taken in 1790. There was little African immigration after that date, and the percentage of the United States' population that was black declined steadily until well into the twentieth century.

Immigration was not fully open even in the late nineteenth century. Chinese immigration to the United States was excluded entirely after 1881, and the Chinese were also banned by Canada in 1923. In a parallel vein, the modest number of Japanese who immigrated were not allowed until after World War II to apply for citizenship. More broadly restrictive laws passed in the 1920s severely limited where immigrants could originate. Except for a large quota allotted other Western Hemisphere countries, most migrants had to come from Europe.

Both countries passed far more liberal immigration laws in the 1960s. With much less European interest in moving to North America, most immigrants now come from non-European sources. Mexico, the Philippines, Vietnam, and the Dominican Republic provided the greatest number of U.S. immigrants in the mid-1990s, followed by China and India. In 2006, three-quarters of all persons obtaining legal permanent resident status were from Asia or the Americas. In Canada, most immigrants are now from Asia, and the non-European share is increasing.

Most recent non-European migrants are settling near their points of entry or in larger cities. For example, the influx of people and capital associated with Hong Kong's transfer from Great Britain to China in 1997 augmented already significant ethnic Chinese populations in Vancouver, Seattle, and Portland. Today very large numbers of people from the Asian rim countries live in Los Angeles and the San Francisco Bay area, more than 2 million in 2006, with another 4.7 million Hispanics in Los Angeles county alone. And Miami is the home for migrants from throughout the Caribbean Basin and Latin America, not just Cuba. The larger cities of both the United States and Canada now contain many new ethnic neighborhoods representative of these recent migrants' rich cultural diversity.

SETTLEMENT EXPANSION

Except for Spain's modest northward expansion along the Rio Grande, Europe's settlement of North America started along the continent's east coast and moved west (Figures 3.1 and 3.2). This is the single most important element in the pattern of European occupation in the two countries. Although interesting variations in the trend exist—places settled earlier or later than might be expected—in general, a place's settlement date can be predicted by knowing its distance from the East Coast and the country's pace of settlement.

North America's first European communities were small. They clung to the coast, looking more toward Europe than toward the land crowding about them. The first tentative settlement push away from the ocean followed the waterways. Rivers and bays offered trade pathways to the ocean and an important link to Europe. Thus, the British settled the indented coastline of the Chesapeake Bay and its tributaries, or they spread a thin band of communities along New England's rugged coastline. The Dutch, and later the English, moved up the Hudson River from New Amsterdam (New York). The French gradually settled the banks of the upper St. Lawrence River. To break away from the water was a break from Europe, and so it was a step taken hesitantly.

Settlement expansion did not pick up speed for a long time. During the first 150 years after the beginnings of permanent European settlement in what came to be the United States, for example, Europeans had moved west only as far as the Appalachians' eastern flanks. But within the next century, the frontier reached the Pacific Ocean. By 1890 the U.S. Bureau of the Census announced that the American settlement frontier was closed entirely; according to the bureaucratic definition, there was no more frontier.

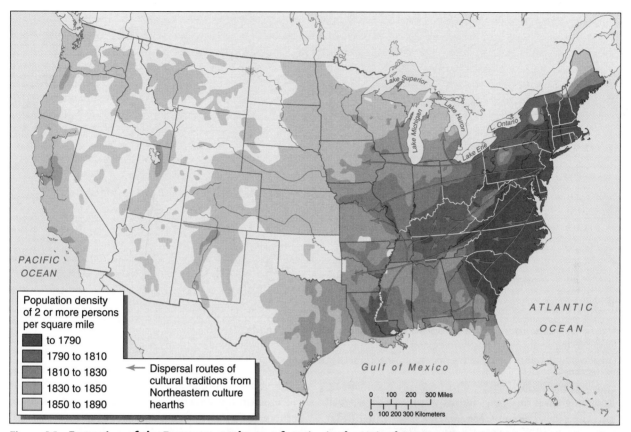

Figure 3.1 Expansion of the European settlement frontier in the United States. For the first three centuries after Europeans began to settle this land, the United States was in a period of active frontier expansion.

The United States' rapid settlement expansion occurred when the country turned away from Europe and toward North America. By the early nineteenth century, Americans viewed occupation of the continent as their manifest destiny. The country's land laws became increasingly pro-expansionist, culminating in the 1862 Homestead Act. Consequently, it became much easier to obtain land along the settlement frontier. In addition, as the population grew from both natural increases and immigration, moving west appealed to many hoping to improve their lives.

Expansionist policies in both Canada and the United States led to an almost constant state of excess labor demand and surplus production. More people were needed to take advantage of available opportunities, and both countries produced more raw materials than their economies could consume. Both the United States and Canada functioned as Europe's economic colonies even after political independence, providing raw materials and buying finished products.

Settlement expanded west across the forested eastern half of the United States in a generally regular fashion with only a few exceptions. Advances were more rapid along certain transportation routes, such as the Ohio River. Early settlers also created a hollow frontier from time to time by passing up some areas in favor of more promising lands farther west. European settlement remained absent in sections of New York's Adirondack Mountains, for example, until the principal settlement line advanced beyond the Mississippi River. A similarly regular settlement pattern developed in southern Canada along the St. Lawrence River and the lower Great Lakes.

Settlement moved onto the interior grasslands more rapidly in the United States than in Canada. The Mississippi River's many tributaries offer easy routes to the United States' interior, whereas the Great Lakes functioned more as a barrier to Canadians' primarily overland movement west. Furthermore, and more important, American settlers found a vast expanse of

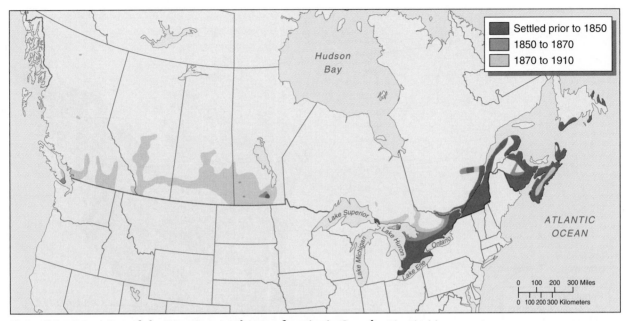

Figure 3.2 Expansion of the European settlement frontier in Canada. The Shield was a major barrier to the north and to westward expansion in Canada. The St. Lawrence lowland and southern Ontario thus developed early as the dominant Canadian settlement focus.

excellent agricultural land with a generally good climate stretching from the Appalachians' western margins well into the Great Plains. Canadian settlers, on the other hand, faced a stretch of land nearly worthless for agriculture that was 2500 kilometers (1500 miles) wide between about 150 kilometers (95 miles) north of Lake Ontario to just east of the site of Winnipeg, Manitoba. This portion of the glacially scoured Canadian Shield has thin, rocky soils and a short growing season. Although most of the United States' Great Plains was settled by non-Indians between 1860 and 1890, the Canadian prairies were not open to settlement until near the end of the nineteenth century. Even then, many early settlers were Americans moving north of the border.

Elsewhere, the regular settlement sequence did not occur. Most of the United States and Canada west of the Rocky Mountains, the Canadian north, and Alaska have low agricultural potential. Much of this huge area was too dry, too hot, or too cold for farming. Rugged topography hampered transportation and further limited agricultural development. Settlement congregated in areas offering identifiable economic potential. For example, small areas with good soil and adequate precipitation, such as in the Willamette valley of Oregon, allowed agriculture. Mineral resources fostered north-central Ontario's mining centers. Other sites developed to service transportation lines ex-

tending across broad empty areas. Whatever the specific reason for growth, the far western and northern settlement pattern was that of a set of points scattered across an otherwise unpopulated landscape.

POPULATION DISTRIBUTION TODAY

Canada, at 10 million square kilometers (3.9 million square miles), and the United States, at 9.5 million square kilometers (3.7 million square miles), have about the same land area but very different population sizes. The 2001 Canadian population census revealed a population of more than 30 million and Canadian officials now estimate a population of over 32 million. By 2000, the United States' population had exceeded 281 million, and in late 2006 the Census Bureau estimated that the country passed 300 million. Canada had a 2001 population density of about 3 people per square kilometer (8.1 people per square mile), whereas the United States had a 2000 density of roughly 29.6 people per square kilometer (75.9 people per square mile).

Neither the U.S. population nor the Canadian population is spread evenly across the respective country's land area (Figure 3.3). Three principal population

Both countries adopted grid systems of land subdivision to facilitate the survey and settlement of their large interior lowlands. The result is the regular, rectangular pattern of fields and roads seen here in Canyon County, Idaho.

zones can be identified. First is the quadrant outlined approximately by the cities of Montreal, Chicago, St. Louis, and Washington, D.C.; this is North America's traditional population core. Most larger cities in the United States and Canada are located in this quadrant: 60 percent of the Canadian population lives in southern Quebec and Ontario; 6 of the 9 most populous states in the United States are here. The earliest population growth in both countries occurred here, and the quadrant has long been the countries' most advanced section economically. Fine natural waterways and many excellent harbors along the Atlantic shore were augmented by one of the world's densest transportation nets. Excellent agricultural lands as well as rich mineral resource deposits are either within the quadrant or around its periphery. Although some other parts of the continent are now growing more rapidly, this region remains North America's vital core.

A secondary population zone lies around the core's southern and western margins and extends west to the Great Plains' eastern sections in the United States. Much of the United States' best agricultural land is in this zone. Population densities in this second zone are generally much lower than those in the core. Cities are more widely and more evenly spaced, and they function primarily as regional economic service and manufacturing centers. As a group, these cities are less significant nationally than the core's cities.

The third population zone is peripheral and extends west from the central Great Plains. In this zone, population and economic growth is limited to sites with special potential. This pattern developed early in the region's settlement history. Although some areas—notably California's San Francisco Bay Area and the Los Angeles Basin and the Fraser Delta-Puget Sound Lowland in British Columbia and Washington State—are now densely populated, most of the zone's population remains sparsely distributed.

The most striking aspect of Canada's population distribution is that more than 90 percent of all Canadians live within 150 kilometers (95 miles) of the country's southern border. To a large extent, this linear East-West distribution is a result of Canada's northern location and the environmental constraints this has on agriculture. In addition, however, Canadian policy has also encouraged settlement from ocean to ocean as a form of insurance against United States expansion northward. As a consequence of the linear population distribution, Canadians have often had closer economic and cultural ties with the areas in adjacent parts of the United States than with other parts of Canada. Rising Canadian nationalism in recent decades, as well as Canadian governmental efforts to improve interregional communications within the country, have lessened this exterior orientation somewhat.

A second consequence of Canada's rather linear population distribution is the pattern's fragmented nature. Southern Ontario and southern Quebec are part of North America's core and have the population density to show it. Separated from southern Ontario by the almost unpopulated Canadian Shield, the Prairie Provinces' farmland settlements might be viewed as part of the secondary population zone in the United States. The Prairie Provinces' population is separated

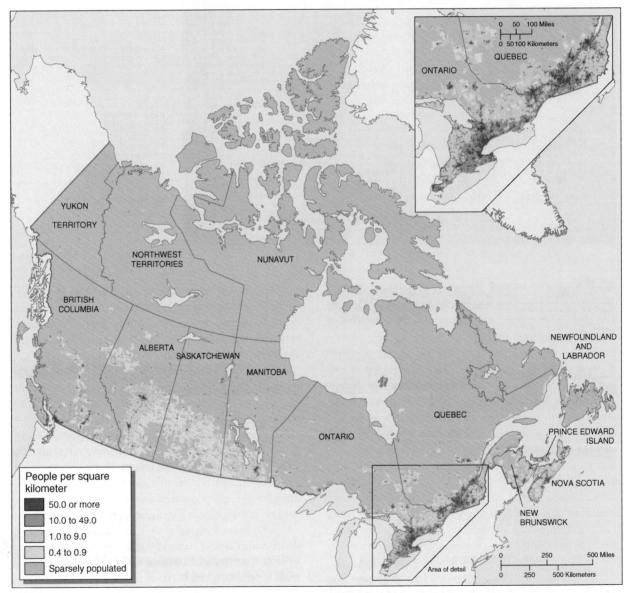

Figure 3.3(a)   Canada population density. Most geographers would argue that we have the technological ability to work within any physical environment, and humans can live anywhere on the earth's surface. Nevertheless, notice how areas of generally lower population density are associated with those of lower precipitation (see Figure 2.3) and colder temperatures (see Figure 2.4).

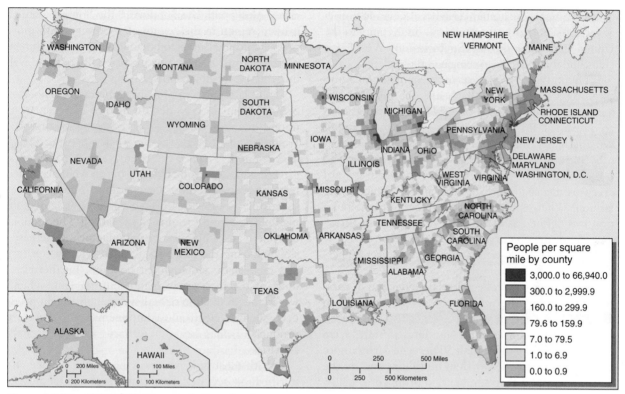

Figure 3.3 (b)   United States population density.

from British Columbia's coastal population by the Coast Ranges and the Rocky Mountains. This far western region, then, is similar to the United States' western peripheral population zone. East of Quebec, the Atlantic Provinces' population is scattered widely along an indented coastline and in a few interior lowlands. The Atlantic Provinces are connected to the rest of the country by a thin, tenuous line of settlement along the southern shore of the St. Lawrence River and overland into New Brunswick. The remaining 5 percent of Canada's population is sprinkled across the country's vast northern regions.

POPULATION REDISTRIBUTION— MOBILITY PATTERNS

North Americans move more frequently than people in any other country. For example, almost 20 percent of the United States' population moves to a new residence in an average year. Because of substantial population redistribution and continued immigration, a high potential for rapid regional change exists.

Americans and Canadians long viewed population growth as a good indicator of economic well being. As long as vast areas of North America were seen as both environmentally hospitable and relatively empty, the view was not seriously challenged. If a region's population is growing rapidly, the region's economy and stature are regarded as sound; if the population is declining, grave concern may be expressed about the region's future health. This view echoes the notion that migration is directed toward places of opportunity.

Under these assumptions, the countries' westward population shift indicates economic growth in their western regions. In 1790, the U.S. population's center—defined as the intersection of a north-south and an east-west line, each dividing the country's population equally—was located southeast of Baltimore. The center moved west each ensuing decade, crossing the Mississippi River for the first time by 1980, and was situated in Phelps County, Missouri, in 2000. The United States' population center was pulled westward until the mid-1960's with people moving within the country. Since then, immigration has also had an impact as most Asian and Latin American immigrants to the United States enter through Southwest and West Coast cities, and many decide to remain in those places.

The mobility history of both countries falls into three distinct periods. First came the east to west movement; then a period of rural to urban migration accompanied, in the United States, by a move out of the South; and finally, the present period, when most movement is long distance and between metropolitan areas. Just as the U.S. population moved westward every decade, so it is urbanized in an equally unvarying fashion. Whereas only about 5 percent of the population was defined as urban in 1790, more than three-quarters of the country's population was urbanized by 2000. Canada's 2006 census also classified over 80 percent of all Canadians as urban.

These statistics reflect a relative decline in rural population, but the number of people engaged in farming was decreasing at the same time. Between 1960 and 1987, for example, the United States farm population fell by more than 50 percent—from more than 15 million to fewer than 6 million, and by the end of the 1990s that number had dropped to below 3 million. Although not experiencing as dramatic a decline, Canada's farm population also decreased substantially.

The population shifts from east to west and from rural to urban locations were in response to perceptions of economic opportunity. First, more and more farmland became available as the settlement frontier of the two countries pushed west. Then a tremendous surge in urban employment was generated by the Industrial Revolution's technological advances. To reemphasize a point, both trends developed together throughout the nineteenth century and extended into the twentieth century in Canada. Westward movement was relatively more important at first, whereas movement to the cities gained ascendancy by the end of the period. Once North Americans were predominantly urbanites and economic opportunities were also urban based, most subsequent population migration occurred between metropolitan areas.

Population statistics suggest that a fourth major mobility period is at hand in the United States. Areas long experiencing no change or even declining in population are now growing. Much of the American South is a prime example. For more than a century, the South experienced considerable outmigration as Southerners were drawn to opportunities in northern and western cities. Along with the Southwest, the South became a growth region in the last quarter-century, with more people entering the region than leaving it. The Bureau of the Census predicts this pattern will continue well into the twenty-first century.

This latest pattern of internal migration is not surprising when we consider the nature of recent economic changes. Many observers suggest that the United States is becoming the world's first postindustrial country. That is, the country's major economic growth is not in manufacturing but in what are called tertiary and quaternary occupations. These are, respectively, occupations providing services and those manipulating and creating information. The number of Americans employed in manufacturing increased slowly during the past two decades, whereas tertiary and quaternary employment boomed. Furthermore, most of the increase in manufacturing employment was in the production of high-value, lightweight products, such as electronic components, and these are products demanded by tertiary and quaternary activities.

With most new jobs available in the service and information sectors, many employers and employees are free to choose work sites for reasons unrelated to heavy manufacturing efficiencies. In the past, the

As North America's urban residents move, they seek homes in newly developed suburbs. In many instances, houses are produced as quickly as possible for waiting customers.

continent's manufacturing core region offered strong economic advantages for new job creation and labor was attracted to the region. Now, high-value light industry, as well as tertiary and quaternary activities, can choose to locate elsewhere, benefiting from nearly ubiquitous high-speed, inexpensive communications and transportation. In addition, noneconomic reasons are more important when people consider where they want to live. The Sunbelt states of the South and West, with mild winters and long periods available for outdoor recreation, provide residential environments many Americans desire.

The socioeconomic nature of the migrants presently leaving and entering the South seems to support this interpretation. The average migrant to the South, whether black or white, is better educated and commands better pay than the average outmigrant. The new arrivals are therefore responding to growth in the number of high-quality Southern jobs relative to opportunities in the North's industrial centers.

Firms that ship packages for overnight delivery use spatial collection and distribution systems that funnel packages to one, or a few, central locations for sorting. The distribution hub, which serves as the company's regional focus, depends on the latest computer scanning equipment to track the thousands of packages that flood the hub each night.

URBANIZATION

The U.S. and Canadian populations are largely urban. By residing in urban locations, North Americans live in environments with a very different character from what was experienced a few short generations ago. Although many national ideals remain rooted in small-town and agrarian culture, these ideals were not the experience of most Americans and Canadians after World War I.

The differences between rural and urban environments are more than technological. Rural life is demanding, but the demands have a different rhythm than is found in urban places. Travel to see others, to shop, and sometimes even to go to the mailbox may require more time and effort than in cities. Opportunities in urban places are more concentrated, whether for face-to-face interaction, for specialty goods and services, or for a new job. In cities, problems arising from too many people in an area are the norm, not problems related to low population density. Rural economic activities also tend to be extractive. By contrast, most urban economic activities involve manipulating materials or ideas, or serving others.

Highly interdependent urban environments are spatially organized very differently than rural areas. Spatial organization is the way activities at separate locations are linked by their interactions. Urban areas possess much more intense and complex spatial organization than rural areas. Extensive rural areas are devoted to one or two extractive activities, such as agriculture or timbering, and the resulting spatial organization focuses on a limited number of marketing and supply centers. In large cities, the variety of interconnected economic and cultural activities is tremendous and ever-changing. As a result, urban spatial organization is very difficult to comprehend and must be lived in selectively by residents.

Most North American urban areas grew significantly during the twentieth century. In a few instances, growth was so great that several major urban areas merged to form city clusters. Geographers and others began to see these clusters as something quite different from anything previously known in urbanization's history. The extensive group of large cities between Boston and Washington, D.C., is the prime example, but others exist as well (see Figure 1.3). Another large urbanized cluster lies around the southern Great Lakes margin. Milwaukee and Chicago are this region's western anchor, and Buffalo and Pittsburgh are its eastern counterpart. Some observers offer part of California, from San Diego to San Francisco, as another likely set of merged urban areas, but California's resource limitations may restrict how fully the northern and southern clusters merge (see Chapter 15). East coastal and central Florida became largely urbanized by the year

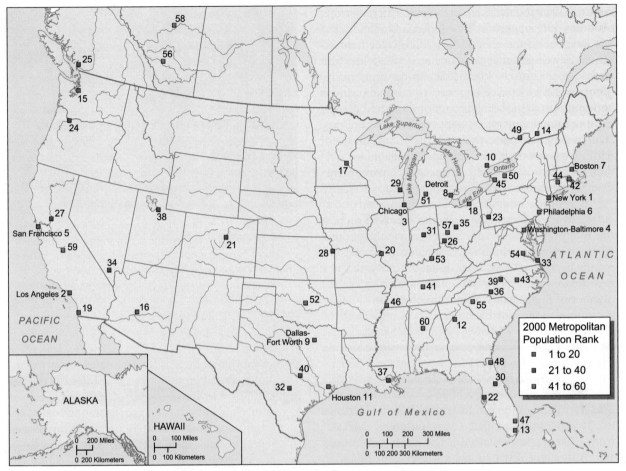

Figure 3.4 **The largest urban areas in Canada and the United States.** Most large urban areas in the two countries are located at highly accessible ocean or inland waterway sites.

2000, and numerous smaller urban clusters repeat the pattern across the country.

Each major urban cluster and each individual town and city as well exists where it does for identifiable reasons. Why are cities located where they are? As a general answer, most large urban places develop where major transportation routes connect with each other. Land and water connections are frequently important. Other factors may also matter. Hinterland quality, proximity to alternative transportation, security concerns, and even the local environment's healthfulness may be important. However, where goods and people transfer from one transportation mode to another (called *break-in-bulk points*), a great many opportunities exist for processing, exchanging, manufacturing, repackaging, selling, and buying goods. In other words, cities usually are founded where there are economic advantages associated with transportation.

Most large U.S. and Canadian urban centers are located adjacent to navigable water (Figure 3.4). Some are on a seacoast or large estuary: Boston, New York, Baltimore, Quebec, Miami, Tampa-St. Petersburg, Mobile, San Diego, Los Angeles, San Francisco, Seattle, Honolulu, and Vancouver, British Columbia, are examples. Others are on naturally navigable waterways, such as Chicago, Milwaukee, Minneapolis-St. Paul, Detroit, Cleveland, Buffalo, Hamilton (Ontario), Toronto, Montreal, Philadelphia, New Orleans, Memphis, St. Louis, Kansas City, Cincinnati, Pittsburgh, Louisville, and Portland, Oregon. Still others are on rivers or channels modified extensively to give the cities water access: Houston and Knoxville are examples. Exceptions to water orientation exist, such as Atlanta, Calgary, Denver, and Dallas-Ft. Worth, but these were on early land transport routes. Atlanta, for example, is located at the southern tip of the

Appalachian Mountains. The city became a key inland railroad transportation center in the South prior to the Civil War.

The reasons why any city grew where it did are often divided into aspects of its site and its situation. Most large urban areas in the two countries are located at highly accessible ocean or inland waterway sites. A city's *site* relates to its immediate physical environment—that is, the landscape characteristics where the city is located. Pertinent information includes the topography across which the city spreads, soil characteristics, drainage, and the depth of the bedrock. Is the terrain flat or rolling? Is the city on a plateau or in a valley? Are large areas of poorly compacted river soils present? Are there good water-bearing sands? Are swampy areas present? Does the place have a river for water supply or water transportation? Answers to these and similar site questions often account for different urban growth rates between competing cities.

A city's *situation* relates to the city's position relative to other places. Situation is sometimes called relative location. Situation factors include proximity to other centers (short distances might make trade easier); location between two areas that produce different products (this may allow the city to serve as intermediary in the two areas' exchange with each other); and the productivity of the land around the city. In geography, situation describes locational features that link one place to other places, thereby forming the spatial relations of this set of places. Situation is defined to a degree by a city's accessibility to regions that contribute to the city's growth. Thus, situational factors are related to the urban growth process, whereas site factors are related to a settlement's initial establishment as well as its subsequent growth.

PATTERNS OF REGIONAL CULTURE

The territory occupied by Canada and the United States is neatly subdivided into political units—2 countries, 50 states, 10 provinces, and 3 territories, with more than 3000 counties in the United States and hundreds in Canada. These units structure our political activity and provide its boundaries at each scale. Political units, however, are unreliable indicators of a culture's spatial expression. Some political subdivisions do

possess substantial meaning as culture regions (Quebec may be the best example), but most do not.

Culture is a concept whose meaning is debated. At a general level, culture is a people's assemblage of beliefs and learned behavior. A common culture bonds people together, whether an ethnic or a social group, or residents of a region or country. Culture also provides a group member's identity and heritage, and in the process lays the basis for the group's meaning and purpose. Culture is transmitted symbolically through language, not through biological means. Thus, a common language is critical in maintaining a culture or culture region. The United States is the world's largest and most populous country joined geographically and socially by a common language. Some have argued that this is one of the country's greatest strengths.

American English may be the United States' language, but there are significant regional differences in the use of this language. Do you live in Providence, Rhode Island? Then you probably call a big bun sandwich a grinder or perhaps a torpedo. New Orleans? To you, the same sandwich is a Poor Boy. Pittsburgh? You probably call it a hoagie. Southern Florida? Cuban Sandwich may sound more familiar. Many in the remaining U.S. regions refer to that same sandwich as a submarine. This dining delight's particular name is not especially important, but what we call it locally is an element of our local culture (Figure 3.5). The point is simply that regional differences in American culture exist. Indeed, most of the regions described in this text are at least partly culture regions.

Regional cultural variation is expressed in many different ways. Indiana, Kentucky, Ohio, and Illinois together produce far more big-time college basketball players per capita than the national average. The vast majority of early country music singers were from the upper South, especially the southern Appalachian Mountains and the Nashville and Blue Grass basins of Tennessee and Kentucky. While the South is now a true two-party region, for decades many white Southerners thought of themselves as "yellow dog Democrats"—given the choice, they would rather vote for a yellow dog than for a Republican. The array of regional differences is immense and gives the United States considerable regional distinctiveness.

An area's landscape is a blend of the natural environment and the local cultural imprint. A rectangular land survey system widely used in the United States during the nineteenth century created striking regularity in much of the Middle West's landscape. German and

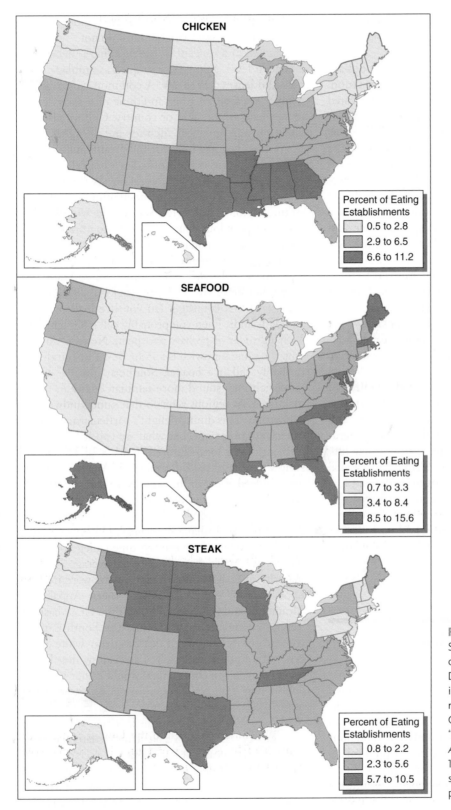

CHICKEN

Percent of Eating Establishments
- 0.5 to 2.8
- 2.9 to 6.5
- 6.6 to 11.2

SEAFOOD

Percent of Eating Establishments
- 0.7 to 3.3
- 3.4 to 8.4
- 8.5 to 15.6

STEAK

Percent of Eating Establishments
- 0.8 to 2.2
- 2.3 to 5.6
- 5.7 to 10.5

Figure 3.5 Eating preferences. Specialty restaurants are not evenly distributed across the United States. Different eating preferences are one indication of the country's regional cultural variations. (Barbara G. Shortridge and James R. Shortridge, "Specialty Restaurants," *The Taste of American Place* © 1998; data from 1992 Census of Retail Trade, subject series, Table 19. Reprinted by permission.)

The boys pictured are Cree Indians, but much of what is going on in this photo is recognizable across cultures within Canada and the United States.

English farmers in southeastern Pennsylvania built large, even massive, cattle and hay storage barns with a second-story extension on one side over the first story. Most students of folk architecture agree that this "Pennsylvania barn" is a key identifying landscape element in the Pennsylvania culture area. Urban ethnic areas may be identified more easily by the names on small neighborhood stores and restaurants.

Much of culture is conservative and seeks consistency with its past, but change itself is part of American culture. Many cultural shifts result from changes in technology or economic conditions. Migration is another key ingredient. Millions of Latin American and Asian immigrants incorporated into both countries during the last two decades are stimulating new cultural patterns. Southern Florida's recent Hispanic influx dramatically altered the fabric of life and thinking in the southern part of the state. Vancouver's large Chinese population added a new dimension to the city's culture. Nearly every large city and many smaller ones are sharing in these changes.

AMERICAN RELIGIONS

Religious choice is deeply personal and individual. Even so, a distinctive geographic pattern of religious denominations exists in North America (Figure 3.6). The pattern suggests that one's choice of denomination is most often based on the religion and cultural heritage of one's parents.

The larger Christian churches were brought to North America by European migrants from countries in which one or two religions predominated. The resulting denominational distribution today closely matches migrant settlement source areas. For example, German and Scandinavian settlers carried the Lutheran Church to the northern Great Plains and northwest portion of the agricultural core region. The wide distribution of Roman Catholicism in the United States was accomplished by Hispanics who brought it into the Southwest, southern and eastern Europeans who introduced it in the Northeast, Middle West, and most large cities outside the South, and the French Acadians (Cajuns) who moved to south Louisiana from eastern Canada.

The United States also is a place that has been hospitable to new denominations. One way of explaining the appearance of new religions is the constitutional separation of church and state (new religious practices are protected) and the population's relative wealth (operating an independent denomination can be expensive). Another element could be a desire for national or regional identity. Several denominations were created at the end of the American Revolution in reaction against Great Britain's continued dominance. The Episcopal Church, formerly part of the English Anglican Church, is an example. Similarly, the American Civil War led to a north-south split in Presbyterianism when each subdenomination felt that the other did not adequately reflect its particular interests.

The freedom to develop a new religion is accompanied by considerable individual and congregational creativity. Individuals established their own churches, or congregations or groups of congregations left a denomination to form a new one because of disagreements over biblical interpretation or church administration. Many such nascent American churches follow a conservative theology and were founded in the South. Although the number of practitioners may be large, the churches are nowhere numerically dominant and hence are not shown in Figure 3.6.

One church founded in the United States is the Church of Jesus Christ of Latter-Day Saints, commonly known as the Mormon Church. Founded in upstate New York, in the mid-nineteenth century, it was carried west by its followers searching for an isolated

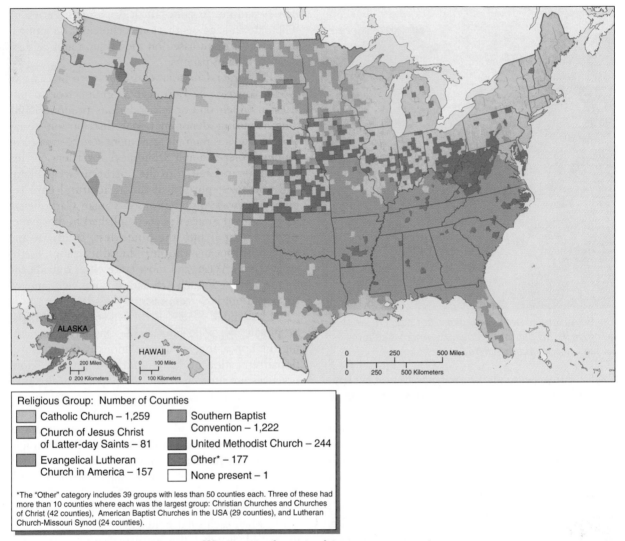

Religious Group: Number of Counties

Catholic Church – 1,259

Church of Jesus Christ of Latter-day Saints – 81

Evangelical Lutheran Church in America – 157

Southern Baptist Convention – 1,222

United Methodist Church – 244

Other* – 177

None present – 1

*The "Other" category includes 39 groups with less than 50 counties each. Three of these had more than 10 counties where each was the largest group: Christian Churches and Churches of Christ (42 counties), American Baptist Churches in the USA (29 counties), and Lutheran Church-Missouri Synod (24 counties).

Figure 3.6 **Predominant religious affiliations in the United States.** International migration into the United States (Catholics and Lutherans), internal migration (Mormons), and the development of regional identity (Baptists) are among the influences on the religious geography of the country. (Adapted from D. E. Jones, et al. Religious Congregations and Membership in the United States, 2000. Copyright Association of Statisticians of American Religious Bodies, 2002; published by Glenmary Research Center, Washington, D.C.)

place to settle and follow their beliefs (see Chapter 13). They eventually chose Utah. Today, most residents of Utah are Mormon. Mormons believe in large families, with the result that average completed family size in the United States (the total number of children ever born to an average family) is highest in Utah. The United States averages about 2.0 children per family whereas Utah's average is 2.2 children per family.

Early European migrants seeking freedom of worship brought the Baptist religion to North America as a nonestablished church. Although founded in the South, it spread widely across the new country. Baptists did not become dominant regionally until after the Civil War when the religion's northern and southern branches split. During the last third of the nineteenth century, the southern Baptist Church became a religious expression of Southern culture and easily the South's dominant church. Although the South's religious diversity is growing, this denomination remains most prevalent across the region.

PERCEPTION CASE STUDY

Manifest Destiny

Senator Thomas Hart Benton was on the bandwagon of popular racist sentiment. Standing before the U.S. Congress in May 1846, Senator Benton justified white supremacy:

It would seem that the White race alone received the divine command to subdue and replenish the earth! for it is the only race that has obeyed it—the only one that hunts out new and distant lands. . . . Civilization or extinction has been the fate of all people who have found themselves in the track of the advancing whites.[1]

With an image of America's divinely approved expansion, flavored by racism and fueled by economic need, Senator Benton's rhetoric played into a strongly destinarian conception of the United States. Political eyes were looking west then, and, in the flowery language of the time, the land there was deemed vigorous and fresh from the hand of God. Like others, Benton viewed its settlement by white Americans as part of a providential plan in which the United States would smite the tyranny of kings and carry peace and civilization to the unenlightened—or else.

In the early 1840s, the conviction was growing that it was America's destiny to expand all the way to the Pacific Ocean. Politicians and influential intellectuals alike shared a sense of restless expansionism. In penny journals and political documents, they talked of repelling contaminating monarchies from soil consecrated to the rights of man; of settling land with the free spirit of new born democracy; of widening horizons of opportunity. Even a young Walt Whitman wrote of adding "new stars to our mighty firmament." By the mid-1800s, Americans only needed a phrase that captured this ebullience, and New York Democratic journalist John L. O'Sullivan coined it. He wrote in the *Democratic Review* about the country's "manifest destiny to overspread and possess the whole of the continent which Providence has given us for the development of a great experiment of liberty and federated self-government."[2]

Although advocates of Manifest Destiny differed in their conceptions of its scope, their numbers grew until the idea was a force to be reckoned with. Some who supported it had their eyes on Oregon and California. Blaming the post-1837 depression on Americans' inability to secure markets for agricultural surplus, they saw Oregon and California as doors to trade. One Missouri Democrat noted that Asia's ports were as convenient to Oregon as the ports of Europe were to the eastern part of the country. An Alabama Democrat pointed out that California's large, safe harbors invited the Orient's rich commerce.

Others spouting Manifest Destiny thought more than trade was at stake. They viewed expansion as an opportunity to preserve America's agricultural character and, thus, safeguard democracy. Essentially Jeffersonian, they equated urbanization and industrialization with class strife and social stratification—inherently British characteristics they wanted the young United States to avoid. In their view, the United States must expand, or it would echo Britain's misery and bloated wealth.

Democrats thought expansion complemented their opposition to centralized banking and high tariffs, both of which they believed fostered the factory system. Their reasoning was that expansion combined with low tariffs would give farmers land and access to foreign markets; and if Americans stayed agrarian, the country's foundations were secure. These variously edged messages, trumpeted in the press, appealed to the working poor, many of whom were Irish immigrants. To them, expansion meant economic opportunity. Indirectly, Manifest Destiny was a way to disperse the multitudes who were inundating America.

Another fuel for Manifest Destiny's acceptance was what historian Frederick Merk called "the foreign devil game."[3] The Texas annexation, for instance, was approved by a reluctant Senate only after rumor flew that Great Britain was plotting to turn Texas into a satellite. The country risked war with Britain over the Oregon Territory. In his inaugural address in 1845, James K. Polk pressed an American claim to the whole Oregon Territory between California and 54°40′ N, asserting that America's claim to it was clear and unquestionable.

Manifest Destiny, as used by O'Sullivan in the *Democratic Review*, gave Andrew Jackson's ideology of expansion political and intellectual presence. Expansionism also drew on the ideas of early Republic leaders like Thomas Jefferson and John Quincy Adams, who proclaimed the American people's right to displace any people, European or other, who stood in the way.

The doctrine had devastating effects on Native Americans. In their inexorable march west, Euro-American settlers, steeped in their own needs and beliefs, settled on Indian lands, taking over village sites and intruding on hunting grounds. Native resistance was met by federal suppression; Indians who did not die from starvation or were killed were forced onto reservations. In attempts to "civilize" them, federal officials and missionaries set up schools and Christian churches. Native languages, arts, and ceremonies were discouraged. In the currency of Manifest Destiny, expansion and annihilation were two sides of the same coin.

---

[1] Thomas Hart Benton, *Congressional Globe*, May 28, 1846.

[2] Quoted in Anders Stephenson, *Manifest Destiny: American Expansionism and the Empire of Right* (New York: Hill and Wang, 1995), p. 42.

[3] Frederick Merk, *Manifest Destiny and Mission in American History* (New York: Vintage Books, 1963), p. 41.

Review Questions

1. What were the major source areas of North America's European immigrants from the colonial era through the mid-twentieth century? Can you identify any generalizations between these source areas and their settlement destinations in North America?

2. What general statements can you make about the expansion of the settlement frontiers in the United States and Canada?

3. Where are the U.S. and Canadian population cores located? Identify some of the factors that help to explain this distribution pattern.

4. Discuss the spatial patterns of religion in the United States. Is there a geography of religion in Canada?

5. What is Manifest Destiny? How did the concept contribute to the expansion of the United States? How did the concept have relevance for Canada?

CHAPTER 4

MEGALOPOLIS

Preview

1. Megalopolis is the most urbanized region within North America and is dominated by a chain of large, densely populated metropolitan areas that have coalesced over time to produce even larger conurbations.

2. Urban growth in Megalopolis benefited from access to good ports and rivers for transportation and trade and from an abundance of rivers to provide water power for manufacturing.

3. Urban landscapes are characterized by high spatial interaction, functional complexity, a need for extensive public services, accessibility, and constant change.

4. Over time, urban populations in Megalopolis expanded outward into suburban and rural areas and changed in racial and ethnic composition.

5. Crowding, efficient transportation networks, and urban sprawl are all geographic problems that Megalopolitan cities have struggled with in recent decades.

merged to form the extensive, many-centered region that is Megalopolis.

The primary theme of Megalopolis is its urbanness. In varying degrees, urban activities like municipal utilities and fire and police protection provide for the millions who live there. And urban forms—manifested by dense patterns of streets and buildings, industrial centers, retail and wholesale clusters, and governmental complexes—are never far away. Indeed, the extent to which people's lives are dominated by urbanness is apparent on a drive along any segment of the 500-mile stretch of Interstate Highway 95 between Boston and Washington, D.C. Office and apartment buildings, small shops and mammoth shopping centers, factories, refineries, extensive residential areas, gas stations, and fast-food eateries by the thousands assail the senses.

But Megalopolis is not just a collection of coalesced urban forms linked by a major transportation line. It also contains green spaces like parks or other

In 1961 the French geographer Jean Gottmann published a monumental study of what he called "a very special region" located in the northeastern United States.[1] Gottmann argued that the massive and complex urban region extending from southern New Hampshire to Washington, D.C., and its northern Virginia suburbs (Figure 4.1) differed from any region that had existed in the world before. Although it was worthy of attention because of its importance in North America, Gottmann predicted it also represented a future in which similarly evolving urban regions would develop elsewhere in the world. Professor Gottmann named this North American region Megalopolis.

Megalopolis formed along the northeastern coast of the United States by the gradual coalescence of large, independent metropolitan areas such as Boston, New York, Philadelphia, Baltimore, and Washington, D.C. As these cities grew, the effects of their growth spilled over to surrounding, smaller places. Large suburbs also grew to make their own contributions to urban sprawl as pockets of open countryside became transformed into urban landscapes. Eventually, the outer fringes of each spreading metropolitan region

[1] Jean Gottmann, *Megalopolis: The Urbanized Northeastern Seaboard of the United States* (New York: Twentieth Century Fund, 1961).

Intense, complex urban activities define the landscape in which most North Americans live. Nowhere is this definition more powerfully present than in Megalopolis, represented here by Times Square in New York City.

Figure 4.1
Megalopolis.

land available for recreation as well as substantial agricultural areas. Tucked in ever-shrinking pockets, farmland in the six states (Massachusetts, Connecticut, Rhode Island, New Jersey, Delaware, and Maryland) that are almost entirely contained within Megalopolis comprised just 20.7 percent of the total land area in 2002. Only 9.2 percent of the land in Rhode Island was farmland in 2002, down from 9.7 percent in 1997.

Megalopolis is a complex region. At least five large but distinct metropolitan areas, with many smaller cities and rural pockets between and around these urban zones, fall within its borders. As Professor Gott-mann put it,

> In this area . . . we must abandon the idea of the city as a tightly settled and organized unit in which people, activities, and riches are crowded into a very small area clearly separated from its nonurban surroundings. Every city in this region spreads out far and wide around its original nucleus.[2]

---

[2] Ibid.

In spite of Megalopolis's mixed character, its massive urban presence makes this region extremely important in the United States. Ten of the country's 46 metropolitan areas exceeding 1 million people in 2000 were located in Megalopolis. The counties defined as Megalopolis by Gottmann contained 48,426,552 people in 2000, almost 40 million of whom lived in the six largest metropolitan areas. The region held more than 17 percent of the total U.S. population in only 1.5 percent of the country's area.

But Megalopolis is important for reasons besides merging cities with dense populations. The region also possesses an unusual mix of economic qualities. Average per capita income is high. Of the more than 22.5 million persons employed in Megalopolis in 2002, a higher than average proportion worked in white-collar and professional occupations. Transportation and communication activities are also prominent, partly because of the region's coastal position. One-sixth of all North American export trade passes through the region's six main ports. The statistics could go on, but the point is this: In spite of the growth in other parts of North America, Megalopolis has been and continues to be what Gottmann called a "very special region."

THE LOCATION OF MEGALOPOLIS

Megalopolis is unique in North America, even though other coalescing urban regions in the United States and Canada have mimicked Megalopolis's form.

Why has this particular portion of the continent developed as it has? Whenever geographers ask why a place developed its set of distinguishing characteristics, the first feature they usually consider is the region's location. In some cases, location is of secondary importance, but often, as with Megalopolis, it is primary. Clues to Megalopolis's origins and growth are found in two aspects of its location: site and situation.

Site Characteristics of Megalopolis

The *site* characteristics of a place refer to the physical features that make up the immediate environment of that place's location. Many of Megalopolis's site characteristics are visible in the region's outline. Since it occupies a coastal position in North America, its eastern margin is deeply convoluted. Peninsulas jut into the Atlantic. Islands are scattered along the coast, some

There are many fine harbors in Megalopolis, but none can match New York's. Protected from ocean currents and storms, the confluence of the Hudson River on the left and the East River on the right has long been a focal point for trade, transportation, and interaction.

large enough to support communities. Throughout the region, bays and river estuaries penetrate the landmass in a mirror image of the land's penetration of the ocean. This interpenetrated coastline brings more land area close to the ocean and provides more opportunities to take advantage of cheap water transportation than if the coast were straight. This site characteristic, a coastline of abundant indentations, was important to Megalopolitan growth.

Proximity to the ocean, however, does not always equal access to oceanic travel. Geographic research suggests that accessibility is the actual, not just theoretical, ability to go from one location to another. This means quality harbors have to be present, and Megalopolis straddles some of the best natural harbors in North America. To be sure, cities such as San Francisco, Seattle, and Mobile are built around excellent natural harbors, but they are far from each other. Not only are Megalopolis's harbors such as those at Boston, New York, Baltimore, and Providence, Rhode Island, excellent, but they are also close to one another. Philadelphia, on the Delaware River, does not strictly have an ocean harbor, but dredging maintains that port's access to the ocean. In addition, numerous smaller harbors, like those at New Haven and Groton, Connecticut, New Bedford, Massachusetts, and Portsmouth, New Hampshire, are located along the coast of Megalopolis. Thus, the abundance of good, reasonably close harbors provided European colonial settlers along this stretch of North America's coastline the ability to export and import goods across the land-water boundary.

That so many good harbors are located close to one another is no coincidence. The reason for this geographic feature lies in the region's geologic history. During the Pleistocene, which ended 10,000 to 12,000 years ago, the northern half of Megalopolis was covered by ice. This most recent Ice Age contributed to the excellence of Megalopolis's harbors in several ways. Sea levels were much lower during the Pleistocene because a tremendous amount of the earth's water was held frozen in the massive ice sheets. When the ice cover began melting, large river spillways formed, and the erosive power of the rivers cut deeply into the flat coastal plain across which silt and gravel-laden water rushed. With sea levels rising, the lower river valleys were drowned to form *estuaries*, and the ocean margin shifted toward the land's interior. Chesapeake Bay and Delaware Bay were formed in this manner from the drowned lower courses of the Susquehanna and Delaware rivers. The navigability of the lower Hudson River was enhanced as well. These glacial river valleys, cut deeply by the meltwater, formed some of the harbors that later proved useful in the development of Megalopolis.

The other major contribution of the Pleistocene is more specific to one or two locations. The large amounts of soil, stone, and other debris scraped up by the expanding glaciers were deposited as *moraines* when the ice front retreated. One such terminal moraine is the southern arm of Cape Cod. The rise in sea level and deposition of sandy soil by ocean currents sweeping north along the East Coast formed the distinctive north-reaching forearm of Cape Cod. Retreating glaciers left an even longer series of ridges just south of what is now the coast of Connecticut. These moraines became an island when the seas rose, and it, too, was widened by deposition from the ocean. The island was not made so wide, however, that it could be called anything except Long Island.

Long Island has enhanced the quality of New York's harbor in two ways. First, one end of Long Island is tucked into the harbor mouth formed by the Hudson River without actually blocking it. This feature increased considerably the length of coastline available for port facilities, which was already significant along the Hudson. Second, Long Island provided good land needed to accommodate urban growth around a large, fully developed harbor. Greater New York's growth was restricted to the west of the Hudson in New Jersey by tidal marshes and the erosion-resistant ridge of the Palisades. To the Hudson's east lies only a narrow finger of land, Manhattan Island, bounded on the north by the small Harlem River and on the east by Long Island Sound and its extension, New York's East River. But beyond the East River is Long Island, a flat to slightly rolling land without the barrier marshes of New Jersey. New York's boroughs of Brooklyn and Queens developed early at the western tip of Long Island, and the island offered a great deal of room for additional urban expansion toward the east. Both the harbor and the port city that grew around it benefited from the presence and orientation of Long Island, a landform owing its existence to glacial deposition.

Besides Megalopolis's many high-quality harbors, few other site characteristics contributed so positively to the region's urban economic development. The climate is not exceptionally mild, although the summers are generally of sufficient length and wetness to support farming. Soils are variable, with the good soils inland from Baltimore and Philadelphia better than most of those found closer to New York. From New York to Boston and beyond, the thin and stony soils are generally only fair, with the exception of the notably superior soils in the Connecticut River valley. The agricultural resources of Megalopolis, though adequate, are hardly exceptional.

The general topographic features of Megalopolis south of New York provide additional urban site benefits. Traveling inland from the Atlantic Coast, a sequence of distinct physiographic regions is crossed (see Figure 2.1). A very flat coastal plain is succeeded by a rolling, frequently hilly landscape called the Piedmont. Farther west is the more mountainous terrain of the Appalachians. The mountains do not parallel the coast, but approach and eventually merge with the coast in New England.

Although the differences between the Piedmont and the mountains are dramatic and topographically obvious, the subtle break between the Piedmont and the coastal plain is more significant to the development of Megalopolis. The Piedmont's irregularly rolling relief is underlain by very old, very hard rocks. This surface is resistant to erosion, and the level of the Piedmont is maintained above that of the coastal plain. Where rivers flow off the Piedmont, therefore, a series of rapids and small waterfalls form. Because these falls are found along a line tracing the physiographic boundary, this boundary has been named the *fall line*.

Early European settlers found the fall line to be a hindrance to water navigation but an obvious source of water power. Settlements developed along it, located as far inland as possible but still possessing access to ocean shipping. In addition, because the fall line was often the head of navigation, goods brought inland or those to be exported had to be unloaded at fall line locations for transfer to another mode of transport. The goods had to be handled anyway at this point, so transporters found it advantageous to break large, bulky shipments brought by ocean freighters into smaller loads. By the same token, transporters moving goods from the interior for export used fall line sites as places to consolidate small shipments. In many cases, manufacturing took place at these sites as well. Thus, fall line cities such as Washington, Baltimore, Wilmington, Philadelphia, and Trenton trace the Piedmont-coastal plain boundary. Only New York among Megalopolitan cities has water access beyond

the fall line via the glacially eroded Hudson River valley.

Situation Characteristics of Megalopolis

Geographers take a broader view of a region's location when they consider aspects of the region's *situation*, or location relative to other places. One of the most important aspects of Megalopolis's situation is its position relative to Europe. The coastal portions of the New England and Middle colonies were readily accessible to trading ships sailing between Europe and the New World. It was not coastal position per se that was critical here—that is a site characteristic. The crucial situation factor was that this portion of North America was on, or very near, the most direct sea route between Europe and the most productive plantations of the Caribbean colonies and southern North America, at least on the homeward voyage. Looking at a map with the North Atlantic as its focus, you can see the nearly straight line, or *great circle route*, between the Caribbean Sea and the British Isles (Figure 4.2). The dominant wind patterns and ocean currents supported this shipping route. The ports around which Megalopolis later grew were convenient stopping places that contributed directly to Europe's rapidly expanding transoceanic trade during the eighteenth and nineteenth centuries.

Ships stopped at New York, Boston, and the other ports because there was business to be transacted. Goods were purchased for local consumption and manufacture, and goods were sold that could be found in few other places. Sugar and molasses were brought from the Caribbean to be distilled into rum. Cotton was transported from South Carolina to be woven into textiles. The obvious profitability of sea trade (or sea piracy as it sometimes happened) encouraged the growth of shipbuilding. This, in turn, led to exploitation of nearby forest resources, another of the region's situation factors that contributed to its early growth.

The location of the core cities relative to the interior of the continent was also critical to the growth of Megalopolis. Philadelphia and Baltimore grew rapidly because each was the focus of a relatively good-sized and good-quality agricultural region. Southwestern New Jersey, southeastern Pennsylvania, northern Maryland, and much of the *Delmarva* (*Dela*ware-*Mary*land-*Virginia*) Peninsula possessed good soils—among the best in Megalopolis. Access routes to the interior were constructed early and helped support the growth of

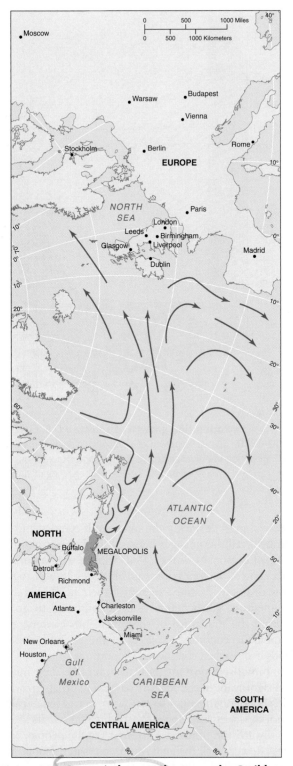

these cities' trading functions. Inland from Boston, in New England generally, the soils are too shallow and rocky and the terrain too rolling for this land to be considered good farming land. Agriculture did not contribute as much to Boston's growth as it did to Philadelphia's and Baltimore's, although Boston did have the resources nearby to support fishing and ship-building industries. The New England hills were covered with hardwood and pine forests that were nearly ideal for ship construction. Also nearby were the productive fishing banks off the New England, Nova Scotia, and Newfoundland coasts and farther south in the rich Chesapeake Bay.

The importance of a city's access to the continental interior, however, is most apparent in the case of New York. New York's local agricultural resources are better than those of Boston but not as fine as those of Philadelphia or Baltimore. Long Island had some good land for farming, but little of exceptional quality was found nearby on the mainland. New York's chief advantage lay in its position at the head of the best natural route through the Appalachian Mountains to the interior. The Hudson-Mohawk River system, later augmented by the Erie Canal, railroads and highways, provides access to the Great Lakes. The Great Lakes, in turn, provide access to the broad interior of the continent. When settlement density and economic activity on the interior plains increased during the early nineteenth century, some of the goods produced were shipped down interior rivers to St. Louis or New Orleans, but large amounts were carried east over routes connecting the interior plains to the urban cores of Megalopolis. The city in Megalopolis that benefited most from this growing trade was the one with the greatest natural access to the interior: New York.

Megalopolis as the Continental Hinge

Megalopolis and the city cores that dominate it possess real advantages for growth. Although none of these cities has great quantities of what are traditionally thought of as essential resources, such as high-quality agricultural land or mineral deposits, they do share a set of *accessibility resources*. Such resources are naturally occurring physical features of a place that facilitate movement into or out of that place. As such, accessibility resources are a subset of both site and situation features. Good harbors, good routes to the interior, a location between regions conducting trade—all are accessibility resources, and all are definitive features of Megalopolitan centers.

Figure 4.2   Great circle route between the Caribbean and Europe. The high quality harbors of what came to be Megalopolis were located close to the most direct sea route from the Caribbean to northwestern Europe.

During the colonial period, trade grew between Europe, the Caribbean area, and the North American mainland, and cities formed around the best harbors along the East Coast. Small-scale manufacturing began appearing in the larger port cities from Baltimore north. The mutually reinforcing relationship between manufacturing and trade stimulated growth in both activities. As urban industry grew, the demand for labor increased. This drew immigrants from northwestern Europe or diverted large numbers of workers from farming, swelling the cities' populations. Banks and other financial institutions underwrote investments in local manufacturing and shipping. Service activities, wholesale and retail businesses, centers of information and control all grew and supported further urban expansion. The greatest growth occurred within the four largest port cities in Megalopolis: New York, Philadelphia, Boston, and Baltimore. Each was the focus for an expanding hinterland.

The richness of each city's hinterland and the respective city's access to it were critical to the way growth proceeded. Boston, for example, grew earliest. It was the largest city of Megalopolitan core cities through 1750. Its advantages were those that could dominate early development—proximity to the densely settled colonies of southern New England, easy access to the rich fishing banks nearby, resources for a shipbuilding industry, and the trade that accompanied growth in shipping. Philadelphia and, to a lesser extent, Baltimore enjoyed growth stimulated by early development of the rich agricultural possibilities in their hinterlands. The cultural and political activities focused on Philadelphia during the last half of the eighteenth century also contributed to its growth, so much so that by 1760, Philadelphia's population reached 18,756, exceeding Boston's at 15,631. As land west of the Appalachians began to be developed, however, New York's preeminent site and situation advantages began to be felt. With only 13,040 people in 1756, New York experienced a population increase to 98,373 by 1810, surpassing that of every other city in the young nation (Figure 4.3).

The most unusual feature of Megalopolis, however, is not the fact that these cities grew; the site and situation advantages were there. Rather it was that four such large cities—later to become five with the addition of Washington, D.C.—continued to grow in close proximity to one another. Urban economic activities often cluster; there are advantages in being near each other. As a result, weaker or smaller urban places lose new growth to stronger and larger neighbors. However, so great was

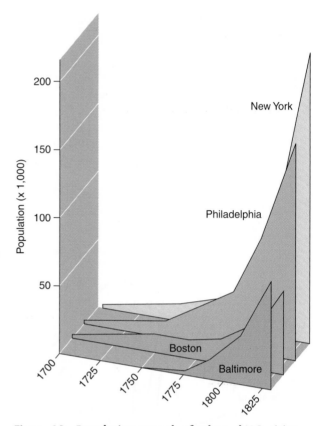

Figure 4.3 Population growth of selected U.S. cities, 1700–1830. The early growth of Boston and Philadelphia indicated their importance before the American Revolution, but New York surpassed both in size by the beginning of the nineteenth century.

national growth during the nineteenth century and so strong were the linkages between the interior and the ports of Boston, New York, Philadelphia, and Baltimore, that none of the four ports was able to wholly absorb the flow of goods destined to any of their neighbors and competitors. All benefited and all continued to grow. During the first third of the nineteenth century, the combined economic resources of these cities' hinterlands had reached continental proportions.

Initially important local resources became much less significant in the region's continued urban growth. Goods, people, and ideas flowed into Megalopolis from overseas and were absorbed, changed, and sent on toward the interior. Different goods, people, and ideas flowed back through Megalopolis in the opposite direction. This dual set of flows led Professor Gottmann to describe Megalopolis as the hinge of the continent. This axis of urban places functions much like a long pivot for North America's exchange with Europe, Africa, and western Asia.

THE URBAN ENVIRONMENT

Megalopolis is a coherent region, in spite of its complexity, for several reasons. In a general way, the function of acting as a continental commercial hinge is shared among sections of Megalopolis, even if the various cities within the region do not coordinate their commercial efforts. In spite of their relative proximity, the major parts of this region have all grown partly because the tremendous volume of goods transferred across the hinge has been much greater than the capacity of any single port. These port cities also have shared in the region's growth because each has sought competitively a larger share of national export and import trade, while none has an absolute advantage of position.

Throughout Megalopolis, however, it is the urban forms and urban functions that give the territory its most significant regional unity. The almost continuous and occasionally overwhelming urbanism of Megalopolis is its primary characteristic. Tall buildings, busy streets, crowded housing, and industrial plants accompany an array of cultural opportunities, such as theaters, symphony orchestras, art museums, and large libraries. Also apparent are impressions of deterioration in dilapidated structures, traffic congestion, and air pollution. All of these, and more, are present in the metropolitan cores of Megalopolis.

These characteristics of urbanism are also found in major cities elsewhere in the United States and Canada and in most large cities around the world. What makes Megalopolis distinct is the fact that the urbanism has spread out so far from the core cities that they have begun to merge with one another in a process of metropolitan coalescence. This process occurred in Megalopolis earlier than elsewhere, partly because the degree of merging is more extensive, but primarily because it encompasses so many large cities within a relatively small geographic area. In this way Megalopolis became a kind of gigantic laboratory in which intensively urban patterns and peculiarly urban problems could be observed developing on a very large scale. It is the most urban region on a continent with a largely urbanized population. Examination of its components helps set the tone for study of the other regions of Canada and the United States either through comparison or contrast.

Land is in great demand in the largest cities in Megalopolis. The rare and startlingly open green spaces of Central Park are taken advantage of by many New Yorkers when time allows.

Population densities in Megalopolis remain high, averaging about 348 persons per square kilometer (901 per square mile) in 2000. This average density continues well above Megalopolis's 1960 density of 266 per square kilometer (688 per square mile), even though national population redistribution led to greater than average growth in other parts of the country.

Unsurprisingly, population densities within Megalopolis are not the same from place to place. Some peripheral counties have population densities that are only 10 to 20 percent of the region's overall average. Settlement density increases as a city is approached, until very high densities are reached near the city core. In New York City, for example, densities in the year 2000 averaged 10,192 persons per square kilometer (26,404 per square mile). This general pattern of lower density periphery and higher density core occurs with considerable spatial regularity. Variations are evident from city to city in the rate of density increase and in the height of the density peak, but the pattern repeats itself again and again.

When experienced up close, urban landscapes often seem chaotic, without sense or order and the consequence of millions of independent and conflicting decisions. This disorder is even greater in the many-centered region of Megalopolis. There are pattern regularities, however, and these regularities help us understand Megalopolis and urban landscapes more generally.

Most geographers would agree that modern cities result basically from the locational consequences of economic decisions. It is true that cities have political foundations, and they also constitute a form of social clustering. Some cities have also been founded or have grown because of religious factors. However, when someone decides to move or to relocate his or her business into one city rather than another or into a nonurban setting, traditionally the fundamental economic advantages of such a choice dominate the decision. So pervasive are these advantages that large numbers of people live very close to one another, often closer than they would prefer, and tolerate negative consequences of dense clustering in order to participate in the benefits the city offers.

An increasing number of urbanites have tried to minimize some disadvantages of city living by moving their residences to suburban locations. Others have moved even farther, well beyond the suburban fringe, into a zone referred to as *exurbia*. From small towns, areas of converted vacation homes and rural estates, exurban commuters travel each day (or each week) to distant workplaces or have developed ways to work from home. But this spreading population has not eliminated the disadvantages of clustering; it only shifted who faces them. The workplaces for those living in the region are also more widely spread than in the past; a smaller proportion of the sprawling metropolitan population than previously still find it necessary to enter the central city for work. The population spread has also created a variety of new problems, many of which are geographic and rooted in the defining characteristics of Megalopolis's urban landscape.

Components of the Urban Landscape

Interaction

In general, the cost of moving an item is directly proportional to the distance it must be moved. The farther

As cities grew during the late nineteenth century, communication became intense. Telecommunication technology was still crude enough to lead to scenes like this one on Broadway in New York City in 1889. The landscape gradually changed after 1900 as wires began to be laid underground.

it must be moved, the more it will cost to move it. Activities cluster in cities so that movement costs are minimized. *Spatial interaction*, which refers to the movement that occurs between places, is basic to the existence of a city. Thus, its imprint is an apparent feature of urban landscapes, and the visibility of interaction depends on what is interacting. The lines of interaction for human movement can be seen in streets, subways, bridges, tunnels, sidewalks, and parking lots. In older cities, such as those in Megalopolis, the downtowns were laid out when travel was by foot and horse-drawn carriage. Therefore, only about 35 percent of the total central areas of these cities is devoted to features such as streets, sidewalks, and parking lots. In newer cities, developed largely after the rise of the automobile, the proportion is much higher. In Los Angeles, for example, almost two-thirds of the city's downtown is devoted to automobile uses. Devoting such high proportions of urban land exclusively to facilitate interaction demonstrates the critical importance of the ability to move from one place to another in urban regions.

Other forms of interaction may be equally important but less visible. The easy movement of information and ideas is supported in cities by intensively developed telephone, telegraph, and teletype systems and, more recently, by fiber optic computer linkages. A century ago, telephone lines were not wound into cables and placed underground. The central business districts of all large cities were crowded with crisscrossing spiderweb networks of telephone lines testifying to the intensity of communication demands. Telephone land lines are no longer part of the visible landscape, and it is just as well because intraurban information flow multiplied many times during the twentieth century. Today many people use cell phones, computer services and various other hand-held wireless devices, accelerating interaction but without visible linkages.

The movement of people and information are two obvious examples of intensely concentrated interaction near the city core. But many more things exhibit the same general pattern of movement. Also hidden, and taken for granted unless something goes wrong, are items like electricity, a fresh water supply, and sewage and garbage disposal. Until the volume of garbage reached high levels because of urban residential densities, it could be fed to animals, buried, or discarded in nearby dumps without becoming a health hazard. In rural areas, individual septic tanks are often sufficient to handle a single family's sewage safely, and each dwelling can have its own well without overdrawing the underground water supply. Whereas some typically urban municipal services have been extended into nonurban areas, the extension is relatively recent. Even electricity was not extended into most of rural America until well after the federal Rural Electrification Act was enacted in the 1930s.

An example of the tremendous intensity of urban interaction was provided by a story in the *New York Times* in 1973. In the process of constructing a new subway line, work was slowed by the intricate networks of utility lines buried beneath the pavement of Second Avenue in New York City. In one block, the crews found "one 30-inch and two 10-inch gas mains; a four-foot water main and a 12-inch water main; a box sewer three and one-half by two and one-third feet; 16 three and one-half inch ducts for electrical wires along with 43 three-inch ducts; 64 three-inch ducts for telephones and 30 four-inch ones. In addition, they found footings for the old El; house connections for sewers, water, and gas; cables for fire alarm-system boxes, cables for police call boxes and Transit Authority cables."[3] Remember, all this was found in only *one* block more than 35 years ago and, ironically, in the process of constructing a new interaction line!

The intensity of intraurban interaction reflects the diversity of activities as well as their density. Movement between spatially distinct locations does not just happen. The cost of movement must be borne, and, by implication, a return must be expected to make the movement worthwhile. Perhaps the prime motivation for interaction is the geographic separation of supply and demand. Economically, this means that an item or a service is needed or desired at a location that cannot meet that need; the good or service must be obtained from somewhere else. When two or more places are deficient in items that can be obtained at the other places, they are said to be complementary, or to have *spatial complementarity*. When locations are complementary, interaction will occur unless they are too far apart, some closer place to obtain the good or service can be found, or some other barrier to movement exists.

Functional Complexity

The implication of densely packed yet complementary activities for city landscapes is that a city is a place of considerable functional complexity. Some kinds of activities—termed *functions* by geographers—tend to cluster together, whereas others are scattered more generally across the urban area. Because a wide variety of activities are carried on in a city, interaction is stimulated between functions or within zones of the same function. When the activities are mapped, the result is a greatly mixed pattern of land use.

Philadelphia's land area, for example, is dominated by residential, commercial, and industrial uses. For two centuries the Delaware River waterfront was the primary industrial region of the city. Low-cost water transportation and Philadelphia's role as a hub of

[3] The *New York Times*, February 18, 1973, p. 69.

international oceanic trade combined to generate the location of industry. This land-use pattern also had an impact on the rest of the city. Almost the entire waterfront along the Delaware River was zoned for industrial uses, as was a large area on the lower Schuylkill River. In the past three decades, heavy industrial activity within the city has declined as firms have moved elsewhere or gone out of business. The complex pattern remains, however, as light manufacturing, high tech, and service industries have moved into areas once occupied by heavy manufacturing firms. In addition, the city has begun to transform former industrial zones into recreational areas for residents.

The generalized nature of land-use categories masks a tremendous diversity of function within each land-use type. Philadelphia's densely packed and intensively interacting central business district is in the same land-use category as retail and wholesale clusters located north and west of the city. Large shopping centers fall into the same land-use category as the headquarters of a large insurance company, whereas the activities carried on at each place are complementary and produce interaction flows. Other interaction arises from activities grouped as institutional, such as the city government, various centers of higher education, hospitals and clinics, and Philadelphia's former naval shipyard (now called the Philadelphia Naval Business Center). This wide diversity of functions is interspersed with and often interacts with the resident population of ethnically and socioeconomically diverse communities. This complex pattern is repeated in other major cities of the region.

Public Services

A third aspect of the urban landscape is implied by the dense population, the complex functional mix, and the high level of interaction stimulated by these diverse activities. The concentration of population and urban activities requires an array of supporting functions. Traditionally organized and operated as branches of city government, these public service functions are only indirectly productive in an economic sense, but they are absolutely necessary for commerce and industry.

As discussed earlier, city interiors are more concentrated than urban peripheries. The density of municipal services is also more intense geographically near the urban center. In addition to water, electricity, sewage, and garbage collection, cities provide police and fire protection, construct and maintain public movement facilities, health care and educational facilities, and document vital population statistics.

There is much variation in the quality and quantity of these services from city to city. Some cities, for example, also extend support to cultural activities such as museums, art galleries, and performance spaces. Recently, a few cities chose to turn the operation of selected public services over to private firms as municipal

Sidewalks, abandoned buildings, and vacant lots often substitute for more formal recreational spaces in older areas of Megalopolitan core cities struggling to provide adequate social services. Here, children play around an open fire hydrant in Brooklyn, New York.

governments seek to reduce the size of city budgets. However, whether provided publicly or contracted privately, such services must be present in some form given the large number of people and functions clustered in urban areas.

As the United States and Canada became increasingly urbanized—that is, as a larger proportion of the total population lived in urban areas—many public services were extended far beyond metropolitan limits. In rural areas, nonurban governments (county, state, provincial, and occasionally federal) also chose to provide public services, although to a less intensive degree. Urbanites who move to nearby suburbs or semirural communities often consider the convenience of urban-level services a necessary part of governmental responsibility and may demand more services than were previously provided. As a result, small communities experiencing a significant influx of new residents may be forced to change their tax structure to institute services that are more typical of larger urban places.

These municipal services are so pervasive in large cities that immense governmental structures have developed to administer them. In the New York metropolitan region, to take the extreme case, almost 1600 distinct administrative agencies were in operation in 2004. Clearly, an extensive network of services is an important corollary to functional diversity and the interaction flows typical of urban areas like those within Megalopolis.

Accessibility

A fourth major component of the urban landscape is its high level of *accessibility*. Accessibility is created and maintained as a public service, but it is demanded by the great amount of interaction and shaped by the spatial arrangement of land uses. A location's accessibility also may affect a decision to locate a new activity at that site. And of course, limited access constrains the amount of interaction possible between places.

Easy access within an urban area was not always as dominant a concern in city organization and urban structure as it has become. The original street plans of most cities in Megalopolis, for example, followed the simple rectangular grid pattern popular in seventeenth- and eighteenth-century Europe. Grids were laid down as fully as local terrain allowed in New York, Philadelphia, Baltimore, and, with modifications, Washington, D.C. Most smaller cities and towns along the eastern seaboard were also modeled on this pattern. Even in Boston, where the harbor's configuration

and the winding rivers that enter the harbor make use of a square grid street pattern awkward, portions of the city were laid out in this fashion.

During the cities' early years, when the functional mix and the interaction it stimulated were less intensive than they are today, the grid pattern was satisfactory. But as the cities grew, access problems inherent in this street system became apparent. A square grid, for example, possesses very frequent, right-angle intersections. Because flow is interrupted at each intersection, a greater traffic volume leads to longer pauses at each one. By 1900, Baltimore and Boston had each exceeded populations of 500,000, Philadelphia had reached nearly 1.3 million, and New York was approaching a population of 3.5 million. The main impact of the automobile was then still in the future, but serious congestion was already present in these cities' centers.

As long as all sections of the city were subjected to the same general limited accessibility, pressures to change the form of access were not great enough to have much effect. When the population had to live within walking distance of work or within reach of the trolley tracks, none suffered significantly poorer access than any other. However, a series of changes gradually accumulated so that following World War II cities' areas expanded rapidly. One change was that an increasing proportion of the cities' workforce began living at distances and in residential densities that made it uneconomic for mass transit to reach them. Another was that the speed and flexibility of truck transport accelerated the diversion of short-haul freight from rail to road carriers. Traffic planners responded by recommending construction of peripheral circumferential, or ring, highways and high-volume, limited-access expressways. The goal was to separate local movement, which remained on the grid, from crosstown and through traffic. Partly successful, these changes as well as many others also increased demand for access within the city center, between the center and the periphery, and eventually between sections of the periphery. The entire street and highway pattern became more complex and difficult to manage.

Intensity of Change

All of these components accentuate yet another feature of the urban landscape, namely, change. All landscapes change, but urban landscapes shift and rearrange their patterns in ways made more vivid by the density, intensity, and complexity of urban functions. The pace of change is faster, more diverse in its expressions, and more obvious to large numbers of people. Tens of thousands of new residents enter a large city like Philadelphia or New York each year. Even greater numbers leave, some to distant cities and some only to the metropolitan fringes. Among the effects of this ebb and flow of people, structures are destroyed and new ones are built. Street patterns are altered, the pattern of functions changes, and flows of people, goods, and ideas shift to fit the new mosaic. Such changes can be observed in any major region in North America, but in some ways, the changes actually created Megalopolis. After 1940, this region became less and less a compact series of large, essentially separate cities performing somewhat similar functions for the continent. Increasingly, the sprawling metropolitan areas merged to form a single urbanized region. The transformation to one sprawling super city is far from complete and may never occur because of yet other changes. But the process is more advanced here than anywhere else on the continent. The impacts and the problems these changes have raised for Megalopolis provide insight into metropolitan patterns developing elsewhere.

CHANGING PATTERNS IN MEGALOPOLIS

Perhaps the most fundamental and far-reaching change in Megalopolis during the last 50 years has been the great areal expansion of each of the major metropolitan areas (Figure 4.4). Greater New York, which had both the largest initial population and the most intensive economic concentration, has extended its population the farthest. But the port cities of Boston, Philadelphia, and Baltimore have also grown greatly. Washington, D.C., also burgeoned as the federal government rapidly expanded its operations. The

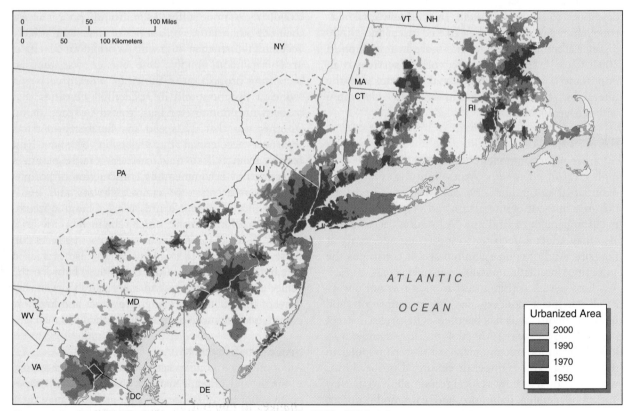

Figure 4.4   Spread of urban population in Megalopolis, 1950–2000 (after C. E. Browning). The sprawling metropolitan regions in Megalopolis have grown at different rates, but the tendrils of urban densities from each center reach far from the original cores.

District of Columbia was no longer large enough to absorb the swelling civil service population and the growing number of people needed to feed, clothe, and otherwise serve them. As a result, urban development spilled into Virginia and Maryland.

Changes on the Land

The spread of urban population far beyond city limits had a strong impact on rural activities in Megalopolis. As pointed out earlier, substantial areas within Megalopolis are not yet urbanized, but the cities surrounding these rural pockets have had a profound influence on them. There are two types of such influence: those related to land use and those related to land value.

As city populations grew, a greater number of people not directly engaged in food production had to be fed by foodstuffs shipped in from rural areas. The tens of millions of people in urban Megalopolis consume agricultural products from all across America and beyond. Many of those farming the agricultural land close to the cities, however, chose to specialize in higher-priced foods and in products with high perishability. The distance to market was short, and the price

All large cities in Megalopolis have responded to the ever-increasing demand for interaction by building numerous limited access highways. At times, as in this photo from Boston's massive "Big Dig" project, the result appears as complex as the problem.

of the final product did not have to be raised much to cover transport costs. Dairy products, tomatoes, lettuce, apples, and a wide variety of other intensively produced table crops came to dominate farm production in sections of rural Megalopolis.

Accompanying this shift in agricultural emphasis was the second influence of expanding urbanization. Land prices on the urban margin are driven up by activities that use the land with greater intensity than farming does. Good farmland purchased at a few hundred dollars per acre in the 1920s or 1930s might be sought for residential subdivision at many times the price. For example, a 125-acre farm purchased decades ago for $20,000 as agricultural land might eventually be sold to a real estate developer for $2 million. The developer, in turn, subdivides the land into 250 home lots of 0.5 acre each and then, after putting in utilities and streets, sells each lot for $40,000, or a total of $10 million.

Even if a farm family is able to resist the profitability of such a sale, the taxes on the land rise sharply toward urban levels as nearby areas begin to be used for urban activities. Unless local land-use controls are put in place to keep land in agriculture, the only way a family can remain in farming is to pursue intensive agricultural production devoted to high-value products. As a result, high-value farm-product operations cluster between the expanding urban regions of Megalopolis.

Urban sprawl and the corresponding shifts in agricultural practices in Megalopolis were pulled along the main lines of interurban access. Strong flows of traffic developed early between the cities of Megalopolis. Each center stimulated growth in the others through complementarity as well as competition; transportation and communication lines were needed to carry the interaction. When people who continued to work within the major cities relocated their residences, a high proportion chose sites that allowed them easy access to their workplaces. The arterial highways and, to a lesser extent, the interurban railroads and main feeder roads became the seams along which metropolitan populations spread first and farthest. As a consequence, the urbanized areas merged first along these main interurban connections.

Changes in Population Composition

As the populations of the separate urban areas grew, the composition of the populations also changed. Prior to

As suburbs spread, farmland is converted to urban uses. Who benefits from such conversions, and who does not?

1910, the cities of Megalopolis absorbed large numbers of immigrants from Europe. During the 1840s and 1850s, millions of immigrants arrived in the New World from northern and western Europe, with 70 percent of these from Ireland and Germany (Chapter 3).

Throughout the latter half of the nineteenth century, large numbers of Italians, Poles, Russians, and others in small groups and families arrived from other parts of Europe. These migrants passed through one or another of the large Megalopolitan ports, usually New York. Those who did not continue west into the farming areas or urban centers of the Midwest and Great Plains settled in dense clusters within the cities of Megalopolis, usually forming communities of each nationality. Communities within the very large urban areas offered cultural and linguistic support for new arrivals. Many such communities have had enough social cohesion to remain spatially identifiable today.

When World War I broke out in Europe, the flow of immigrants stopped suddenly and a new set of migration flows began to filter into Megalopolis. After 1910, the slow trickle of African Americans moving from the South began to grow, with most migrants settling within the cities where there were already small African-American populations. As the migration continued through midcentury, black residential densities increased and spread out from the original areas.

This pattern of black settlement expansion also illustrates the increased social and economic differentiation that occurred within ethnic and racial groups in these cities. As portions of each group found better-paying jobs and merged into the larger culture, families moved toward the city's edge or beyond. Faced with individual and institutional discrimination, African Americans and other non-European minorities for many years did not move as far or in as large numbers as other urban groups. But they, too, fit the trend. The general urban pattern of intense core and diffuse periphery was reinforced. The most intense ethnic and racial separation near the city center gradually changes to loss of ethnic identity as one travels away from the center.

Changes in Population Redistribution

For many years after the mid-1950s, it appeared that central-city populations would continue to decline, with much of the population leaving the large core cities to relocate in one or another of the nearby suburbs. However, during the 1970s and 1980s three entirely new aspects of urban change appeared in the metropolitan regions of the United States. These latest changes are national in scope, but the largest and oldest cities, such as Megalopolis's port cities, demonstrate the changes most dramatically.

First, sometime during the late 1960s and for the first time in U.S. history, people began to leave the largest metropolitan regions—considering both central city and suburbs together—in greater numbers than those who moved in. Suburbs still grew, but their growth no longer exceeded central-city decreases. Even more distant smaller cities and towns and the

Most cities in Megalopolis contain old housing units that have been renovated by professional people wanting to live near their work and their city's abundant nightlife. Creating such gentrified neighborhoods raises the city's tax base while dislocating the area's previous and typically poorer occupants.

rural areas between them were the primary recipients of these population shifts. This pattern is especially prevalent in communities and areas near metropolitan clusters but too distant to be considered part of them. As a result—and in spite of the significant shift in national population away from the Northeast—many of the smaller places within Megalopolis continue to grow. One trend in population redistribution, therefore, is reducing the number who live in Megalopolis as a whole, while another is rearranging the pattern of population within the region.

A second recent aspect of urban change is an eruption of high-rise office clusters at various locations in the metropolitan areas. The appearance of new metal and glass office skyscrapers since the mid-1970s has transformed the downtown skyline of many North American cities. This feature has not been limited to the old city cores. Even more dramatic than downtown office clusters has been the emergence of massive office buildings within the suburban rings surrounding the central cities. Not only are these suburban clusters new, but they are very large—with many exceeding the square footage of office space of a major city downtown. For example, office space in Montgomery County, Maryland, adjacent to Washington, D.C., is triple that of downtown Baltimore and nearly as much as the entire Baltimore metro area. This cluster is only one of more than a dozen such new office zones around the District of Columbia. One of these, Tysons Corner, has over 25.3 million square feet of leasable office space. The phenomenon is repeated throughout Megalopolis and across the country, at varying scales. This change appears to affect most significantly the location of white-collar jobs and the pattern of travel to them rather than the location of residences.

Third, the outflow of population from residential areas near city centers began to moderate during the 1980s. The attractiveness of a central location and shifts in the type of workforce desired downtown since the late 1970s led to the thorough renovation of residential structures in parts of the interior city. As real property renovations attracted higher income residents and created a higher tax base, existing lower income residents were dislocated. The process of real property investment, called gentrification, usually began in small areas within the central city but often spread outward. It represents a counterflow to the movement from the central city to the suburbs, but it is too early and the patterns too complex to see the long-term result of this facet of population redistribution.

It is difficult to foresee the impact of national population redistribution on Megalopolis as a whole. This region comprises the greatest single collection of old, large metropolitan areas and, therefore, may find it difficult to adapt, but it is also a region of tremendous economic energy and potential. The core-city populations have shared in the national trend, as we have seen (Table 4.1), and many suburbs showed decline as well when 1980 census data were released. But the geographic advantages that stimulated commercial, financial, industrial, and cultural activities during two centuries of urban growth are still present and are indicated by the growth of office clusters. This is shown by the lower rates of population decline in the central cities during the 1980–1990 and 1990–2000 periods than earlier, and population growth in cities like New York and Boston during the period 1990–2000, indicating that these cities continue to be a draw to migrants from other parts of the United States and from overseas.

TABLE 4.1

Megalopolis Central City Population Change, 1970–2000				
		Percentage Change		
City	**2000 Population**	**1970–1980**	**1980–1990**	**1990–2000**
New York	8,008,278	−10.4%	−0.5%	+9.4%
Philadelphia	1,517,550	−13.4	−8.6	−4.3
Baltimore	651,154	−13.1	−8.5	−11.5
Boston	589,141	−12.2	−1.6	+2.6
Washington, D.C.	572,059	−15.6	−10.0	−5.7

Source: U.S. Bureau of the Census.

Clearly, urban areas are landscapes of change, and the changes in Megalopolis suit the unusual character of the region. The changes have been continuous; they have been drastic; and until very recently they have occurred on a scale unmatched anywhere else in the world.

There are clear reasons for the region's dynamic nature. Each aspect of change is a response to demands and pressures that led to a particular resolution. Some changes came about because of conditions hundreds or thousands of miles from Megalopolis, conditions that led to immigration, for example, or conditions that altered the cost of oceanic shipping. Some changes occurred in response to developments inside the region. In each case, new problems were created as previous pressures were resolved. The new problems often appeared to be more complex and more difficult than those faced earlier, although this may be largely because they were new.

PROBLEMS IN MEGALOPOLIS

The problems faced by the residents of Megalopolis are not merely the problems of any large city multiplied four or five times. Megalopolis is not a super city with super problems. Although many of the difficulties of urban organization and growth in Megalopolis are those found in most large metropolitan areas, this region contains numerous large urban places competing economically with each other while their territories merge in a complex, uncoordinated manner.

Most of Megalopolis's problems can be viewed in the context of three characteristics of urbanness discussed earlier: density, accessibility, and spread. The first two include sets of problems that are fundamentally urban, occurring in the cities of Megalopolis on a somewhat larger scale than elsewhere. The third characteristic, however, also includes a set of problems that are multiurban in nature.

Disadvantages of Density

High population density and spatially intense economic activity are typical urban patterns, and both have problems associated with them. When cities experience a more rapid influx of migrants than can be accommodated by housing openings, residential areas become overcrowded. The core cities of Megalopolis

contained overcrowded residential areas for many years as migrants from Europe poured in. Not only were new housing structures constructed more slowly than the increase in population, the new building sites were usually far away from where immigrants could afford to settle. The magnet of social or cultural cohesion and the constrictions of low migrant incomes and restrictive property tax structures exaggerated the extreme overcrowding in sections of the city with some of the oldest buildings. These same factors reduced owner incentives to maintain those buildings. Meanwhile, great opportunities for profit arose from land development around the expanding urban fringe.

Intense concentrations of economic activity also produced problem conditions exaggerated by high urban population densities. Many of these difficulties are related to the issue of municipal waste management. The problem is not that waste is produced in large quantities but that it is produced faster than its negative aspects can be neutralized. When this happens, the excess spills over into the environment, the environmental balance deteriorates, and the result is pollution.

Industrial activity frequently is singled out as the chief polluting activity in urban areas, and in many cities industries do contribute greatly to environmental deterioration. It would be a mistake, however, to lay the chief blame for urban pollution on industry. In general, cities in Megalopolis possess much lower proportions of heavy industrial activity than do many cities along the Great Lakes, yet their air and water pollution problems are severe. Washington, D.C., for example, has almost no heavy industry, but its residents can testify to frequent smoggy days every summer.

More important than industrial pollutants are what are euphemistically called municipal sources. These sources are the urban dwellers who generate increasing levels of waste in such high amounts that the waste cannot be absorbed locally. Hundreds of thousands of vehicles are driven in and out of each core city every day, many carrying only the driver. Thousands more travel between suburban locations. Millions of homes and millions of acres of office space in each metropolitan area are heated every winter and air conditioned every summer with electricity or natural gas supplied by public utilities. Millions of gallons of waste water are expended from washing machines, dishwashers, garbage disposals, and flush toilets every day. And all of this is concentrated within a few hundred square miles. In the New York metropolitan region, for example, 5 million cubic yards of treated

Almost everyone wants to travel someplace else within cities, at least several times each day. With so many origins and destinations and with so many people moving within a small area, trips are slowed to a crawl during the busiest times of the day.

sewage residue (or sludge) is produced each year. The accumulation of this sludge spread evenly over Manhattan would bury the entire island to a depth greater than 3 feet in just 12 years. In Megalopolis, the pattern of urban concentration repeats the problem again and again within a small portion of the continent.

The problems of pollution, when they are addressed at all, usually elicit one of two approaches. One approach is technological—developing and applying better mechanical and chemical ways of neutralizing the negative aspects of waste. The other is ecological—finding a better balance with nature's capacity to absorb waste by producing and consuming less and by reusing what was previously discarded, thereby reducing the output of waste. A third choice, a spatial approach—reducing waste density by spreading population and economic activity over a greater area—might be discussed more often, but there are already well-recognized problems associated with an increasingly dispersed population.

Accessibility and Density

High population densities create difficulties in getting from one place to another. Congestion occurs when activity sites become so densely packed and the level of interaction between them so intense that the routes for interaction become overcrowded. Real access is reduced while demand for access continues to increase. The phenomenon is familiar as a daily cycle to all who have lived in, or visited, the core areas of major cities.

Morning and evening rush hours find movement so difficult that "rush" is not an appropriate description of these hours of the workday.

The problems of intracity accessibility are made even worse in Megalopolis by two features. One is the obviously large number of very densely populated urban areas, including the continent's extreme case in New York City. In 2000 the average time to commute downtown in New York City during a workday was 40 minutes, second only to Palmdale, California, among all metropolitan areas in the United States.[4] The second feature is related to these cities' sites. All were founded and grew because of their easy access to ocean shipping and the excellent harbor facilities provided by the sites. The coastal configurations, however, create severe problems for land movement. Boston and Baltimore have no land approach to their cores from the east and southeast, respectively. New York and Philadelphia cannot be reached directly by land from the southwest and the south, and Manhattan is an island. The rivers within these cities and in the District of Columbia congest movement by funneling it across a limited number of bridges and through tunnels. When they are all open to traffic, there are 14 bridges (including 3 for railroads) across two rivers in the District of Columbia, 22 bridges into or within Philadelphia, and 20 major routes of access for land movement onto Manhattan Island in New York. Baltimore, Boston, and many smaller cities, such as New York's New Jersey suburbs; Wilmington, Delaware; Providence, Rhode Island; and Connecticut's Bridgeport, Hartford, and New Haven, also possess transportation networks that are constricted and limited by the expensive need to bridge across or tunnel under water.

The cities of Megalopolis have approached this problem in several ways. The first method typically used when existing movement facilities become overcrowded is to increase their capacity. Four-lane roads are widened to 8 lanes, and 8 lanes to 12 lanes—or even 16 lanes. Limited-access expressways are constructed parallel to existing roads to enable faster movement of traffic through populated areas. Additional telephone, water, sewage, and electricity lines are installed. Sooner or later—often by the time they are opened—these expanded routes are carrying traffic volume up to, or in excess of, their designed capacities. This repeated experience shows that traffic congestion will not be alleviated solely by building more and

[4] U.S. Census Bureau, *Census 2000.*

wider roads. Some analysts maintain that building new roads actually encourages traffic to increase.

A second method for alleviating congestion is frequently proposed but, for a variety of reasons, is seldom successful. Rather than increasing movement capacity by building parallel routes, each existing line can be used more efficiently. An average automobile at rest occupies almost 20 linear feet of road and can contain up to four or six adults in moderate comfort, although only through carpooling are such high passenger levels reached. A bus, on the other hand, may occupy 50 linear feet of road but can carry six to eight times as many persons as a car. The ability of mass transit to move much larger numbers of passengers per unit of road space than the private car appears to offer a clear opportunity to increase movement capacity by making more efficient use of existing lines of access.

A third approach to alleviating traffic congestion is to combine private vehicle convenience with the multipassenger movement character of mass transit. On larger arterial expressways, an additional lane or two is reserved exclusively for vehicles carrying two or three persons, not just the driver. These HOV (High Occupancy Vehicle) lanes must be monitored, but they do increase the number of people carried in and out of the city without increasing the number of vehicles or roadway space.

The apparently greater efficiency of mass transit is not so obvious a solution, however. First, mass transit systems already carry a large proportion of commuter traffic into some Megalopolitan cores but have been in serious financial difficulty for decades. More than 52 percent of working New York City residents reported in the 2000 U.S. Census that they traveled by public transportation to get to work, but longer commuting times, aging bus and subway fleets, and increasing prices keep others away. In 2005, the average New Yorker spent just over 30 minutes getting to work every day. The U.S. average is about 24 minutes.

Second, and strongly related to the financial problems of urban mass transit, is a long-term decline in transit ridership. Movement by private automobile gives at least the impression of greater comfort, convenience, and flexibility of travel and greater personal safety to the passengers, as well as the illusion of lower costs per trip. Only when fuel is less available (as occurred briefly in 1973 and again in mid-1979) has public transit ridership increased significantly. Employer and municipal attempts to encourage carpooling have met with modest success.

Federal, state, and local governments have responded to private transit demands by constructing hundreds of miles of new expressways, building new bridges or increasing the capacity of existing ones, and in other ways expanding capacity for private movement in and around Megalopolitan centers. Meanwhile, public transportation has declined in quality and capability, with a few notable exceptions such as the Metro subway system in Washington, D.C.

One aspect of urban landscape change underway in Megalopolis is alleviating the congestion resulting from centralized activities. As suburban populations have grown, many stores and services dependent on

Many neighborhoods in Megalopolitan cities, such as this area of Providence, Rhode Island, were once suburbs. With small yards and modest houses by today's suburban standards, these areas were often among the first opportunities for people seeking detached dwellings for their homes.

TABLE 4.2

Distribution of Jobs in Selected Sectors, New York City, 1968–2002 (in thousands)						
						Net change
Selected Sectors	1968	1978	1988	1998	2006	1968–2006
Industry						
Manufacturing	840.0	538.9	369.4	261.8	104.6	−735.4
Construction	102.5	69.2	124.6	92.9	118.3	+15.8
Finance, insurance, and real estate	437.1	418.6	539.5	472.5	464.2	+27.1
Services	747.9	818.1	1136.7	1309.1	1703.9	+956.0
Government	526.0	500.7	592.2	523.5	557.4	+31.4
Total New York City employment	2653.5	2345.5	2762.4	2659.8	2948.4	+294.9

Sources: Employment and Earnings, States and Areas, 1939–1974, Bulletin 1370-11 (Washington, D.C.: U.S. Bureau of Labor Statistics, 1975); and *Employment and Earnings,* Monthly Reports (Washington, D.C.: U.S. Bureau of Labor Statistics, November 1979; December 1989; May 1998; August 2007).

proximity to potential customers have relocated out of the central city. As discussed earlier, new urban workplaces have also located in suburban and exurban locations rather than in metropolitan core locations. With more activities available outside the central city, fewer trips are taken into what had formerly been the main focus of metropolitan congestion: the city's central business district (CBD). A *New York Times* survey[5] of suburban residents three decades ago, for example, found that half made fewer than five nonbusiness trips per year into New York City, and 25 percent said that, except for business, they never went into the city at all. These proportions have undoubtedly increased further since the survey was taken.

For years, it was believed that the trend toward urban decentralization would have less effect on total traffic congestion in this larger region than in other urban areas. Occupations emphasized in Megalopolis's distinctive employment mix—that is, white-collar occupations—require daily decision making or are activities that support those making the decisions. The advantages of face-to-face interaction in the decision-making process help to concentrate white-collar jobs. Finance and insurance (as activities controlling investment), publishing firms, major magazines and newspapers (indicating the flow of information), and cultural leadership for much of the nation (through the theater, opera, symphony orchestras, and such activities as clothing design) are all much more important components of Megalopolis than of comparable areas elsewhere in the United

[5] *New York Times,* November 12, 1978, p. 131.

States and Canada. These activities employ, or are a response to demand from, white-collar workers.

Even Megalopolis's central cities, however, have recently experienced an outflow of decision-making jobs in unprecedented numbers. Among the nation's top 1000 corporations, for example, more corporate headquarters are located in New York's suburbs than in the city, a trend accelerated by the September 11, 2001 attacks. Improvements in computer technology and telecommunications have made information exchange more rapid without requiring the former level of centralization, and such improvements promise to continue. But to some degree, headquarters relocations have occurred as the disadvantages associated with central-city location accumulated.

The overall occupational structure of central-city employment changed dramatically during the last 35 years. For example, New York City has experienced a great decline in manufacturing employment (Table 4.2), but a large gain in service sector jobs. In 1947, industrial jobs in New York City exceeded 1.1 million, but now that number is just over 139,000, a decline of over 85% in a half century. This shift signals the increased importance of the New York metropolitan area as a financial and information hub for the world. Manhattan alone has approximately 100,000 workers in the securities and commodities sector and 250,000 workers in business service fields. But the transfer of jobs from one sector to another is not without problems. Areas that once had manufacturing jobs, such as waterfront locations, are not always the same districts that gained service jobs, so city leaders still

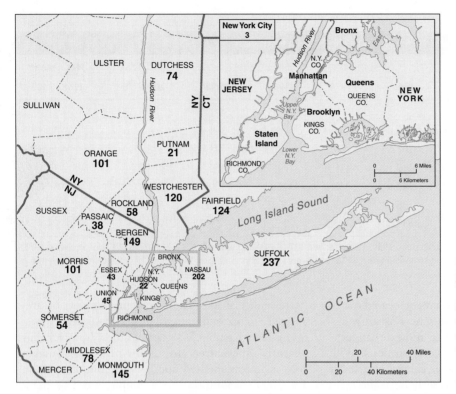

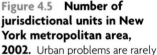

Figure 4.5   Number of jurisdictional units in New York metropolitan area, 2002. Urban problems are rarely contained within formal administrative units. Governmental operations can be very difficult in a metropolitan region with so many separate jurisdictions.

must struggle with redeveloping depressed areas while managing growth in other zones.

Some Repercussions of Sprawl

As urbanized areas expand, many of the problems found in the metropolitan cores spread toward the peripheries, initially in the older, interior suburbs and later in what earlier were entirely separate cities and towns. Even a generation ago, more than one-third of the suburbanites interviewed in the 1978 *New York Times* survey[6] believed their communities were becoming "more like a big city" with all its attendant problems.

Two somewhat contradictory trends follow from the slow aging of the interior suburbs. One is additional urban sprawl as inner suburban residents relocate farther out toward the metropolitan periphery. An opposite trend is reflected in the inflow of population as central-city residences are rehabilitated in ways that attract middle- and upper-income families and individuals, as part of the gentrification described earlier. In spite of the disadvantages, the large cities of Megalopolis possess great economic and social possibilities. These newly renovated

inner-city locations are attractive to a significant number of America's metropolitan population.

An equally problematic aspect of Spread City—as the sprawling metropolitan regions have been inaccurately called—is that not just one city but a great many independent municipal areas are involved. Everyone familiar with the great cities of the United States has an image of a large city surrounded by a multiplicity of smaller suburbs. In fact, the process of suburbanization is so advanced that many of the smaller cities near the major center are hardly suburban. The urban character that typifies the original core city now extends uninterrupted far beyond its legal boundaries and includes hundreds of independent urban territories. The extreme case, undoubtedly, is the New York Metropolitan area, which contains over 1600 distinct political units when national, state, county, local, and even sublocal governing bodies are taken into account (Figure 4.5). More than four decades ago, one observer described this situation as "a governmental arrangement perhaps more complicated than any other that mankind has yet contrived or allowed to happen."[7] These

[6] Ibid.

[7] Robert C. Wood, *1400 Governments* (Cambridge, Mass.: Harvard University Press, 1961), p. 1.

jurisdictions include municipal and township govern-ments, school districts, and a variety of special districts, some with taxing power and some without. This pattern is repeated at varying scales throughout Megalopolis and much of the rest of urban North America.

The implications of this fragmented pattern of jurisdictions are staggering. The economic geography of the metropolitan region does not coincide with its political geography. Most people live in and pay the bulk of their local taxes to a city different from the one in which they work. Neighboring cities com-pete for preferred businesses or industries. A desir-able industry is nonpolluting, pays a high average wage, and contributes significantly to the municipal-ity's tax base. A desirable resident is middle or upper income, so that the city can accumulate tax revenue to pay for the public services it is expected to provide. The populations and industries most able to support these public services through taxes are also the ones most able to move to a new location if local condi-tions deteriorate. As the core city continues to lose its tax base through population and economic relocation, the burden increases for those who remain. New York's brush with bankruptcy in 1976, Cleveland's in 1979, and Philadelphia's in 1991 are reflections of these tensions.

The central cities' problems are not isolated diffi-culties peculiar to the older sections of large metropoli-tan regions. Central cities are economically and socially woven into the spatial fabric of the entire urbanized area of which they are a part. It is coordination and coopera-tion between the hundreds of politically independent municipalities in each region that have been lacking. Attempts are made from time to time to improve region-wide coordination by establishing metropolitan study and planning regions that cross county and state boun-daries. Boston, New York, the Delaware valley area (Philadelphia), the Chesapeake Bay area (Baltimore), and the national capital region (Washington, D.C.) all established the beginnings of planning commissions whose jurisdictions extend far beyond the geographic limits of each large central city. The real impact of each commission, however, is variable and has never been

Many cities in Megalopolis have large retail and office clusters located just outside the central city but well within the suburban fringe. Tysons Corner, Virginia, pictured here, has more leasable office space than many U.S. cities.

very strong. Furthermore, the trend within Megalopolis and other large metropolitan areas in the United States lately has been toward a decrease of influence of these multi-municipal agencies in regional coordination and planning.

A logical question at this point would be, has there ever been any equivalent regional planning agency for all of Megalopolis? The answer is no, there has not, and it is unlikely that one would be effective even if it were in existence. Megalopolis is a massive region of coalescing metropolitan territories, with each territory containing a distinct urban core. But Megalopolis is *not* a super city. What binds the many parts of Megalopolis together is still much less than that which dominates each of its major urban areas.

Although not a super city and with its preemi-nence in North America challenged during the last two decades, Megalopolis remains a complex urban-ized region of tremendous importance to the rest of the United States and Canada.

URBAN FRINGE CASE STUDY

Queen Anne's County, Maryland

Ask farmers or fishermen whose livelihoods are rooted in tradition what they think about changes to Queen Anne's County on Maryland's Eastern Shore since the Chesapeake

Bay Bridge gave easy access to the area and most say they don't like it. As Boyd Gibbons observed in *Wye Island*, "With little difficulty you can find people on the Eastern Short who will tell you, without a trace of humor in their voices, that they would gladly blow up the Chesapeake Bay Bridge, if that would return the Shore to its former tranquility."[8]

When the Chesapeake Bay Bridge opened in 1952, Maryland's Eastern Shore was one of the rural oases in the creeping urbanization flowing out from major urban centers between Boston and Washington, D.C. Queen Anne's County, like the rest of the Eastern Shore, was sparsely populated with rolling fields, scattered forests, waterfronts, and tranquil rivers. Because it was relatively isolated, strong traditions in farming and fishing methods, architectural styles, cooking, and even social relations evolved. It was, as journalist and politician Frederic Emory wrote, "at once a delightful place of residence and an exceptionally favorable location for farmers, stockraisers, and truckers."[9]

Indeed, until the Chesapeake Bay Bridge was completed, the expanse of water between Annapolis and Kent Island literally kept the world at bay. Ferries running regular routes between the western and eastern shores brought summer vacationers and hunting and fishing enthusiasts from the growing cities of Baltimore and Washington. Some stayed in resorts near the ferry docks, and some built luxurious shore-side houses as weekend retreats. The more adventuresome, wealthy, or less pressed for time headed to the ocean beaches lining the Delmarva coast. But with their numbers small and their stay short, these so-called outsiders had scant effect on the self-sufficient, independent natives and the tidewater's traditional lifeways of farming and fishing.

Certainly, the rural character of Queen Anne's County was unaffected for most of the nation's history. Like those living in much of the Delmarva Peninsula of which it is a part, farmers grew grains and produce in fertile, easily worked soil to help stock the food shelves of nearby cities. Like much of Delmarva, the county and its people were psychologically and physically isolated (Figure 4.6). Bounded by the Chesapeake Bay on the west and the Atlantic Ocean on the south and east, the world went around Delmarva. Ships chugged up the Chesapeake to unload goods and people at the port of Baltimore or headed to the politically important cities of Philadelphia and Washington, D.C. Others headed for New York, Boston, or Richmond. Overland transportation routes skirted the Delmarva Peninsula to avoid the natural barrier posed by the Chesapeake.

There was a gloomy aspect to life on the Eastern Shore, however. By the 1950s, the economy had suffered for decades—particularly in Queen Anne's County. Property values were declining, as were the number of farming and fishing jobs. Unemployment was a chronic problem. Those lucky enough to find jobs scraping boats or pumping gas barely made a living at subsistence levels. Shanty towns dotted the region, providing squalid homes for the laborers at canning and seafood processing houses in places like Kent Narrows. In search of better opportunities, many young people drifted away from the area looking for jobs. So did their parents.

To begin to address the economic problems of the Eastern Shore, a 1950 study by the Maryland State Planning Commission focused on the economic consequences of building a bridge spanning the Chesapeake.[10] The commission forecast an upswing in the Eastern Shore's economy based on the increased vacation trade that a trans-Chesapeake bridge would create. They estimated that perhaps 2 million vehicles would cross the bridge the first year it was open. With a regular increase in traffic each year, the money the tourists would drop in Eastern Shore gas stations, restaurants, and motels would boost the area's failing economy.

The commission's expectations have been met and exceeded. By improving access to nearby urban populations, economic improvements accompanied by population growth began slowly but became obvious even to casual observers by 1980 (Table 4.3). As a result, Queen Anne's County today is caught up in a dilemma typical of areas on the margins of Megalopolis's urban sprawl.

The county's rural appeal is waning as more and more settlers are drawn from the Baltimore and Washington, D.C., metropolitan areas. Based on commuting patterns, Queen Anne's County was added to the other six jurisdictions constituting the Baltimore metro area. Each morning, thousands of people travel U.S. Route 50 across the Chesapeake Bay Bridge to their jobs in the city and retrace the route back to the Eastern Shore at night. The irony, of course, is that the rural tranquility these commuters seek is destroyed by their very presence.

On Kent Island, the Queen Anne's County access point for the Chesapeake Bay Bridge, this slow destruction of rural

TABLE 4.3

Growth in Population, Queen Anne's County and Maryland's Eastern Shore		
Years	**Queen Anne's County (percent)**	**Eastern Shore (percent)**
1950–1960	13.6	15.6
1960–1970	11.2	6.1
1970–1980	38.5	14.8
1980–1990	33.1	15.9
1990–2000	19.5	14.7

Source: U.S. Bureau of the Census.

[8] Boyd Gibbons, *Wye Island* (Washington, D.C.: Resources for the Future, 1987), p. 113.

[9] Frederic Emory, *Queen Anne's County, Maryland* (Baltimore: The Maryland Historical Society, 1950), p. 4.

[10] Maryland State Planning Commission, *Probable Economic Effects of the Chesapeake Bay Bridge on the Eastern Shore Counties of Maryland*, Publication No. 62, April 1950.

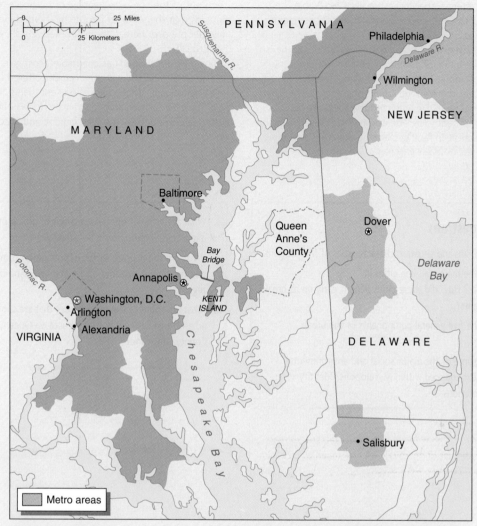

Figure 4.6   Queen Anne's County and urbanization on Maryland's Eastern Shore. The spread of urbanization onto the Eastern Shore was slowed for decades by Chesapeake Bay. But with the Bay Bridge providing easy access, commuters to Baltimore and Washington, D. C., now find residence in Queen Anne's County feasible. Adapted from maps © by Rand McNally from Commercial Atlas and Marketing Guide, 2003. R. L. 04-5-69.

character is particularly apparent. Although it still has the aura of countryside in places, the island is under siege from urbanization. Housing developments are replacing farm fields. Shopping centers, not combines, cluster along the highway's edge. Interspersed between and among the shopping centers are the usual array of fast-food eateries and gas stations. Traffic jams on the island were gargantuan until a new flyover bridge was built above Kent Narrows, particularly on summer weekends when vacationers headed for Delmarva beaches competed for access with commuters returning home. Amid this, county officials grapple with how to pay for larger schools

and increased services for a ballooning population. No longer a sleepy place of farmers and fishermen, Kent Island is becoming, in the words of Queen Anne's County's former planning director Joseph Stevens, a bedroom community for the Western Shore's metropolitan areas.

Robert Walters, a 75-year-old life-long county resident who has spent most of his life plying the Chesapeake's waters for clams or oysters, gave his impression of the change. "The way it is here now, if you haven't been some place in three months, you'll see something else," he said. "It used to be you could ride up and down the roads and just see fields and farms. Now there's just one shop right after another. It's getting like Glen Burnie," he added, referring to a Baltimore suburb with a 2000 population of just under 39,000.

Unlike many other Kent Island natives, Walters said he isn't bothered by the influx of people. "Guess I'm different, but different people don't bother me," he said. "The way I see it, I couldn't live over there either. I don't blame them for wanting to come here." He admitted that because he lives in a tranquil part of Kent Island, he doesn't feel closed in by the growing congestion. In fact, because he now runs a framing shop out of the converted garage of his home, Walters said he finds that the influx of people has helped him. "There's new people over here, building homes, and they need what I have to offer."

But other locals aren't as charitable. Most give the short answer an 85-year-old waterman crisply offered when asked what he thought of the changes on Kent Island. "Not much," he said. Typically taciturn among strangers, locals—like the elderly waterman—keep to themselves, lament, and watch the changes unfold.

Review Questions

1. Explain the site and situation characteristics of Megalopolis.

2. What are the major fall line cities in Megalopolis? How did these cities benefit early on as break-in-bulk points?

3. Describe the five general components of the urban landscape.

4. What are some of the major social and environmental problems experienced within Megalopolis? Identify some possible solutions.

5. What are some of the repercussions of urban sprawl? How might these repercussions be addressed?

6. What are the major changes in population composition and residential patterns within the dominant cities of the region?

7. What are some of the major issues that are showcased in the urban fringe case study that focuses on Queen Anne's County, Maryland?

CHAPTER 5

THE NORTH AMERICAN MANUFACTURING CORE

Preview

1. The North American manufacturing core extends across the territories of several regions.

2. The location of the manufacturing core is a function of the quantity, quality, and location of natural resources as well as transportation assets that aid the movement of resources.

3. The growth of railroads in the nineteenth century, and especially steel rails after the American Civil War, spawned rapid growth in the manufacturing core.

4. Although relatively poor in natural resources, East Coast industrial cities, such as Boston and New York, benefited from abundant access to water power, highly skilled labor, and good ports.

5. The movement of mineral resources such as iron ore and coal were crucial to the development of interior manufacturing cities such as Pittsburgh, while Chicago rose to prominence as the hub of a major rail network.

1000 companies, 469 were headquartered in the 16 states that are included in the manufacturing core region. In addition, 10 of the United States' 25 largest metropolitan areas and 10 of Canada's top 16 Census Metropolitan Areas in 2000 were in this area.

The international boundary this region straddles has little obvious effect on its shape. Loosely defined on three sides by the Ohio River valley, Megalopolis, and the southern Great Lakes, the region's western margin blends gradually with agricultural landscapes across southern Indiana, Illinois, and beyond. Although manufacturing is found in important clusters west of the Mississippi River, such as in California's fabled Silicon Valley, its impact is less concentrated across the United States and Canada outside what geographers call the continent's *manufacturing core*.

M anufactured products are everywhere around us—in clothing, in food, in residences and office towers, in transportation and communication, in computers and televisions, and more. Many of these items are manufactured in Mexico, Europe, and Asia, especially cars, small appliances, electronics, and textiles. Yet industry located in the United States and Canada still produces most of what people in both countries consume. Medium-sized towns and large cities often continue to have at least some local employment in manufacturing. Manufacturing has been so fundamental to the economic development of these two countries that we simply assume that sufficient volume and a range of choice of any item will be available whenever we need or want it.

Manufacturing is found throughout the more populated regions of North America, but its distribution is not even. Excluding northern New England, the northeastern United States and southern Ontario together comprise the continent's single most significant manufacturing region (Figure 5.1). Although occupying only about 3 percent of the total land area of the United States and Canada, the region contains factories that produce over half of both countries' steel and over 71 percent of all cars. Of the 2003 Fortune

Industrial landscapes can have a stark beauty that often masks the harsh reality of working conditions. Here, an Illinois foundry worker examines molten metal.

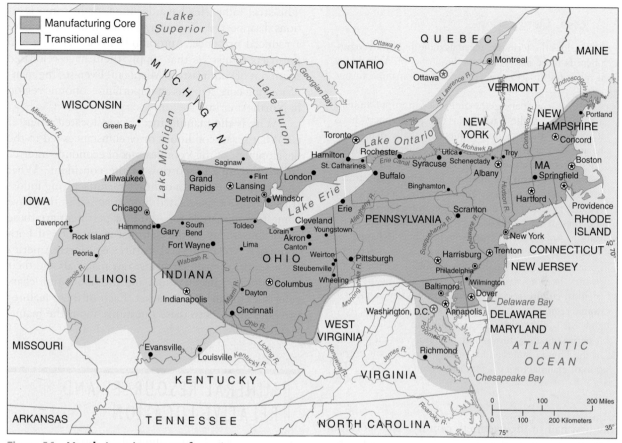

Figure 5.1 North American manufacturing core.

The North American manufacturing core's territory overlaps the territories of several other regions. Megalopolis (Chapter 4) extends across its eastern portion. Although Megalopolis is basically an urban region, the metropolitan areas within it contain significant manufacturing and are strongly linked to other centers in the manufacturing core. Because of this, most of Megalopolis is also part of the manufacturing core. Similarly, part of the area occupied by North America's agricultural core (Chapter 11) overlaps the western portion of the continent's manufacturing core (see Figure 11.2). This complex zone of overlapping core areas has led some observers to declare somewhat loosely that the combined areas of the manufacturing core, the agricultural core, and Megalopolis define a core region for the entire continent.

Manufacturing's role in defining a continental core suggests how important this form of economic activity is in defining the character of the United States and Canada. Most of the important ports, main centers of communication, and primary financial centers for both countries have been within or very near the manufacturing core. Although outside the core's boundaries as drawn here, the political capitals of both countries are on the region's immediate margins. The intensity of urbanization is much higher in the manufacturing core than in comparably sized territories on the continent and encompasses the two largest clusters of coalescing metropolitan areas: Megalopolis (Chapter 4) and the group of large urban regions between Milwaukee and Chicago on the west and Cleveland and Pittsburgh on the east.

In many respects, the vitality and productivity of the Midwest's farm population created the resources and demand for industrial production. Success in agriculture supported the manufacturing core's early market centers, and the gradual mechanization of agriculture demanded diversified manufacturing support. Mechanical reapers, winnowing machines, and cultivating implements were required by the tens of thousands during the late 1800s. Tractors, hay balers, pumps, other supporting equipment, and increasingly specialized farm machinery became important local sources of industrial demand during the first half of the twentieth century. Transportation lines were improved and expanded to carry the tremendous volume of agricultural products produced on the Midwest's farms. Thus, agriculture stimulated early urban and industrial growth, and successful manufacturing and rapidly growing urban centers continued to support intensification and improvement in agriculture.

Within the manufacturing core, a set of landscape features reflect the area's highly complex, distinctive flavor. It is a region with many large cities and even more cities of medium size. It is a region with numerous industrial concentrations and the transport networks demanded by them. It is a region of smokestacks and factories, some of which are massively drab on the outside but vibrant and energetically alive inside. It is a region of great ethnic diversity. Its industrial cities grew as waves of migrants left central, eastern, and southern Europe for new lives in the United States and Canada. The cities' cultural tapestries were enriched by subsequent African-American, Hispanic, and Asian migrants as well.

Changes in manufacturing and its concentration in the traditional core have been underway for decades.

The steel industries have declined, or in some locations disappeared altogether. Chemical industries have a reduced presence. Automobile manufacturing and the parts industries that supply them have relocated south out of southeastern Michigan. Even so, the manufacturing core remains an important economic region in North America.

Why is the manufacturing core located where it is? What set of conditions or circumstances led to the development of this complex mix of economic interrelationships on this portion of the continent? What encouraged the growth of heavy manufacturing industries and all of the related human activities that have come to dominate here? And how pertinent are these conditions today as the national economies of both countries evolve? Several features of North America are essential to manufacturing: raw materials and their distribution, transportation, and the population characteristics of labor and market. Each of these features provides answers to these questions about the manufacturing core.

MINERAL RESOURCES AND RELATIVE LOCATION

North America's metallic mineral resources are located in three broad zones of metamorphic rock: the Appalachians, the western mountains, and the Canadian Shield (see Chapter 2). The mineral fuels, on the other hand, are found more widely, but especially beneath the sedimentary lowlands west of the Appalachians and between the Gulf of Mexico and the Arctic Ocean. The quality, quantity, and variety of minerals available to those who sought to exploit them were extremely good. In many cases, too, deposits could be mined with no more than average difficulty. The United States and Canada are large countries well blessed with natural resources that drive heavy industry; it is not surprising, then, that manufacturing came to be significant in their economies.

Two additional geographic factors critical in defining the size, character, and location of manufacturing clusters in North America are relative location and accessibility. Mineral deposits are found along a northerly arc covering the northeastern quarter of the continent (the Shield) and in two linear areas, one extending north-east-southwest (the Appalachians) and the other northwest-southeast (the Rockies). With

River barges on the Ohio River are usually pushed by a single tugboat in groups of two or three abreast and as many as six units long. This group of coal barges is moving slowly through Cincinnati toward one of the region's many industrial sites.

respect to each other, the eastern and western mineral zones form a broad V cut off at the base; the Canadian Shield fits into the northern open end of the V. Consequently, the broad interior plains are nearly enclosed by these zones of metallic minerals. The interior plains are also underlain by large deposits of high-quality mineral fuel, especially in the eastern section where the Shield and the Appalachians are only narrowly separated by the St. Lawrence lowland. In terms of the mineral requirements of heavy industry, then, the relatively small triangular region bounded by the Appalachians, the Canadian Shield, and the Mississippi River contains most of what is needed.

No doubt, the three factors present—resource quality, resource quantity, and close relative location—would have led to the development of a concentrated manufacturing core in this general region. But other advantages exist in the form of what is called natural *accessibility resources*. Accessibility is a critical concept in human geography, defining the ability to interact between locations and the relative ease and costs of movement. Accessibility can be created or improved through human effort. Roads can be built or widened. Railroads can be constructed and maintained. Airports, canals, and pipelines all improve the accessibility of a place. Investment in transportation facilities, however, involves clear risk of financial loss unless the need for these facilities is overwhelmingly obvious. Navigable waterways, on the other hand, are natural routes of access that do not require as much initial investment

or contain the same possibility of financial loss. The presence of such waterways is a very real resource during the early stages of economic growth when inexpensive transportation is critical.

The interior portion of the North American manufacturing core possesses great accessibility resources. Connecting the mineral-rich Canadian Shield and the fuel-rich interior plains, the Great Lakes represent an internal waterway unlike any in the world. The five Great Lakes interconnect with only two significant changes of elevation. The small drop of about 6.7 meters (22 feet) between Lake Superior and Lakes Huron and Michigan was overcome by locks, first opened in 1855, at Sault Sainte Marie. The much greater elevation change between Lakes Erie and Ontario posed a more serious barrier to water transport, but the Welland Canal (first opened in 1829) was built in Ontario to skirt Niagara Falls. The Erie Canal was constructed by 1825 in New York to permit some freight to avoid Lake Ontario altogether.

With these exceptions, the lakes offered more than 800 kilometers (500 miles) of inexpensive transportation to early European settlers of North America. This transport encouraged the shipment of agricultural products to eastern markets. The same cheap transportation was of critical importance to industry in the later nineteenth and early twentieth centuries as the Shield's iron ore was shipped to sites near places where coal was mined in Illinois, Indiana, Ohio,

West Virginia, and Pennsylvania. The location of the industrial capacity that developed along the southern margins of the Great Lakes—Chicago, Gary, Detroit, Cleveland, Hamilton, Toronto, and many others—can be largely attributed to this natural accessibility resource. Although closed to traffic during the winter months, the Great Lakes are a transportation route of incalculable importance.

Within the interior core, additional accessibility resources are also available. Flowing west from deep within the coal-rich Appalachian region, the Ohio River crosses the interior plains for hundreds of miles, dividing the Eastern Interior coal-field before joining the Mississippi River. Dozens of tributaries supply the Ohio River with water and also provide further access, either directly because they are navigable or less directly because their paths offer easy routes through the tributaries' valleys for overland movement. Beyond the manufacturing core's western margin, the Mississippi River and its tributaries provide access from the south and west. The Mississippi also physically connects the industrial capacity of the core and the tremendous agricultural potential of the interior plains.

So unique is the combination of spatial and mineral resources located between the Appalachian Mountains and the Mississippi River that the manufacturing core in the United States is often thought of as this interior territory alone. References to the industrial Midwest or, more dramatically, to America's industrial heartland may fire the imagination, but they are geographically incomplete.

The inland, interior portion of the manufacturing core is a direct western extension of the large ports of Megalopolis. Generally, the Appalachians form a barrier to easy east-west transportation, except along the Hudson-Mohawk valley connecting New York City to the interior. But as early as colonial times, people began to construct land routes to neutralize this barrier. One consequence is that the North American manufacturing core includes both an interior portion and Megalopolis, the urban region through which the interior has its primary linkage to international commerce. Although the essence of Megalopolis is urbanism and exchange, Megalopolis contains massive industrial development and is a fundamental part of the continent's manufacturing core region. This overlap of Megalopolis and a portion of the manufacturing core is a preview of how North America's regions are thematically complex and in many cases not territorially separate.

Growth of the Region

The manufacturing core is a product of the late nineteenth century. It was not until the 1870s that manufacturing industries started appearing in intensities that encouraged extension of the tentative label "manufacturing core" to the area west of the Appalachians. Manufacturing was not a major concern during the early years of European expansion and settlement. Initial settlement paths followed the natural lines of access—the waterways. Although some families came overland, most moved into the area via the Ohio River and its tributaries or from the margins of the southern Great Lakes. By 1880, even though the overwhelmingly dominant economic activity was agriculture, average population densities of 18 persons per square kilometer (46 per square mile) existed almost everywhere in the area that was to become the manufacturing core.

Prior to 1830, U.S. urban and industrial development was limited almost entirely to Atlantic Coast ports and the ports' immediate hinterlands. European settlement west of the Appalachians was still limited to scattered subsistence agriculture and a few urban outposts. Between 1830 and the outbreak of the Civil War, however, population density in the interior increased. Agriculture intensified, and farmers started producing a regular surplus which, coupled to a growing population, sparked demand for functionally efficient centers of exchange.

The region's growing economic complexity was reflected in the gradual spread of railroad lines across the interior plains (Figure 5.2). This was also a time of rapid increase in water freight transport on the Ohio, the Mississippi, and the Great Lakes. The Erie Canal was complete and provided a direct link between the interior and New York City's rapidly growing port. The other great ports of Megalopolis could not match the support capabilities of the Erie Canal, but they attempted to incorporate portions of the productive interior into their hinterlands by improving overland routes. Baltimore, with the Baltimore and Ohio Railroad, and Philadelphia, with the Pennsylvania Railroad used gaps in the Appalachians to make connections with the interior plains of the Midwest. Not to be outdone, New York City was also connected via the Mohawk valley with the construction of the New York Central Railroad.[1]

---

[1] Students puzzled over the railroad's name, New York *Central*, should note the location of the Mohawk valley within the state.

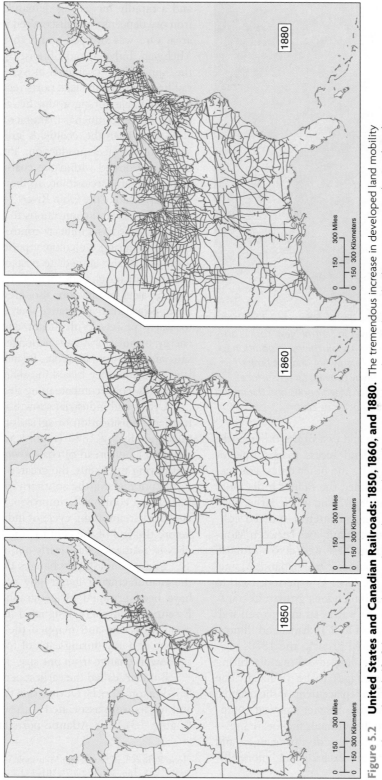

Figure 5.2 United States and Canadian Railroads: 1850, 1860, and 1880. The tremendous increase in developed land mobility across the northern half of the United States during the period is in sharp contrast to the more limited rail construction in the South and in Canada. (Reprinted with permission, Association of American Railroads.)

The valley of the Mohawk River in central New York is an excellent passage through the Appalachians. In addition to the river itself, a railway, two highways, and the limited access New York State Thruway provide direct surface transport connections between Megalopolis and the inland cities of the continent's manufacturing core.

Boston, located north and east of the other ports, found its main lines of land access to the interior United States preempted by the other cities' connections and therefore intensified the railnet in its more immediate New England hinterland much earlier than the other cities did. Canadian interest spurred construction and consolidation of rail lines between Montreal and Toronto and from there west across Ontario to Michigan, as well as southeast to Portland, Maine. The last decade of the pre-Civil War period, 1850 to 1860, foreshadowed the tremendous growth in transportation that followed the war as an interwoven web of rail lines flung out across Ohio, Indiana, and Illinois.

After a decade of slowed growth, the 1870s produced several changes that allowed the manufacturing core's full range of advantages to be realized. The development of steel led to replacement of the original iron rails, thereby permitting heavier railroad equipment, higher speeds, and longer hauls than were previously possible. Standardization of rail gauges and rolling stock allowed interline exchange and more efficient movement.

Accompanying the railroad's growing efficiency and a rapidly increasing demand for steel, extensive iron ore deposits were discovered in the Lake Superior area. Ore was shipped south to lake ports in Ohio, Michigan, Indiana, and Illinois. Bituminous coal from the Appalachian coalfields moved north over the new rail lines to the same lake ports receiving iron ore. The industries that sprang up in the lake port cities began to define the continent's manufacturing core.

Cities near the coalfields grew rapidly with the expansion of steel production. Pittsburgh is a prime example. Located within the coal-producing region, it enjoyed natural accessibility to the west by occupying a site at the head of the Ohio River. Thus, Pittsburgh commanded good land connections to the east and west. It had also served as a military control point for westward expansion during the colonial period and continued to grow during the immediate postcolonial period. Cities without Pittsburgh's advantages and less closely tied to coal or steel production also grew as coal-generated steam provided an alternative source of industrial energy to water power. With steam engines providing the energy, manufacturing was no longer tied to water power sites and waterfront locations. Inland settlements that had lagged because they lacked the ability to generate power locally now could compete using steam-driven power.

Urban and industrial growth was accompanied by further intensification of agriculture. Rail lines multiplied rapidly by 1880 (Figure 5.2), reflecting these changes. The spread of coal-powered electric generators during the 1880s, the greater flow of coal moving to the ports along the southern margins of the Great Lakes, and continually improving transportation created a self-reinforcing cycle of industrial growth. Well before the end of the century, the manufacturing core of North America was clearly outlined.

The technological changes that directly affected the manufacturing geography of the United States have been grouped by geographer John Borchert.[2] Essentially, Borchert designated five categories of city population size and mapped the relative growth or decline of a city during each of four historical epochs to show its change from one size category to another.

Borchert called the earliest period, 1790–1830, the Sail-Wagon Epoch. During this time, almost all cities and towns were associated with water transportation (Figure 5.3). The Atlantic ports and towns that had

[2] John R. Borchert, "American Metropolitan Evolution," *Geographical Review* 57, no. 3 (July 1967): 301–332.

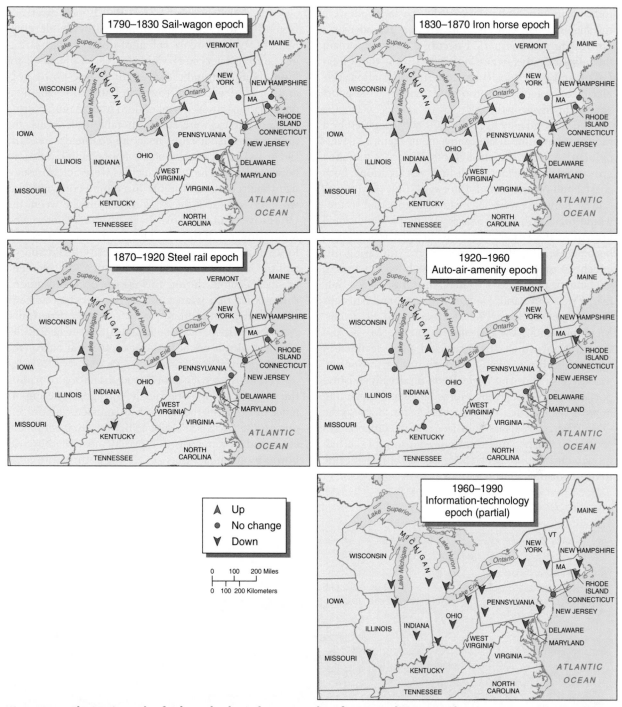

Figure 5.3   Change in rank of selected urban places: Epochs of metropolitan growth
(after Borchert, to 1960). The relative growth of many cities in the U.S. manufacturing core illustrates
the shifts in the importance of each city's site and situation advantages as the country's economy and
technological base developed after 1790.

their beginnings along some of the coastal rivers were the major urban centers. The greatest inland urban growth during this period occurred along the main inland waterways—the Mohawk River, the Great Lakes, and the Ohio River. Land transportation, primarily by wagon, was too rudimentary to support equivalent city growth.

The second epoch, 1830–1870, was triggered by a radical innovation in land movement—railway development. At first, the Iron Horse Epoch stimulated growth in the already established port locations, both coastal and on the inland waterways. Indeed, the new railway networks were so constructed that their routes would focus on the port cities. Aside from additional growth of the larger port cities in what was soon to become Megalopolis, the greatest growth during this epoch occurred in inland cities such as Pittsburgh, Cincinnati, and Louisville (all on the Ohio River); Buffalo, Erie, Cleveland, Detroit, Chicago, and Milwaukee (all on the lower Great Lakes); and St. Louis, Memphis, and New Orleans (all on the Mississippi River). Many other cities grew and a few declined in population. But the key to growth during this period was expansion of the hinterlands of already-existing port cities by constructing railroad lines and expanding in areas behind the frontier of settlement. Because the Ohio River, the Mississippi River, and the Great Lakes continued to be important, the most rapid urban growth during this epoch took place within what came to be the manufacturing core.

The third epoch, the Steel-Rail Epoch, stretched from 1870 to 1920. It was stimulated by the development of steel, the replacement of iron rails with stronger and heavier steel rails, increased demand for bituminous coal, and the spread of electric-power generation. The greatest growth in national urban areas occurred in cities peripheral to the manufacturing core, but there were several notable exceptions: the numerous smaller cities near the coalfields, near the Great Lakes, or on one of the major rail connections between larger cities. These cities were able to establish themselves because the interconnecting rail network crisscrossed the region densely between the Ohio River and the Great Lakes. Akron, Canton, and Youngstown, Ohio, are clear examples; they are located between the coal-and-steel city of Pittsburgh and the iron-ore port and steel city of Cleveland. The greatest decline during this epoch occurred among those cities that were most tightly tied to waterway navigability. Within the broader core area, Erie Canal cities (Syracuse, Utica-Rome, and Albany-Schenectady-Troy), Louisville on the Ohio River, and St. Louis on the Mississippi were major losers.

Perhaps an even more crucial factor in forming the manufacturing core was the change in the importance of two costs related to the manufacturing process during the Steel-Rail Epoch. Before 1870, shipment costs and power costs comprised most of the costs borne by manufacturers. By 1920, both had become less dominant in location decisions.

Looking northeast across downtown Pittsburgh, it is easy to understand the strategic nature of its location at the head of the Ohio River, formed by the confluence of the Allegheny River (to the north) and the Monongahela River (to the south). Note also the competing demands of water movement and land movement, as demonstrated by the city's many highway and railroad bridges.

The cost of shipping raw materials was an important share of total costs at the beginning of the Steel Rail Epoch because a significant proportion of the raw material was used up or discarded during the manufacturing process. Therefore, industry tried to minimize these costs by locating near the raw materials or in places where they might be shipped cheaply. That is, industry tended to locate in port cities on navigable waterways. This locational tendency was reinforced at some river locations by the continuing importance of water power as an energy source.

Toward the end of the 1800s, cheap and efficient railroad transportation removed some of the advantages of waterway location for industry. The spread of electric power also reduced manufacturers' dependence on port and river cities. Although some port cities and some industries retained their locational advantages, the general effect of the railnet was to make it more worthwhile for manufacturers to locate near their markets. In the Canadian section of the manufacturing core, southern Ontario developed more rapidly than southern Quebec. Ontario shared the important Great Lakes waterway and also was much closer to the growing markets of the interior core in the United States. Toronto, Hamilton, and Windsor all have the advantages of both accessibility and market proximity.

Borchert's fourth epoch, the Auto-Air-Amenity Epoch, spans the period from 1920 to 1960. Transport innovations such as the automobile and the airplane effectively minimized the impact of shipment costs in the production process and increased individual mobility. Migration within the United States was directed at those places perceived as having amenities, such as California, Florida, and Arizona. Industry's market and labor orientation, begun in the previous epoch, was strengthened as shipment and communication costs declined. Industry was drawn to places of greatest population growth, which were primarily the amenity areas outside the traditional manufacturing core.

Growth in the manufacturing core was steady during Borchert's fourth epoch, but there were clear prospects for future decline. Large markets and labor pools existed. Manufacturing and the linkages between industries were well established. The main exceptions to the general stability were slower than average growth in the steel city of Pittsburgh and more rapid than average growth until the 1960s in the automobile production centers of Detroit, Flint, and Lansing, Michigan.

The two countries entered yet another historical epoch after 1960, one that has extended into the twenty-first century and might be called the Information Technology Epoch. As the economies of both the United States and Canada become more dependent on the production and exchange of information, the means of processing and transmitting this information encourages the growth of industries that do not need cheap bulk transportation or large population clusters. The factors supporting growth in the large manufacturing core cities during the first two-thirds of the twentieth century no longer provide those cities with special development advantages. Even so, the skilled labor forces, large markets, and established air transportation patterns of those cities make some of them strong competitors for future growth.

THE ECONOMIC CHARACTER OF THE REGION'S CITIES

Until recent decades, little industry in the manufacturing core was located beyond the immediate vicinity of cities. The striking parallel between urban growth patterns and manufacturing industries in much of the United States and Canada confirms their interdependence. However, the forces stimulating growth in one city may be very different from those stimulating growth in another, so urban places have very different economic characters. As a result, the visible economic landscape of cities in North America's manufacturing core varies.

The economic character of the cities falls into two broad regional categories: (1) that associated with cities in the eastern portion of the manufacturing core, that is, the Atlantic coastal cities and the narrow region east of the Appalachian Highlands, and (2) that associated with the cities in the interior part of the core between the Ohio River and the Great Lakes.

Eastern Cities of the Manufacturing Core

With Boston, New York, Philadelphia, and Baltimore based early and firmly on commerce and the financial exchanges it stimulated, these port cities and their satellites lured people long before manufacturing became dominant in the national U.S. economy. Although manufacturing was drawn to the East Coast by the promise of matchless local markets, tremendous labor supplies, and easy access to water transportation, the economies of most cities in Megalopolis also continued to attract what came to be known as office industries.

New England was exceptional within the eastern core because manufacturing industries developed there when its ports were growing. So early was New England's industrial development that this region can be called North America's manufacturing hearth.

More densely populated than other portions of British colonial America, New England had relatively poor agricultural opportunities and nonagricultural labor became available there earlier than elsewhere. Shipbuilding industries thrived along the coast and generated countless subsidiary manufacturing operations needed to supply this complex industrial undertaking. When factory industry grew in importance elsewhere in North America, New England had several advantages that let its regional manufacturing keep step. Experience and continued immigration from Europe were important, but the steady availability of power churned from New England's small but abundant rivers was primary. Water power, sufficient labor, and proximity to coastal ports encouraged the rise of textile industries that helped establish the foundation for New England's manufacturing development. Although many of the textile industries eventually shifted to South Atlantic states (Chapter 9), the highly skilled residual population drew in enough replacement industry to maintain a slow regional population growth, and a high proportion of that population worked in manufacturing (Table 5.1).

Boston, as New England's regional capital, characterizes many of the changes in this portion of the manufacturing core. Boston's apparel and leather industries as well as shipbuilding in nearby Connecticut are remnants of an earlier period. Most growth since World War II has occurred in electronic components and machinery, very loosely called the "high-tech" industry. The production of these goods demanded greater shipping speed and flexibility than older and heavier industries, and well-connected suburban locations provided those advantages.

The growth of new industry on the city's periphery also emphasizes the relative decline of Boston's port. Although the harbor and its facilities are excellent, industrialists in New England (and Boston) now ship most products by land, either to markets in the rest of the United States or south to New York City for export through Megalopolis's primary port.

New York's primacy among North American harbors is unparalleled. As might be expected, manufacturing industries found proximity to this node of international commerce and the population cluster

TABLE 5.1

Percentage of 2007 Nonfarm Labor Force[a] in Manufacturing by State in Core Region		
	Percentage by Region	Percentage by State
New England	10.1	
Maine		9.6
New Hampshire		11.7
Vermont		11.6
Massachusetts		9.1
Rhode Island		10.2
Connecticut		11.4
Middle Atlantic	7.9	
New York		6.4
New Jersey		7.8
Pennsylvania		11.4
Maryland		5.2
Delaware		7.5
Interior	14.6	
Ohio		14.4
Michigan		14.6
Indiana		18.7
Illinois		11.3
Wisconsin		17.3

[a]U.S. national average: 10.2.

Source: Bureau of Labor Statistics, May/June 2007 monthly data.

around it advantageous. This pull was so strong that New York's industrial mix became extremely diversified. Many industries were located on Manhattan well into the beginning of the twentieth century.

The increasing demand for space by space-intensive office businesses, shown in the rapid construction of skyscrapers, gradually pushed heavier industry to lower Manhattan's margins or beyond the island's confines into the New Jersey tidal marshes across the Hudson River. More recently (see Chapter 4), many manufacturing jobs left the New York area altogether and were replaced by nonmanufacturing employment.

The New York metropolitan economy has been dominated for some time by office industries housed in the numerous, tightly packed skyscrapers that sprang up after the turn of the century. Forming concrete canyons extending for blocks are headquarters for dozens of companies and corporations, a banking and insurance cluster, publishing houses, and the myriad of service and control centers. These office industries require a worldwide information network and the facilities to transmit requests and responses rapidly. New York is industrial by virtue of the large number of people employed in office industries. Even with the recent relocations of several corporate headquarters, it remains the economic decision-making capital of the United States.

Philadelphia and Baltimore, initially so different in industrial inheritance and urban character, are apparently becoming more alike. Philadelphia's manufacturing base was almost as diversified as New York's, although it had almost twice New York's emphasis on food-processing industries, from the truck farms of southern New Jersey, the lower Delaware River valley and southeastern Pennsylvania, and Eastern Shore Maryland. Because New York lies only 120 kilometers (75 miles) to the north, Philadelphia's early industrial growth suffered somewhat from the presence of New York's better harbor and superior access to the interior. But better access to the coal and steel regions of western Pennsylvania, respectable port facilities, and a heritage as the United States' early political and cultural center maintained Philadelphia's growth.

Baltimore, on the other hand, has always been located on the manufacturing core's periphery. Like Philadelphia, its port possessed good rail connections with the interior's coal and steel regions. Baltimore's industrial mix, which included transportation equipment and shipbuilding and repair, reflects these connections.

Metals fabrication and chemical industries also are significant in the metropolitan economies of both Philadelphia and Baltimore. Steelworks exist at Sparrows Point east of Baltimore. Similar steelworks are near Trenton, north of Philadelphia. Those, along with the chemical industries in Philadelphia and around Wilmington, Delaware—located between the two larger cities—emphasize the coastal connections of these cities with the heavy industrial interior.

Because Baltimore and Philadelphia share these locational advantages, they will probably become even more similar in the future. Increasing U.S. dependence on imports of iron ore from Labrador, Venezuela, and Africa and continuing importation of petroleum from Venezuela, Nigeria, and the Persian Gulf states will increase the importance of their coastal locations and their access to inland demand centers.

Cities of the Interior Core

The larger portion of North America's manufacturing core, the industrial Midwest and southern Ontario, lies west of the Appalachians. Major cities in this part of the core derived their character from their location near the continental interior's rich mineral and agricultural resources. Almost all of the large cities in the manufacturing core's western portion are located along the Ohio River and its tributaries or along the shores of one of the Great Lakes (Figure 5.1). The Ohio River valley and the southern Great Lakes comprise two of the three zones of manufacturing in the interior core. A third zone lies between these two and is made up of numerous, generally smaller centers.

Most important in the development of urban centers in the manufacturing core's interior has been the movement of metallic mineral ores from the margins of the Canadian Shield to the coalfields of western Pennsylvania and West Virginia, along with the smaller movement of coal in the reverse direction (Figure 5.4).

Iron ore is mined at the Mesabi Range of northern Minnesota, at the Steep Rock deposit in western Ontario, and at the Gogebic, Marquette, and Menominee Ranges in northern Michigan and Wisconsin. Mesabi ore is now processed into pellets at the mining site to increase the iron content of the material shipped. Large ships designed specially for Great Lakes travel carry pelletized Mesabi ore to the southern shores of Lakes Michigan and Erie. Iron pellets and ore are carried to the southern shore of Lake Michigan, to Hammond and Gary, Indiana, where the iron is met by coal transported north by rail from the large Illinois coalfields. Even larger shipments of iron ore are destined for Lake Erie ports. From there, the iron is either carried south, primarily to the steel cities along the Ohio River, or converted to steel in lakeside cities using coal carried north on the return rail trip from the Appalachian fields. Iron is also shipped beyond Lake Erie to the Canadian port of Hamilton on Lake Ontario, supplementing Canadian ore brought up the St. Lawrence from Labrador deposits.

Of the interior core cities, Pittsburgh is the one whose name became synonymous with steel. Located at the junction of the Allegheny and Monongahela rivers where they join to form the Ohio River, Pittsburgh

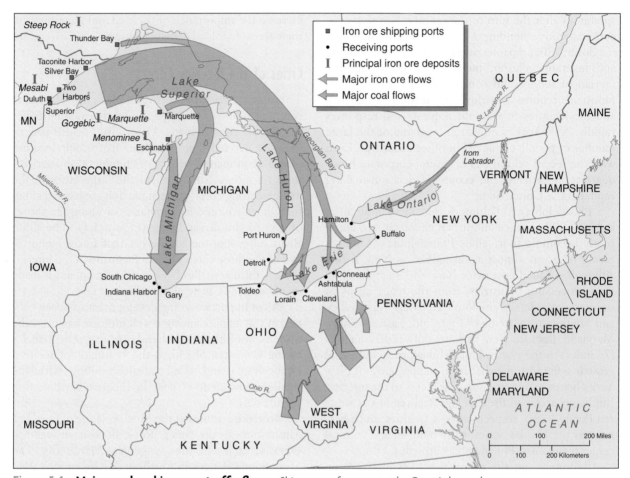

Figure 5.4   Major coal and iron ore traffic flows. Shipments of ore across the Great Lakes and both coal and iron ore between the Ohio River and southern shores of the lakes have been fundamental to the development of urban industrial concentrations in the interior section of North America's manufacturing core region.

is in an excellent position to take advantage of access to both raw materials and downriver markets. The Allegheny and the Monongahela rivers drain the coal-rich margins of the Appalachians, and the Ohio flows along the southern margins of the agricultural core and into the Mississippi. The beginnings of the United States Steel Corporation spurred Pittsburgh's development as a steel-producing center. As Pittsburgh grew, industries dependent on steel crowded onto the narrow river bottoms to take advantage of proximity to the steel mills and to low-cost water transportation. Metal-fabricating industries, machine parts manufacturers, and other industrial consumers of large quantities of steel located their plants in and around Pittsburgh. Nearby cities also benefited from the powerful pull of steel. Youngstown,

Canton, and Steubenville in Ohio; Wheeling and Weirton in West Virginia; and New Castle and Johnstown in Pennsylvania shared in the industrial growth of this region and developed their own steel and steel products industries (Figure 5.5). With the decline of the U.S. steel industry and the closure of many of the mills, however, these cities have experienced economic hardship (Figure 5.6).

Urban industrial growth did not happen solely at the source region of coal. Iron ore shipped across the Great Lakes had to be transferred to railroads at points along Lake Erie's shore for final movement to the Pittsburgh area. Railroad cars returning north to the lake carried coal rather than moving empty. This coal provided fuel that was not available locally and helped

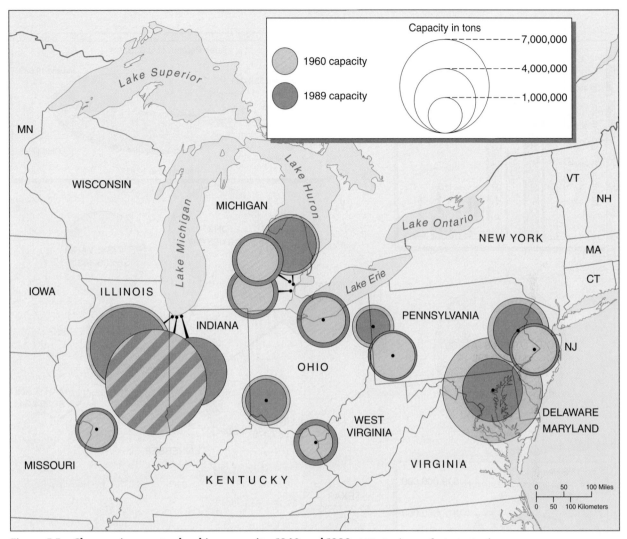

Figure 5.5   Changes in raw steelmaking capacity, 1960 and 1989. U.S. steel manufacturing in the region began to decline during the period from 1960 to 1989, a trend that continued into the twenty-first century.

metal products industries and other industry take advantage of the *break-in-bulk* function at the lake's edge.

Cleveland was the largest of Lake Erie's port cities. A canal connecting the narrow and winding Cuyahoga River with a tributary of the Ohio River stimulated Cleveland's initial growth. Although Cleveland quickly outgrew this small initial advantage, it was enough to give the city a head start on nearby urban competitors. The diverse industrial base that resulted there took advantage of the accessibility offered by the lakes and

by the major east-west railroads in the United States connecting New York with Chicago and the agricultural core to the west. Cleveland's growth also spilled over into adjacent Ohio ports, such as Lorain, Ashtabula, and Conneaut. Growth pushed as far east as Erie, Pennsylvania, and as far west as Toledo, Ohio, as well as to inland centers such as the rubber and plastic goods-producing city of Akron.

Buffalo sits at the eastern end of Lake Erie, and its location has had major effects on its industrial character. Buffalo was the premier flour-milling center on the

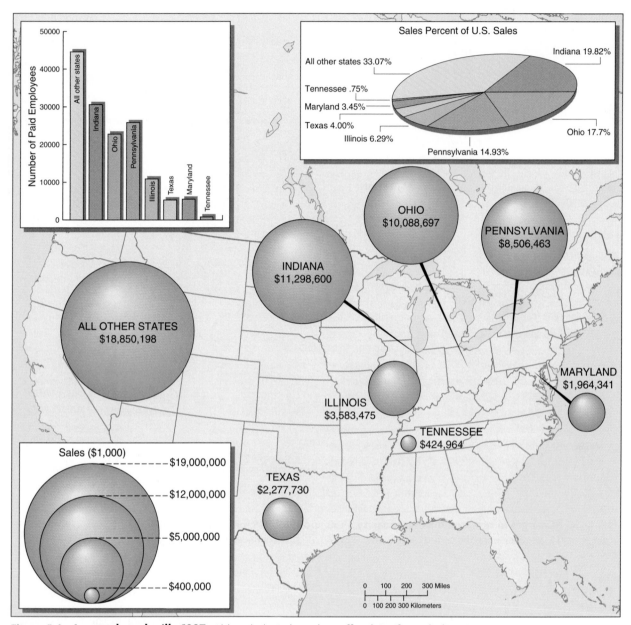

Figure 5.6 **Iron and steel mills, 1997.** Although the industry has suffered significant declines in recent decades, iron and steel production remains concentrated in the interior portion of the manufacturing core region.

continent until after World War II. Although the volume milled there has declined since the war, wheat from the Great Plains still arrives at the western Great Lakes to be carried by water in bulk to Buffalo for refining before sale in Megalopolis's large eastern markets. The decline occurred partly because Buffalo's locational advantages as a transfer site were weakened when the St. Lawrence Seaway opened. Buffalo also grew because the same factors generating steel and metals manufacturing elsewhere along the lakeshore helped ensure that a significant portion of the city's manufacturing would be connected to this type of industry. Another locational advantage was hydroelectric power derived from harnessing Niagara Falls. This drew chemical and aluminum industries, among others, to the Buffalo area because their high electric-power requirements could be met.

Of Ontario's two major port cities in this part of North America's manufacturing core, Hamilton has the strong mix of steelmaking, metals-fabricating, and automobile industries typical of the eastern Great Lakes. Toronto, much larger than Hamilton, developed a more diverse industrial base. Why Toronto grew more than Hamilton is complex (see Chapter 6). It has few obvious locational advantages over Hamilton, yet Toronto has a full mix of light industry rather than specialization in one or two heavy industrial sectors.

On the narrow water passage between Lakes Huron and Erie, Detroit and Windsor initially grew because this site controlled traffic between these lakes and was a natural crossing point between Ontario and southern Michigan. Despite such apparent locational advantages, the cities did not reap significant development benefit during manufacturing's early period in North America. Detroit did not grow rapidly until the first part of the twentieth century because it is located more than 80 kilometers (50 miles) north of the primary New York–Chicago rail connection. At first, the more direct but politically more difficult rail route across southern Ontario via Buffalo helped stimulate some growth. However, Detroit did not develop the urban character it is best known for until the rise of the industry that fostered the railroads' chief land transport competitor—motor vehicles.

As successful automobile manufacturers concentrated in Detroit and nearby cities and the demand for automobiles skyrocketed, a wide variety of suppliers were drawn to southeastern Michigan. These industries attracted a large workforce, and the result of this was that Detroit became the fifth largest city in the United States for several years. It held this rank until the 1970s when a decline set in. Today, Detroit is the country's tenth largest city.

Although Detroit was as specialized in its industrial mix as Pittsburgh and as susceptible to similar fluctuations and trends in demand and competition, its range of complementary industries was greater because motor vehicles require an extremely large number of components. After both the steel and motor-vehicle industries became less concentrated spatially, Detroit experienced a slower initial decline in

Manufacturing core cities along the Great Lakes contain many industrial clusters, but The Flats along the Cuyahoga River in Cleveland are one of the most intense concentrations of heavy industry in the region.

manufacturing and service employment than Pittsburgh. For example, metropolitan Detroit's non-farm payroll employment increase, at 17 percent, almost kept pace with the national increase of 18 percent between 1985 and 1995. But metropolitan Pittsburgh's equivalent employment declined by more than 18 percent during this same period. In recent years, Pittsburgh has weathered the economic downturn better than Detroit. Pittsburgh's nonfarm employment in 2002 was 6.9 percent higher than in 1995, but Detroit had only increased nonfarm employment by 4.0 percent. Likewise, at the end of 2006, Pittsburgh's unemployment rate was 4.9 percent compared to 7.2 percent for the Detroit metro area.

The two other metropolitan centers located along the southern Great Lakes margin are Milwaukee and Chicago. Milwaukee's industrial character was typical of Great Lakes port cities, but the city had more agriculture-related industry. In addition to the usual industrial mix of heavy industry and motor-vehicle manufacturing, Milwaukee for years was one of the leading centers of the continent's brewing industry. This was a legacy of the large number of German immigrants who settled in Wisconsin during the late nineteenth century. Milwaukee also has a significant food-processing industry, for the city is a major focus for Wisconsin's middle Dairy Belt.

Chicago is easily the dominant city of the interior manufacturing core. For many years it was called the "Second City," in recognition of its importance and its large population (2,896,016 in 2000), second only to that of New York. Los Angeles, with a larger metropolitan area, grew rapidly enough to pass Chicago to take over second place by the 1980s and reach 3,694,820 people in 2000. Even so, Chicago remains the informal capital of the Midwest and persists as the strongest urban focus of the United States' interior.

Chicago had to overcome a series of early site disadvantages to grow to its present size and economic importance. The city occupied an undesirable swampy lake margin. Although ideal for mosquitoes and similar pests, its early human inhabitants viewed it with little favor. Chicago's drinking water also had a distinctive flavor that was often unappreciated. The Chicago River was too narrow and shallow before improvements were made to provide access to the interior agricultural region. And, finally, after Chicago began to

The Gary Works in Gary, Indiana, seen here from Dunes State Park, provides a good example of large steel mills that dot the southern margins of the Great Lakes. This is the area of the United States where rail-carried northbound coal meets southbound iron ore transported by Great Lakes ore carriers.

establish itself as a regional center, it was nearly consumed in the great 1871 fire.

The disadvantages of Chicago's site were overbalanced by an array of situation advantages. Located along the southwestern shore of Lake Michigan, Chicago occupies the optimum location for people and goods transfer between lake shipping and the rich agricultural region to the west and southwest. The Illinois and Michigan Canal, snaking through the heart of the city, was completed in 1848 to link the Great Lakes and the Mississippi River system—the two greatest waterways of the continent. Chicago's strength as a focus of inland water transportation was supplemented 4 years later when it was connected to New York by rail, allowing Chicago to become the Midwest's primary regional rail focus as well.

With the support provided by its position as a major transportation hub, Chicago grew rapidly, absorbing hundreds of thousands of European immigrants throughout the later nineteenth century. The city spawned a radial network of rail lines into Illinois, Wisconsin, and the agricultural states beyond. A large meatpacking industry developed around the city's extensive stockyards. Lured by growing local markets and good access to markets farther west, other industries such as furniture and clothing manufacturers located in the city. The steel industry was introduced to the Chicago area after the turn of the century. The smokestacks that can still be seen south of the city along the lakeshore at East Chicago, Gary and Hammond, Indiana, were easily accessible to Chicago's unparalleled railroad network.

A city of 1 million by 1890, Chicago doubled in size before 1910 and exceeded 3 million by the mid-1920s. The self-reinforcing cycle of population increase (labor and local markets), economic growth, more population growth, and so on was clearly operating before the turn of the century. Today the volume of Chicago's manufacturing activities (more than 387,000 employed in the city in June 2007) is matched only by the immense diversity of products manufactured. With its volume and diversity, Chicago is at least partly an effective regional counterbalance to the set of intense economic nodes in Megalopolis. And because of its location, Chicago's industrial base was formed on both the typical Great Lakes port pattern of steel and metal products manufacturing and an agricultural orientation provided by the rich lands farther west. Although changes in meat production techniques and shipping have

Chicago's intense demand for space and the accommodation made for water traffic is apparent in this image of the city's lakefront area. Chicago's metropolitan area reaches to the horizon and beyond lies the vast interior plains.

closed the city's stockyards, Chicago's vitality and the implications of its location were caught effectively in Carl Sandburg's 1916 description of the city as "Hog Butcher, Tool Maker, Stacker of Wheat, Player with Railroads and Freight Handler to the Nation."

The manufacturing clusters and major urban centers within the continent's manufacturing core not located along the southern Great Lakes fall into one of two other general zones. One is the Ohio River complex that includes Pittsburgh and its immediate neighbors, the industrialized Kanawha River valley, the older port city of Cincinnati, and many smaller centers in between. The core's other zone of manufacturing is defined by the diverse and scattered centers located between the Great Lakes and the Ohio River and between the Appalachian coalfields and the dominantly agricultural region toward the Mississippi River.

Almost all of the Ohio River clusters are related to the accessibility the river provides in moving coal and other Appalachian raw materials at low cost or in the return flow of materials from the Mississippi River and beyond. Small, but representative cities such as Ironton, Ohio, and Ashland, Kentucky, are sprinkled along the eastern half of the Ohio River. Most had small steel plants or metals-fabricating factories at one time, although many have now closed. When plants close in

these small one-industry towns, whether temporarily or permanently, the effects can be devastating on the local workforce and require major efforts to attract alternative employers. The Kanawha valley industries in such places as Nitro and Belle, West Virginia, on the other hand, are more diversified, using coal as a raw material as well as a fuel and specializing in chemicals, synthetic textiles, artificial rubber, plastics, and glass. Cincinnati also has changed its industrial mix. Formerly a port city for outbound agricultural products, it now has a more balanced manufacturing structure, but one that bears the typical Ohio River stamp of metals industries.

Cities located in between the Great Lakes and the Ohio River—Indianapolis, Fort Wayne and South Bend, Indiana; Dayton, Lima and Columbus, Ohio; Battle Creek, Jackson and Flint, Michigan; and many others—have manufacturing activities that are partly a spillover from the larger centers surrounding the core's interior, partly a response to the intensely successful agriculture of the region (see Chapter 11), and partly a consequence of the network of interaction set up between the larger centers. Cities like Fort Wayne, Lima, and Flint, whose economies were strongly tied to heavy industries, have suffered along with these industries. But other cities, such as Indianapolis and Columbus, are developing rapidly as their service and market-oriented industries have grown.

THE DYNAMICS OF NORTH AMERICAN MANUFACTURING

Just as there are changes over time in the locational reasons why one company succeeds and another in the same business does not, there are also changes in the relative importance of locational advantages that helped form the manufacturing core. New England lost much of its textiles industry to the Southeast but substituted electronics. Chicago lost its stockyards and much of its meatpacking industry to a myriad of smaller sites and new ways of marketing meat. Towns along the Ohio River lost part or all of their steelmaking operations. These and other changes have been stimulated largely by technological innovations.

Behind the losses and gains of individual firms and particular industries are broad shifts in economic activity. These shifts are reflected in the proportion of the labor force working in each sector of the economy. Agriculture's proportion of the total national economic effort, for example, declined gradually during the twentieth century in both countries, while manufacturing maintained its proportion until the last several decades. The drop from nearly 50 percent of the labor force engaged in agriculture in 1880 to 2 percent in 2006 was matched by an increase in service, or tertiary sector activities. The tertiary sectors of the United States and Canadian economies, comprised of activities that could loosely be described as serving the population (retail and wholesale, transport, recreation, finance, and many others), grew as income became more generally distributed among the countries' inhabitants. As the primary, or extractive, sector (agriculture, mining, fishing, lumbering) and the secondary, or manufacturing, sector (steel, chemicals, consumer products, and so forth) became more efficient, productivity increased and fewer people were required to produce the same amount of goods. More of the labor force became engaged in providing services for people in the primary and secondary sectors.[3]

[3] Some geographers argue that quaternary and quinary sectors of the economy also exist. Quaternary sector activities are involved in information creation and transfer; control functions make up the quinary sector. Examples of quaternary activities are education, portions of the communications industry, and research and development. Quinary activities are represented by decision makers in business headquarters and government policy forums. Others argue that this sectoral analysis of national economies is useful only in a general, mostly conceptual manner. Activities do not fall neatly into one sector or another; most jobs involve more than one sector in their day-to-day activities.

While the balance among the economic sectors gradually shifted in both the United States and Canada, the geography of these sectors changed. New technology applied to agriculture permitted increases in total production even as the farm population declined. The tremendous industrial growth in the manufacturing core after the Civil War created a demand for labor that attracted migrants from Europe and from rural America, especially the South. The core's sprawling industrial agglomerations grew to city size and filled with multiethnic populations heavily dependent on secondary sector employment. Then after 1970, industries began to relocate as population shifted within each country, the use of computers and telecommunication grew, competition from foreign manufacturers increased, and access to an educated workforce became more important. This relocation shifted economic growth to centers outside the core in spite of many factors resisting such a change.

The automobile industry provides an example of these changes. The geographic pattern of motor-vehicle assembly was affected in the 1980s by a number of innovations. The introduction of robotics on the assembly line changed the requirements and importance of the labor supply in locational decisions. In addition to this technological leap, the adoption of highly coordinated "on-time" component delivery systems significantly reduced inventory costs because parts arrived as they were needed. "On-time" delivery also meant that final assembly sites no longer needed to be located near the places where components were manufactured. And redesign of the truck trailers that carry completed vehicles meant that delivery time between the assembly plant and the dealer was also reduced. These and other innovations have helped disperse the industry in the United States away from its traditional location in southeastern Michigan.

Much has been made recently of higher relative economic growth in the southeastern and southwestern United States and in western Canada, but the total locational inertia of the manufacturing core region is immense and manufacturing continues to be strongly present there. The persistence of manufacturing employment in core region states, especially those in the interior portion of the core, is evident from workforce patterns over a long period (Table 5.2). By using the percentage of employment in manufacturing for the entire United States as a baseline and comparing each decade's

TABLE 5.2

Percentage of State Employment in Manufacturing Relative to National Average, 1920–2002 (US Average = 100)					
State	**1920**	**1950**	**1980**	**1996**	**2005**
Massachusetts	167	144	119	95	90
Rhode Island	191	170	143	122	105
Connecticut	175	164	139	114	110
New York	127	115	94	82	64
New Jersey	156	146	113	81	76
Pennsylvania	135	137	129	114	112
Ohio	135	141	130	135	141
Indiana	110	134	134	157	181
Illinois	108	124	116	112	110
Michigan	136	158	128	146	145
Wisconsin	111	118	120	151	167

Source: Calculated from U.S. *Census of Population* (1920) and *Statistical Abstracts of the U.S.* (Washington, D.C.: U.S. Government Printing Office, 1951, 1981, 1997, and 2007).

state percentages to this standard, state-by-state shifts in manufacturing employment can be compared without the comparison being affected by changes in the national emphasis on manufacturing.

Within the manufacturing core, changes in relative state employment concentration in manufacturing reflect the two broadest zones within the region. Generally, in 1920 states east of the Appalachians were engaged more heavily in manufacturing than states of the interior core. But the eastern portion has experienced a decline in manufacturing employment since 1920 relative to the trend for the interior states, with

New York and New Jersey manufacturing employment now well below the national average. Though general, these figures, together with those in Table 5.1, suggest that the enormous economic mass of the interior manufacturing core is resisting rapid change even while specific industries within the region experience difficulty. The figures also suggest that absolute shifts in manufacturing employment will continue to alter the economic composition of other regions in North America.

Review Questions

1. How might you define the relative location of the continent's manufacturing core?

2. What are some of the general circumstances that contributed to the growth of an American manufacturing core?

3. How can Borchert's Model be employed to better understand the evolving nature of the American urban system?

4. What are some principal motivations for industrial agglomeration?

5. What are the significant attributes of Pittsburgh's site and situation?

6. Examine the stability of the U.S. Manufacturing Core in terms of the costs of locational inertia.

7. Given the disadvantageous nature of its site, how did Chicago emerge as the dominant city of the interior manufacturing core?

8. How can you explain the tremendous ethnic diversity found in the major manufacturing cities of the core?

CANADA'S NATIONAL CORE

CHAPTER 6

111

Preview

1. Canada's national core contains a large proportion of the country's population and economic activity and is entirely within the provinces of Quebec and Ontario.

2. Much of Canada's historical settlement pattern was linked to the export of raw materials (fur, fish, and forest products) to Europe.

3. The poor soils and drainage of the Canadian Shield limit human settlement. Thus, Canada's population in the core clings to a narrow strip of land south of the Shield along the country's southern border.

4. Cultural conflict between English-speaking and French-speaking Canadians has been a recurring reality in the core.

5. Farming exists in the core and was crucial to the area's early development, but the region's cities and industries are its most significant features.

Canada's urban population and industrial capacity are also found here. Its industry is associated with advantages of accessibility provided by the Great Lakes and the St. Lawrence River or with nearby mineral-rich hinterlands.

French settlers spread into the St. Lawrence's riverine lowlands in the century and a half after 1608; subsequent English settlers mostly went inland from the shores of Lakes Ontario and Erie in the late eighteenth and early nineteenth centuries. By 1867, when the new Dominion of Canada was formed, more than 90 percent of Canada's population resided in the southern portions of Ontario and Quebec and around the margins of the Maritimes to the northeast. Only during the last 125 years have Canadians established permanent settlements in much of the vast interior.

Canada's national core is entirely within the provinces of Quebec and Ontario, lying along the northern shores of Lake Erie and Lake Ontario in a band reaching northeast along the banks of the St. Lawrence River to Quebec City (Figure 6.1). This national core is in the country's historic hearth. Most of the core and the Atlantic Provinces' margins were settled by 1850, while the rest of the country remained a frontier (see Figure 3.2). Early on, the small, nascent region contained the seeds of Canada's continuing national dilemma: how to resolve internal cultural divisions while remaining a single country?

Canada is larger than the United States. The country covers more than 9.8 million square kilometers (3.8 million square miles) compared to the United States' roughly 9.3 million square kilometers (3.6 million square miles). But despite Canada's greater extent, it contains a much smaller share of North America's manufacturing core. Canada's national core extends beyond the manufacturing core's boundary to include cities like Ottawa, Montreal, and Quebec that are culturally and politically important in Canada.

Given its settlement history and situation, Canada's national core contains a large share of the nation's population and economic activity. About 60 percent of Canada's people live in southern Ontario and southern Quebec. Disproportionate shares of

Quebec City's landscape invokes images of Europe and reminds us of Canada's French and British heritage. Here, the Château Frontenac rises above the city.

These historic patterns suggest three themes explaining the essential character of Canada's national core. The first theme deals with the national core's relatively small size and the factors restricting growth within its boundaries. Ordinarily, the development hearth of a rapidly growing and prosperous country like Canada increases in area as the national economy grows. But in Canada the core's extent is no greater than it was in 1867.

The second theme deals with the core's marked cultural division. Typically, a core region's character, and the economic, political, and cultural activities within it represent the country's distilled purpose. In Canada, the core is culturally divided between French and Anglo Canadian spheres. It is politically fragile along the same lines, but historically and economically the core region is one territory, not two.

Third, the character of Canada's national core is affected by its location as part of North America's larger core. Thus, the relationship between Canada and the United States affects the integration of Canada's national core into the continental core.

Canada's national core is located far from the center of the country, a consequence of the historical sequence of European settlement and environmental patterns. The initial European entry into this portion of Canada was from the northeast via the St. Lawrence River. In 1535, Jacques Cartier reached Montreal Island. Further extension of French interest along the Great Lakes waited until the turn of the seventeenth century. After the French and Indian War ended in 1763, the British encouraged settlement of English Protestants to balance the French Catholic population. Two decades later, displaced families remaining loyal to Britain spilled into Canada across the United States border after the American Revolution. This British settlement spread along the northern shores of Lakes Ontario and Erie.

Because of the St. Lawrence River's importance as an entry route for Europeans, the lands reached by passage up the river were given the administrative

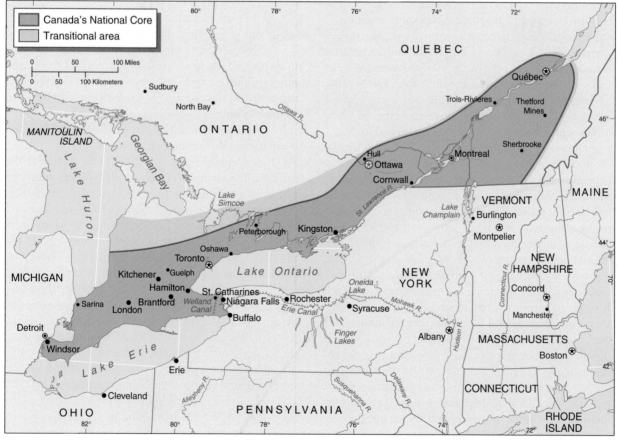

Figure 6.1 **Canada's national core.**

labels of Lower Canada and Upper Canada during the first half of the nineteenth century. Although these names are no longer in use, the land in southern Quebec, occupied by the French Canadian population along the St. Lawrence River, was Lower Canada. The more southerly lands bordering Lakes Ontario and Erie occupied by the English Canadian population were Upper Canada.

Canada's settlement pattern was for a long time linked to production for export to Europe of what are called *staple products*. Staple products are basic economic goods in steady demand. Generally, staple products are bulk goods with limited value added through processing. Thus, when staple products are exported, the originating country receives little more than the value of the raw material.

In Canada, dependence on cod fishing off Newfoundland and fur trapping in the St. Lawrence valley gave early settlers the means to purchase imports from Europe. But as eastern fur trappers exhausted the local supply and fur trapping shifted west, Canadians turned to alternatives such as timber, metals, and raw agricultural products as the bases for their export economy.

From the late eighteenth century on, trade was oriented toward Great Britain. Population increased during the nineteenth century within the St. Lawrence valley and southern Ontario, but the increasing local market did not lead to economic independence. The traditional economic and trade ties with Britain maintained Canada's dependence on exploitation of staple products.

During the decades before World War I, Canada's economy began to evolve beyond a nearly absolute dependence on staples exported from the national core. Although the United States' economy expanded its influence in Canada, and Canadian domestic manufacturing industries grew while settlement spread beyond the national core, these factors did not transform Canada's national economy away from staple production.

Severe environmental conditions prevail in much of Canada. The climate and soils most suited to early European farming are located on the southerly Ontario Peninsula and to a lesser degree along Quebec's narrow St. Lawrence lowland. With Canada's population growth prior to confederation in 1867 supported by the production of staples for export, the pressure on land resources in Canada's national core became considerable. In spite of this, dense settlement did not push much beyond the national core's present limits until after confederation. By the time Canada's population did extend across the continent,

southern Ontario and southern Quebec had accumulated enough economic, political, and cultural inertia to remain the country's leading growth areas. But why did the developing core stay so small? Why was settlement restricted until the late nineteenth century, three-and-one-half centuries after initial French exploration?

LIMITS OF THE CORE

Canada's physiography limits the size of its national core. The Canadian Shield is composed of igneous and metamorphic rock covering much of the continent north of the core. This vast zone's margin forms a long arc from the Arctic coast to the Great Lakes and the St. Lawrence valley (see Figure 2.1). Within Canada's national core, the Shield lies under a thin layer of soil from the mouth of the St. Lawrence estuary southwest to about the capital city of Ottawa, south again almost to Lake Ontario, and then nearly due west to Georgian Bay (Figure 6.2).

Early settlement in Quebec was restricted to land south of the Shield's thin and poorly drained soils. Thus, for years agricultural and commercial populations hugged a narrow band of land only a few dozen miles wide along the St. Lawrence River. The Shield is not quite as restrictive for farming or as visible in southern Ontario as it is in Quebec, but Ontario's agriculture and settlement remained clustered in Ontario's south during the nineteenth century.

It was not the Shield's topography that was the problem for farming. Much of the Shield's southern margin is a rolling, low plateau dotted with small lakes and marshes. Nor were the zones of mixed forests a problem. As elsewhere in North America, where trees stand in the way of farmers' plans, the forests are removed. North of Canada's national core, poor soil and harsh climate kept farming from succeeding.

The Shield's soils are poorly drained, thin, and acidic. During the last Ice Age, mile-thick, blue-tinged glaciers slowly crept south over the Shield. As they traveled, the glaciers bulldozed loose material from the underlying rock surface and then deposited this glacial till in mounds where the glaciers ended. By the time temperatures rose, melting the glaciers 8000 to 10,000 years ago, little soil cover was left on the Shield's crystalline bedrock. Too little time has passed since then for Shield soils to develop more than shallow pockets of thin, acidic layers.

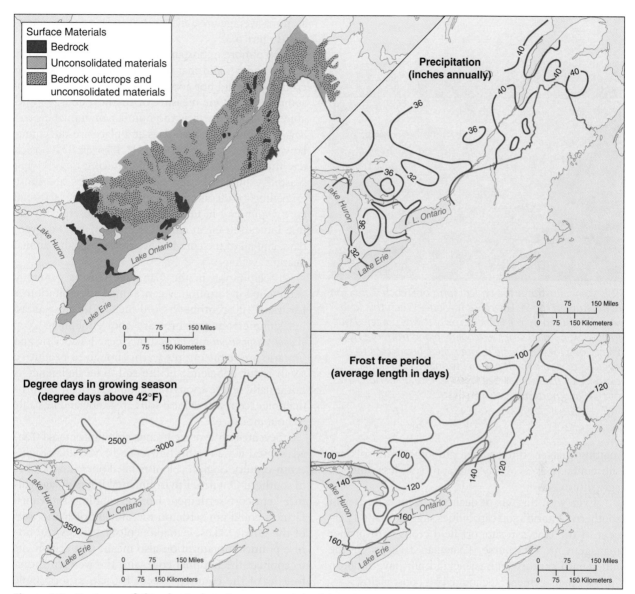

Figure 6.2   Features of the physical environment. Each of these four indicators of the environmental support for farming reflects the importance of southern Ontario and the St. Lawrence valley to the Canadian core's agriculture.

The combination of scouring glacial ice and a resistant bedrock also resulted in the region's poor drainage. Most streams and rivers flow off the Shield to the Great Lakes, the St. Lawrence, or north into Hudson Bay. But their paths are rarely direct and uninterrupted. Most surface water wanders haphazardly from lake to lake toward the Shield's margin. The resulting *hydrography* discourages agricultural use; small swampy pockets and poor drainage are everywhere. This hydrography also discourages commercial development because only a few rivers, like the Ottawa, are navigable enough to move goods.

Soils suitable for farming exist south of the Shield in Ontario and in Quebec's St. Lawrence valley. This land's low relief masks a complex pattern of deep and moderately fertile alfisols (see Chapter 2) which derive directly from glacial till or from outwash material carried by the tremendous volumes of water leaving the

This scene illustrates the typical form of French Canadian agriculture in the St. Lawrence lowland. When this land was distributed by the French crown, it was parceled out in narrow, parallel strips extending inland from the river. Each "long lot," as the fields are called, has access to the river, or to a road built parallel to the river at some distance to the interior. See Figure 6.4 for an aerial view of this land pattern.

melting glacier. Interspersed among the alfisols are substantial pockets of clay and sandy soils less suited to agriculture.

Even with this mix of quality, current estimates of southern Ontario's soil capabilities suggest that 60 percent of the land can support field crops with proper management. Of course, Canadian farmers in the nineteenth century and earlier did not have modern "proper management" technology to boost yields.

Besides soil quality and drainage patterns, climate also limited settlement expansion beyond the Canadian core. When Canadian settlers moved inland, they were farther north and farther from the Great Lakes' moderating influence. These frontier farmers encountered appreciably shorter growing seasons, and their agricultural efforts suffered.

Within the core, the Ontario peninsula's growing season is extended by proximity to Lakes Huron, Erie, and Ontario. The lakes' size and nearness slightly hasten southern Ontario's spring thaw and postpone its autumn freeze. The St. Lawrence River valley and the lowland between the St. Lawrence and the Ottawa rivers also have an extended growing season. Almost all

sections average greater than 120 frost-free days annually (Figure 6.2).

Even more important than short growing seasons, early settlers found that the heat energy necessary for crop growth did not accumulate in the long, but cool summer days. One measure of the heat energy available for crop growth is the number of annual *degree-days*. The annual degree-days at a place are the cumulative number of degrees over 42°F (over 5.5°C) each day this temperature is exceeded during the growing season. Canada's better agricultural areas are those exceeding 3000 annual degree-days. The national core's boundary in Quebec coincides closely with the line indicating mean annual heating of 3000 degree-days, and all of the core in Ontario exceeds this annual heating average.

Precipitation in the Canadian core is generally sufficient for farming. Mean annual precipitation is less than 100 centimeters (40 inches) in most areas, but summer temperatures are cool enough to keep *evapotranspiration* rates low. Only near Lakes Erie and Ontario are temperatures warm enough to produce a small water deficiency in the soil. The deficiency is not usually severe enough to hamper agricultural production except where local soils are sandy and less able to hold moisture.

Seventeenth-century French colonists entered Canada's interior via the St. Lawrence River valley. Eighteenth-century English colonists filled out Canada's core in southern Ontario. Environmental limits on agriculture kept core settlement from expanding. Individual farmers could not settle on the Shield north and west of the Great Lakes. Canada's colonial status and diffuse political organization also meant that arguments to connect the national core with the western three-fourths of the country, except to help export staple products, were unconvincing until late in the nineteenth century. By then, the core's agricultural base was strong, and southern Quebec and southern Ontario were thoroughly settled with two cities approaching 100,000 residents (Montreal and Toronto) and another five exceeding 10,000 (Quebec, Ottawa, Kingston, Hamilton, and London).

Strong economic trade connections with Europe helped solidify the core as a focus of Canada's transportation network. Staple products from Canada's interior flowed through the core on their way to Europe. Eventually, manufacturing industries developed in the region, and raw materials began flowing to Canada's western ports and south into the United

TENSIONS WITHIN THE CORE   **117**

States. By that time, however, the national transportation network was firmly focused on the national core. Canada's centers of finance, commerce, and political activity also remained strong in the core. The region's national dominance was clearly established.

TENSIONS WITHIN THE CORE

The southern margins of Quebec and Ontario share many characteristics. Both provinces are located along the main access route between Europe and the Canadian interior. Both developed urban commercial centers that dominated each province's economic organization. The urban centers competed with each other for dominance of all of Canada. Both provinces' populations gradually shifted from rural to urban residences, a shift typical of industrializing economies. Both possess a restrictive environmental base, and both have economies tied to the United States through investments and subsidiary ownerships. But in spite of their similarities, Quebec and Ontario are strikingly different in one crucial way—the makeup of their cultural landscapes.

Cultural Landscapes

The most pervasive contrasts within the core relate to Canada's heritage of cultural dualism. Most of Quebec's people trace their ancestry to French origins, while most of those living in southern Ontario are of British heritage. Canada also received many migrants from other European countries and, more recently, Asia. Some resettlement between Quebec and Ontario has occurred. Even so, the cultural boundary between French Canadians and those of British heritage continues to follow the provincial boundary (Figure 6.3).

French Canada's cultural base is reflected in linguistic distinctiveness, religion, historical identity, individual attitudes, settlement patterns, and, frequently, in a political posture defensive of French Canadian culture. The sometimes vitriolic and typically passionate adherence to French Canadian culture follows from its minority position in Canada. As of 2001, nine of ten people whose mother tongue is French lived in Quebec. Although one-third of New Brunswick's population and 485,630 of Ontario's identified French as their first language, fewer than one-fourth of Canada's total population reported French as their first language that year. Almost all the rest of Canada's people claim English as their primary language.

French Canada keeps its distinctiveness largely because it does not want to be culturally assimilated. When the British conquered France's Canadian colonies in 1763, most land in the St. Lawrence lowland was occupied by French-speaking farmers. The cities of Quebec and Montreal, with their robust French presence, were rapidly growing commercial centers. What's more, French traders and fur trappers had carried French influence beyond the Great Lakes to the Ohio and Mississippi rivers. Great Britain's seizure of governmental control from France in these parts of North America effectively ended most migration and cultural infusions from France, but it did not destroy Quebec's cultural base. Quebec's economy was simply redirected away from French goals and toward British ends.

Beginning in the 1780s after the American Revolution, British Loyalists moved into eastern Canadian areas less well settled than lowland Quebec and yet possessing greater agricultural potential. As a result, there was little immediate dilution of French culture in Quebec's lowlands. During the nineteenth century, Montreal sustained French cultural dominance in the national core's eastern half despite its growth as a commercial and industrial center within a British-controlled national economy. The city's cultural distinctiveness persisted even though its growth attracted non-French Canadians along with many from rural Quebec.

As French Canada's rural population migrated to Quebec's cities, the migrants helped sustain French Canadian culture in increasingly diverse urban centers. More isolated than urban residents from Anglo cultural influences, migrants from rural Quebec to Montreal and the province's smaller cities and towns kept infusing French Canadian perspectives into growing urban populations until urban French Canada was firmly established. Today, Quebec has almost 80 percent of its population living in urban places. Quebec's nonurban population is also larger today than it was a century ago. The strength and vitality of the province's French heritage seem secure.

Unlike southern Quebec, Ontario had few European settlers until British Loyalists were forced out of the United States after the American Revolution. Some of these Loyalist families moved to Nova Scotia and New Brunswick, but most went to Ontario where they were soon joined by waves of British, and some American, settlers.

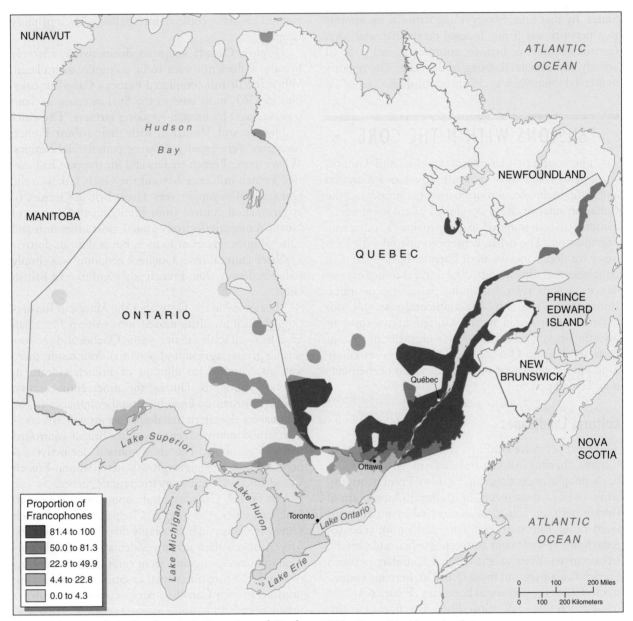

Figure 6.3   Linguistic dominance in Ontario and Quebec, 2001. The political boundary between these provinces does not match the linguistic boundary perfectly, but relatively few people in southern Ontario use French as their first language and relatively few in Quebec consider English their mother tongue.

Colonial governmental policy, partly as a counterweight against potential American invasions, but also to balance the large and growing French Canadian population farming along the lower St. Lawrence, drew migrants to southern Ontario from the British Isles. Many settlers were Scots and Irish, although most were English. By confederation in 1867, southern Ontario was an Anglo Canadian culture hearth and a base for population diffusion beyond the Great Lakes onto the Canadian prairies.

Canada's cultural and regional diversity is served by a *federation*. This is a form of government in which a group of states or provinces unite under a single, mutually acceptable sovereign central authority. Some governmental powers are assigned to the central government, while the states or provinces retain others. Ideally, a federation (or confederation) possesses advantages of a large, unified organization while accommodating local regional differences.

If regional differences persist and are deep enough, however, regional interests may conflict with the central authority's view of national responsibilities. Conflicts between local and national values in the United States, also a federation, were so irreconcilable that the country was torn by civil war and scarred for a century by the war's great destruction and loss of life.

Contentious Biculturalism

Canada's constitutional existence has been tested by regional differences for more than 130 years. Tension between the Canadian central governmental powers and Quebec's provincial powers has revolved around the issue of French Canadian cultural independence and equality within a largely non-French country.

Political concessions have been made to ease tensions in this bicultural country. Officially, Canada is bilingual. This ostensibly means that French and English languages are equally important in the country's governmental operations. In theory, a federal civil servant in British Columbia, for example, could speak French to other government workers and expect to be understood. But in fact, few people in the western provinces speak French; thus, bilingualism is not practiced in these areas in spite of its status.

In recent decades, issues of cultural dualism, cultural integrity, and equality polarized Canadians around the validity of the bilingualism policy. Outside French-speaking portions of Canada, this policy accumulated considerable opposition. Today, either as cause or effect, a correspondent heightening of Quebecois nationalism exists within *La Belle Province*.

Quebec's voters have cast ballots repeatedly to express their opinions about the province's secession from Canada. The 1996 referendum results showed an even division of opinion, with 50.56 percent voting to remain in Canada and 49.44 percent voting for secession. After the close 1996 vote, the Supreme Court of Canada ruled that Quebec secession would require not only a majority vote in the Province, but countrywide as well. It is not inevitable that Quebec will withdraw from Canada and become an independent state, but the political issue remains volatile.

Much more is involved in Quebecois nationalism than language differences, but contention over linguistic English/French parity has been a political issue in Canada for many years. A Canadian census study in the late 1960s aroused French Canadian nationalism when it projected that Montreal's population would be mostly non-French-speaking by the year 2000. To counteract the potential erosion of French, a provincial law was passed requiring that all of Quebec's signs be in French. This provincial law was subsequently ruled unconstitutional by the Canadian Supreme Court, but most signs today are indeed in French. A later law was passed requiring all children permanently residing in Quebec to attend French-language schools unless the parents themselves were educated in Quebec in English-language schools. This second law was intended to reduce migration into Quebec by non-French speakers.

Quebecois separatist efforts and the issue of French Canadian independence have long-term implications for Canada. About one-fourth of Canada's population and about 85 percent of its French-speaking population live in Quebec. Similarly, Quebec's economic productivity represents about 21 percent of Canada's total. Losing Quebec means Canada would lose a significant part of itself.

Quebec's secession would affect residents who live in Quebec but do not want to leave Canada and French Canadians who live outside the province. The referenda on secession show repeatedly that about half of Quebec's voters do not want to leave Canada, undoubtedly for many different reasons. The impact of Quebec independence on the Atlantic Provinces' economies, given the Provinces' location east of Quebec, and on their French-speaking populations (about 22 percent of the total in the Provinces), is also uncertain.

As cultural differences translate into strong political feelings, the Canadian core is threatened by the potential division. The likelihood of Quebec's separation from the rest of Canada may be diminishing as immigrants from other parts of the world settle in the province, but it is too early to be certain of this effect. Should Quebec separate politically from Canada, the remaining region would hardly be a national core. Economic connections between Quebec and Ontario have accumulated for three centuries. Only if bitterness between the two populations is strong following separation are the benefits of close economic association likely to be surrendered along with political union.

Rural Landscapes

Agricultural landscapes are very different in southern Quebec and southern Ontario, although similarities exist. Grain and animal production dominated farmers' efforts for decades in both regions. Near-subsistence farming in the less productive interior was often supplemented by maple sugar manufacture and lumbering during long nonfarming seasons. Even so, each half of the national core developed differently because settlers encountered dissimilar environmental conditions and brought with them distinct cultural histories.

A culture leaves its marks on the landscape, and this occurred in Canada's national core. In Quebec and a few small sections of southern Ontario, France granted land to settlers within blocks, called *seigneuries*. In each grant block, individuals received land according

Although both French and English are the official languages of Canada, French is the language of commerce, and parking, in Quebec.

to the *rang* system. Under this system, farmsteads were located on a riverbank or lakeshore, or on a road constructed parallel to the shore. Farmsteads lay at the head of long, narrow lots extending inland perpendicular to the shore or road. Each lot was called a *roture*, and its dimensions were based on the *arpent*, a French unit of measure equal to 192 feet.

The *rang* system gave settlers equal access to the main transport lines. The system also incorporated differences in land quality on each farmstead; many lots included both the swampy shoreline and better-drained upland terrace.

As demand for land increased, new roads were built farther inland but parallel to the shore or first road. From the new roads, a second, third, and fourth *rang* were demarcated. France laid out southern Quebec's best farmland (and a few areas of southern Ontario) in this manner before colonial Britain's takeover. What is popularly called the French long-lot pattern, the *rang* system still maps out the best farmland in southern Quebec (Figure 6.4).

As Quebec's rural population grew, some French Canadians migrated to the United States or nearby Canadian cities. Many more French Canadians remained rural residents, expanding Quebec's settlement frontier onto the Shield and into the equally marginal hill land south of the St. Lawrence estuary. As the nineteenth century unrolled, other parts of Canada more suited to wheat production became the country's wheat sources. Quebec's farmers shifted their emphasis from wheat to dairying, the latter supported by hay and oats, and specialty crops, such as vegetables, sugar beets, fruits, and poultry. Eventually, marginal land was abandoned and reverted to forest when more factory and service jobs became available after World War II.

The British demarcated Ontario's rural landscape and landscapes in southern Quebec's less well-endowed or less accessible lowlands in squares or rectangles. Individual homesteads are more dispersed under the British arrangement. When French Canadians later settled on land laid out by Britain, the pattern of landholdings rarely changed from the initial British form to the long-lot pattern.

Farming changed in southern Ontario much as it did in southern Quebec, but Ontario's more southerly location and more extensive lowland led to a different result. Most of southern Ontario's farmers planted wheat as a cash crop during the early 1800s to meet increasing food demand in the Canadian core and the

Figure 6.4 Long-lot farms in Quebec. This digital orthophoto quadrangle reveals the unique imprint of the long-lot survey system in the cultural landscape. (Image from the U.S. Geological Survey.)

eastern United States. By midcentury, however, yields declined sharply as the alfisols lost their initial fertility under constant wheat production. Farmers diversified cropping, but wheat remained prominent in Ontario agriculture until the Prairie Provinces were settled in the last third of the nineteenth century.

Mixed farming, producing both feed crops and meat animals, became the primary farm type in Ontario until about World War II. Gradually, innovations in milk storage, transport, and daily farm operation during the 1920s and 1930s encouraged many farmers in the region to undertake dairying. Dairying dominates farming near Windsor, Hamilton, on the Niagara Peninsula, and in southeastern Ontario's non-urban counties. Livestock farming, on the other hand, continues to exceed dairying in value of production near Toronto, London, and Kitchener.

Specialty crop farming is also significant in southern Ontario. Along the middle Lake Erie shore and to the northeast, sandy soils support high-value, flue-cured tobacco farming. The high annual returns for tobacco production offset to some extent anxieties about its long-term demand.

Southern Ontario farmers have also grown fruit, mostly on the Niagara Peninsula. The Niagara Peninsula, located between Lakes Ontario and Erie, possesses soils and a climate approaching the optimum for the growth of some fruits. Set in a relatively southern latitude for Canada, the peninsula's winters are moderated and spring blossoming is retarded because of its position between the lakes. Grapes, peaches, and sweet cherries are the most important fruits grown here.

Fruit production on the Niagara Peninsula does not appear secure, however. The peninsula's urban

The Niagara Peninsula, especially the land along the southwest shore of Lake Ontario, contains many fruit orchards and grape vineyards. Although this region continues to be an important source of apples, peaches, and grapes in Canada, urban activities are encroaching rapidly on the agricultural uses.

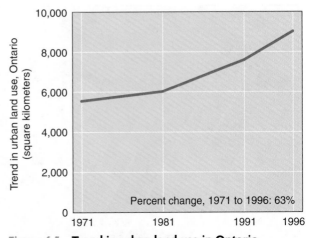

Figure 6.5 Trend in urban land use in Ontario, Canada, 1971–1996. (Data from Statistics, Canada, Environment Accounts and Statistics Division, Ottawa.) The rapid conversion of valuable fruit-producing land to urban uses is a great concern in this section of Canada.

areas are expanding to accommodate growing populations. As early as 1978 a study of land-use change in the Niagara Fruit Belt concluded that the peninsula's best fruit soils were being lost to urban development[1] (Figure 6.5). Attempts to preserve what remains have had mixed success. As in the United States, land changes from rural to urban uses have not been easy to control in Canada, and the long-term success of such attempts remains unclear.

Farmers in Ontario's core region, like those in southern Quebec, are under pressure to apply more capital, mechanize their operations, and become more specialized in what they produce (see Chapter 11 for more on this trend). In both sections of the Canadian core, the trend will continue toward further land abandonment along the region's margins and more intense production within the national core's best farming areas.

Urban and Industrial Dominance

The Canadian national core is defined by three characteristics. It is where the country's distinctive cultural dualism is most strongly represented; where agriculture was productive during the colonial period and is still important; and where industrial and urban development is most complex. Regarding the third characteristic, however, industrial and urban growth occurred in disparate ways within the core's two halves. Growth variations reflect each section's situation, history, and environment. Growth differences also reflect each section's interactions with the United States.

Each half of the core has an urban-industrial center far exceeding all others in their province: in southern Quebec it is Montreal, with 3,635,571 people in 2006, and in southern Ontario it is Toronto, with 4,753,120 residents in the same year. Recent uncertainties over Quebec's possible secession helped Toronto, with a 9.8 percent increase in population, grow significantly faster between 1991 and 2001 than Montreal, whose population increased only 3.0 percent.

Greater Montreal benefited from military and commercial site advantages during the colonial period. The original settlement lies at one end of an island in the St. Lawrence River. The island sits adjacent to the Ottawa River's confluence with the St. Lawrence and the first major rapids on the St. Lawrence. The island

[1] Ralph R. Krueger, "Urbanization of the Niagara Fruit Belt," *The Canadian Geographer* 22 (Fall 1978): 189.

is also at the northern end of the Hudson-Lake Champlain lowland.

Together, these locational characteristics gave Montreal a strong defensive position and military control of two major water routes to Canada's interior. Montreal's situation has even greater economic advantages for trade, and the city's municipal administration exploited these advantages throughout the nineteenth and twentieth centuries.

Montreal's location on the St. Lawrence River, where shipments were required to circumvent the rapids, was important to early commercial development because of the site's break-in-bulk advantages. Montreal surpassed the city of Quebec's population between 1825 and 1830. But Montreal's true dominance in southern Quebec was not assured until the 1850s when the river was bridged, connecting it by railroad to Portland, Maine in the east and to Toronto in the southwest. Access to the southeast also enhanced Montreal's position at the northern end of the Lake Champlain-Hudson River valley and increased its connections with New York City.

Montreal was already a commercial, financial, and transportation center by 1867 when confederation occurred. The city's economy was well along a self-reinforcing growth spiral. Industry attracted labor and a larger market; labor and market growth drew more industry that drew more labor, and so forth.

Montreal came to have a large non-French population. Most of the city's early financiers and industrialists were British, while most of the labor force was French Canadian. Montreal's economic opportunities attracted French Canadians from the countryside, and the city's French-speaking population exceeded its English speakers by 1871. This history has given Montreal its strong bicultural character.

The character of the city of Quebec is much less bicultural than that of Montreal's. Quebec lies at the head of the St. Lawrence estuary and has fewer of the locational advantages that stimulated Montreal's industrial growth. Although Quebec's metropolitan population had grown to 717,600 by 2006, the city still retains the strong French Canadian flavor expected of a cultural, religious, and provincial administrative center.

Significant urban growth in southern Quebec occurred only recently outside of the immediate Quebec and Montreal spheres. The growth of the smaller centers is traceable to resource-based industrial development. Some, such as Trois-Rivières (141,529 in 2006), combine access to Shield resources

like pulp timber and hydroelectric power with imported materials like the bauxite ore processed at nearby Shawinigan. Others, such as Thetford Mines (26,861) and Sherbrooke (186,952), grew by utilizing local minerals (asbestos at Thetford) or abundant local labor and diversified industrial investment. Throughout the province, Montreal dominates by means of its transport and financial connections.

Toronto is southern Ontario's equivalent to Montreal and Montreal's urban competitor within the national core. At first glance, Toronto appears to have few natural site or situation advantages to support its growth. Advantages accrue because the city is located on Lake Ontario, has a good harbor, and lies at the head of the Toronto Passage, a lowland route to Georgian Bay. But other sites along the lakeshore, perhaps at either end of Lake Ontario, seem to provide equivalent urban growth advantages. Toronto had neither Montreal's advantage of location where natural transport routes were focused nor Montreal's early urban growth momentum.

Toronto matched, and in some ways exceeded, Montreal's economic growth, however, because Toronto has several other advantages Montreal does not. Shortly after British rule began in southern Ontario, the government decided that the region's capital would be York (now Toronto) rather than Kingston, Niagara, or London. So Toronto became Ontario's political capital while the city of Quebec, not Montreal, was Quebec's provincial capital. For administrative purposes, the provincial capital of York (Toronto) had to be well connected by road, and later by rail, to the rest of Ontario. Therefore, Toronto's historical political function provided it with a land transportation network that made it the natural focus of all Ontario prior to confederation. Equally important, this land transport network focusing on Toronto provided the city access to a larger and more productive agricultural region than that of Montreal. Its superior hinterland gave Toronto an alternative basis for economic growth and put it on a par with Montreal's strong commercial locational advantages.

Early Canadian urban growth was also stimulated by developments in the United States. As long as most U.S. commercial and industrial activities concentrated along the country's northeastern seaboard, Montreal's proximity to them was advantageous. Although Toronto had good access via the Erie Canal route through the Mohawk and Hudson River valleys, the early edge lay with Montreal. During the late nineteenth century, however, as the southern Great Lakes industrial centers in the United States grew and the North American manufacturing core expanded across the Midwest, Toronto's Great Lakes location and easy accessibility by rail through Windsor to Detroit and Niagara to Buffalo assured the city's urban and industrial expansion.

Toronto's relatively strong hinterland and proximity to major U.S. manufacturing centers are reflected in southern Ontario's urban pattern. Geographers define three zones: an urban-industrial cluster of cities around Lake Ontario's western end; a group of Ontario cities adjacent to southern Michigan's industrial concentration; and an array of cities located along railroad lines connecting the other two zones.

At Lake Ontario's extreme western end, Hamilton (population 692,911 in 2006) developed as an iron-and steel-producing city. Beginning in the late nineteenth century, iron ore was shipped from Lake Superior mining regions through the Welland Canal. Iron ore is now also brought via the St. Lawrence from Shield mines in Labrador. Coal was imported inexpensively from Appalachia by direct rail connections through Buffalo. Hamilton's foundries and rolling mills encouraged further industrial development within Hamilton as well as transport and farm machinery manufacturing in nearby St. Catharines (390,317 in 2006). Hydroelectric power from Niagara Falls, good water access to both the St. Lawrence and the interior Great Lakes, and abundant agricultural production attracted chemical, textile, metals refining, and food-processing industries.

Toronto's immediate hinterland comprises a small crescent-shaped region of especially intense economic

The city of Quebec was founded at the mouth of the St. Lawrence atop a high river bluff to control the river traffic. During the centuries that followed, the city's commercial, cultural, administrative, and industrial activities gradually overwhelmed its original limited purpose. These activities now support over 700,000 people in the city's metropolitan area.

American visitors to Toronto, Canada's largest city, are often surprised by its vitality, modernity, and welcoming openness.

activity and spatial interaction. The land is occupied intensely and productively from the automobile center of Oshawa (141,590 in 2006), through Toronto and Hamilton, to the uncomfortable blend of high-value agriculture and chemical and metal-fabricating industries on the Niagara Peninsula.

This region has been crucial to Toronto's economic rise, an importance recognized in the region's nickname, "the Golden Horseshoe." Here, per capita income levels are Canada's highest. But however fitting the place-name is in evoking the region's crescent shape, the "horseshoe" has a tarnished side. The "golden" label refers only to the region's money-generating potential. As geographer Maurice Yeates puts it, "The word 'golden' can hardly refer to the environment, for much of it is cement, steel, and

macadam, and although areas of beauty exist, there has been little regard for preservation."[2]

The advantages of early growth in Toronto and Hamilton, cheap transportation, good access to raw materials, energy, and markets, along with proximity to labor concentrations and support industry, all stimulated urbanization and manufacturing throughout the Golden Horseshoe. In many respects, this region possesses much of Megalopolis's promise and many of its problems (Chapter 4), although on a smaller scale.

The urban-industrial cluster in southwestern Ontario also takes advantage of good accessibility,

[2] Maurice Yeates, *Main Street* (Ottawa: Macmillan of Canada, 1975), p. 165.

resources, proximity to other manufacturing activity, and, to some degree, local mineral resources. Near Sarnia, petroleum deposits were discovered in Oil Springs shortly after the middle of the nineteenth century. A refinery was built at Sarnia by the beginning of the twentieth century and subsequently connected by pipeline with later, and much larger, petroleum-producing regions in Alberta. As a consequence, this city of 88,793 in 2006 is a center for petrochemical industries.

Complex interindustry linkages from the southeastern Michigan automobile-producing region beyond the Detroit River spilled across the international boundary to benefit Windsor (323,342 in 2006). Windsor's industrial mix is supplemented by food and beverage industries. Location on the continent's major industrial waterway—the Great Lakes—supported both Sarnia's and Windsor's growth within Canada's manufacturing sector.

Southern Ontario's other urban-industrial zone developed between the Toronto-Hamilton area in the east and Windsor and Sarnia in the west. Cities such as Guelph, Kitchener-Waterloo, Brantford, and London are located on major east-west railroad lines across southern Ontario. Unlike cities in the other manufacturing clusters, these cities' industries are limited more to local resource advantages. Industries in this third urban zone include food processing, manufacturing activities dependent on large and accessible wood supplies, and manufacturing, such as electronic machinery production, drawn to an established pool of skilled urban workers.

A summary comparison of industrialization in the two halves of Canada's national core would show that southern Ontario's industrialization is more abundant and widely spread than southern Quebec's. Both portions of the national core possess a single dominant urban center and smaller industrial complexes located near good transportation routes and accessible to local resources. Ontario, however, contains more large cities than Quebec.

At a personal or family level, the consequences of different industrialization levels show in several ways. Quebec ranks poorly among Canadian provinces in average provincial family income (see Figure 7.4). The populations of many of its urban centers, including Montreal and the city of Quebec, include families with very high average annual incomes and others with only marginal incomes. This income disparity exists in rural Quebec as well. Southern Quebec's rural-urban income contrasts are similar to those of other major North American regions. However, Quebec occupies a substantial portion of Canada's national core where incomes might be expected to be higher than the province's actual, moderate average per capita income of Canadian $37,278 in 2006.

Southern Ontario has better access to the United States' interior manufacturing core than southern Quebec. Available to Ontario for decades via the Great Lakes, this relative locational advantage diminished somewhat in 1959 when the St. Lawrence Seaway was completed. With locks and canals large enough to handle oceangoing vessels, the Seaway opened the Great Lakes to direct oceanic shipping. This permitted southern Quebec better access to the continent's industrial interior, thus reducing southern Ontario's apparent advantage.

After more than 40 years of operation, however, the St. Lawrence Seaway's impact on southern Quebec has not been extraordinary. Ironically, southern Quebec may have been hurt by the Seaway. Considerable industry was already located along or near the St. Lawrence River before the Seaway opened. Most of the Seaway's benefits have been enjoyed by Great Lakes ports, including southern Ontario's, for it increased these ports' sources of supply and potential markets. Even though the Seaway benefited both the United States and Canadian portions of the continental manufacturing core, Montreal lost many transshipment activities and the potential industry associated with break-in-bulk freight handling. If the ships do not stop, the increased traffic moving up and down the St. Lawrence has no benefit for southern Quebec.

REGIONAL ORGANIZATION AND CONTINENTAL INTEGRATION

A place's location affects its importance to the country's economy. This simple geographic concept is key to understanding the Canadian core's content and diversity. The city of Quebec, for example, benefited from its location at the head of sea navigation but later suffered because of poor rail connections. Montreal was built on the St. Lawrence where water transport was interrupted by rapids and is a site accessible to New York City. Windsor developed directly across the

Detroit River from Detroit, Michigan. London, located about halfway between Toronto and Windsor, is in the middle of a good agricultural area where two major land transport lines cross.

Similar notes could be made for all large urban centers. Some in Canada, like Guelph, remained small, and significant growth stimuli did not appear until the twentieth century. Others, such as Kingston, went through a growth-decline-growth sequence as changing external factors altered the economic importance of the cities' locations. For geographers, knowing the distribution of a country's economic activity is usually critical in understanding why a city exists.

The Canadian core is partly a culture hearth, partly a political focus, and partly the center of the nation's economic structure. The region's history of cultural diversity and the way this history translates into political action threaten to divide the national core. But so far, at least, the region's economy and interurban connections are holding the national core together.

Another geographic concept is that a region does not exist in isolation. Regions possess specific characteristics and links to other regions. Each region and each place within a region is tied to others by people, goods, and ideas moving between them. These interrelationships are fundamental to economic activities (why each place in a region succeeds or fails where it does) and to the general manner in which regions develop (why each develops as it does). Because of its explicitly spatial nature, the pattern of economic interrelations is said to express a region's *space economy*.

The space economy of Canada's national core reflects the national core's position within North America. Commercial traffic across the Great Lakes and along the upper St. Lawrence River join the two countries as much as the waters separate them. Canada's national core, though divided politically, is integrated economically with the much larger continental core through road, rail, air, and telecommunication connections focusing on Toronto and Montreal as well as on smaller centers across Ontario and Quebec. This cross-border integration was illustrated dramatically in 2003 when the electric power grid across southern Ontario was also shut down when much of the grid in the northeastern United States lost power, all triggered by an apparent failure in northern Ohio.

For a century or more, a less visible form of continental economic integration has also been taking place. American business and industrial firms have invested heavily in Canada's economy. U.S. capital comprised only 14 percent of the total invested in Canada in 1900, with British firms providing most of the balance. The U.S. proportion increased to 53 percent of Canada's foreign investment by 1926 and surpassed 80 percent (and $25 billion) by 1966. In 2006, the total U.S. annual investment in Canada exceeded $241 billion and accounted for 61% of all foreign direct investment. In 2006, the total trade between the countries was more than $500 billion.

This century-long infusion of U.S. capital, which was largely an attempt to circumvent tariffs imposed by Canada on U.S. manufacturers, also reflects the U.S. economy's increasing strength and Great Britain's gradually weakening economy. By establishing branch plants in Canada, U.S. firms gained access to Canadian markets and British markets without paying the import penalty of tariffs.[3] In some cases, investment meant faster development of Canadian Shield resources or construction of new manufacturing plants employing Canadian workers and supplying the Canadian market.

American investment was heaviest in mining and manufacturing industries. More than 50 percent of Canada's mining industry, especially mineral fuel mining, is owned or operated by parent U.S. firms. Fully 40 percent of Canada's manufacturing industry, especially the automobile industry, is controlled from the United States. In total, U.S. firms control nearly 7000 Canadian corporations. Because Canadian manufacturing is concentrated in the southern Ontario–Quebec national core region, these two provinces have been especially attractive to U.S. investment capital. As Table 6.1 shows, over 70 percent of all U.S.-controlled corporations in Canada are located in Ontario or Quebec.

Some Canadians view this high level of U.S. capital investment with alarm. The two countries are usually the best of neighbors, but they are different countries. Much as Quebecois nationalism reflects French Canada's strong concern and resistance to cultural assimilation by the larger Anglo culture, a broader Canadian nationalism from time to time expresses fears of economic and cultural assimilation by larger American interests. Canadian legislation to limit the penetration of foreign (mostly U.S.) capital was seen as too limited

---

[3] This international movement of investment capital may not seem surprising given the two countries' proximity and the relative openness of their shared border. The investment rate, however, was unusual prior to World War II for noncolonial powers.

TABLE 6.1

United States Controlled Corporations in Canada, 2001, by Province							
	Atlantic Provinces	**Quebec**	**Ontario**	**Prairies**	**British Columbia**	**Other**	**Total**
No. of corporations	296	942	4021	934	612	20	6825
Percent	4.3	13.8	59.0	13.7	9.0	0.3	100

Source: Statistics Canada (2001). "Inter-corporate ownership" *The Daily*, 16 March, 2001.

and too late. Also relevant to Canadian concerns is the concentration of significant U.S. economic involvement within the heart of Canada's national economy and in its national core region.

Although Canada strongly desires to remain economically independent of its large southern neighbor, the bonds that bring these two countries together are powerful. The political boundary between them is a selective and semipermeable barrier. Citizens of each country enter the other easily for recreation or business. For most, the international boundary is no more than a paint strip across the road and a few brief questions from a customs agent.

The transparent border between Canada and the United States resulted in an integrated *continental* economic core, and this benefits people of both countries. But a less beneficial consequence is the intense cultural and economic influence exerted on Canada by the much more populous United States. About 19 percent of all U.S. exports in 2006 went to Canada, with the flow of goods and materials from Canada to the United States comprising almost 79 percent of Canada's export total.

In spite of the Canadian core's internal divisions, it is strongly influenced by and integrated within a much larger continental culture and continental economy. These economic realities are likely to remain influential in both southern Quebec and southern Ontario, regardless of future political decisions about Quebec's possible independence.

Review Questions

1. What are some of the factors that have limited the size of the Canadian core area to the relatively small territory it now occupies?

2. What are the implications of the cultural dualism that exists between Quebec and Ontario?

3. Describe the long-lots of southern Quebec. What are the advantages of this unusual survey system?

4. Describe the economic interrelations of Canada's national core.

5. What has been the impact of American investment in mining and manufacturing activities?

CHAPTER 7

THE BYPASSED EAST

Preview

1. The Bypassed East is relatively cold, receives substantial precipitation, and is generally mountainous.

2. Farming in the region, never easy because of rocky soils and hilly terrain, has consistently declined in recent decades.

3. Fishing remains important to the region, but pollution, overfishing, and overseas competition raise serious concerns about future prospects.

4. The region is sparsely populated, with about half the population living in small cities and half in rural areas.

5. The demographic and economic patterns of the Bypassed East are undergoing slow transformation but the future remains uncertain.

the many fine harbors in the Atlantic Provinces, became Canada's major eastern port. Like northern New England, the Atlantic Provinces were settled early, but as Canadian settlement pushed west, these provinces became increasingly isolated from Canada's growth regions.

The Bypassed East, then, lies in a transportation shadow thrown by nearby regions onto which the light of economic development remains focused. This has slowed regional economic growth in the Bypassed East, for geographic accessibility is key to a place's growth prospects (see Chapters 4 and 5).

Southern and northern New England are parts of two separate regions. New England's strong image as a unified region is an historical carryover, even though the image continues to be held almost universally outside the area and is often accepted within it. Certainly, New England was once a region with considerable economic and

North of Cape Cod's distinctive hook, the land between the ocean and the St. Lawrence River estuary becomes more convoluted and broken the farther it reaches out into the North Atlantic (Figure 7.1). There are no large coastal cities north of Boston, and interior cities are generally even smaller than those along the coast. Few major overland routes extend inland. We call this area, comprising the Atlantic Provinces of Canada as well as northern New England and the Adirondacks of New York, the Bypassed East.

Why should this region, so close to Megalopolis and the Manufacturing Core and east of Canada's national core, have a character so different in its economic inertia and political tensions? The answer lies in the region's, and the continent's, geography.

The Bypassed East is near major transportation routes but not on them. Ocean transportation bypasses the region. Ocean-borne trade finds little advantage in the Bypassed East's harbors. Megalopolis's fine harbors farther south along the coast possess easier, more direct access to the U.S. interior.

The region's northern coastline is indented by the St. Lawrence River's wide, deep estuary. As with Megalopolis's harbors, the estuarine indentation augments access to the continent's interior. Jacques Cartier sailed up the St. Lawrence in the sixteenth century hoping to find a sea route to Asia. Instead, he found rapids blocking his path at the present site of Montreal, and he abandoned his search. Rather than find a route to Asia, he found an easy passage deep into the heart of what was to become Canada. Montreal, rather than one of

Weathered landscapes, small towns, winding coastlines, and quaint buildings represent the Bypassed East for many North Americans. This scene is from Prince Edward Island.

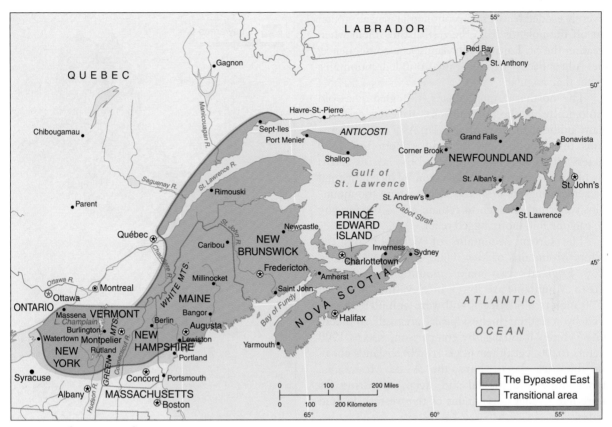

Figure 7.1   The Bypassed East.

institutional uniformity. Today, however, southern New England is a part of metropolitan America. Northern New England, for the most part, is not. The north is much more like Canada's Atlantic Provinces in its characteristic landscapes, problems, and prospects.

The international boundary that places the Bypassed East into two different countries has little bearing on the region's uniformity. There may be no better example in North America of the fact that regional boundaries shift over time and that the shift may not be recognized if the previous regional identities are strongly embedded in a people's awareness and traditions.

THE PHYSICAL ENVIRONMENT— THE JOY AND THE HINDRENCE

Much of the Bypassed East is beautiful; it is also relatively empty. The Presidential Range in New Hampshire's White Mountains contains some of the eastern United States' most rugged topography. The region's extensive shoreline thrusts out into the Atlantic, meeting the ocean's waves with dramatic headlands and many small coves bordered by rocky beaches. Large empty areas, almost totally lacking in settlement, are only hours away from some of the continent's largest cities. This combination of natural beauty and limited population does much to etch the Bypassed East's imagery.

Topography

Physiographically, most of the Bypassed East is part of the Appalachian Highlands' northeastern extension. However, the northern Appalachians' structure bears little surface resemblance to the southern Appalachians' clearly delineated Blue Ridge-Ridge and Valley-Appalachian Plateau sequence (see Chapter 8).

Northern New York's Adirondack Mountains are a southern extension of the Canadian Shield rather than part of the Appalachians. The Adirondacks were severely eroded by the same continental glaciation that cut off the uplands from the rest of the Shield when creating the St. Lawrence lowland. As a consequence, the Adirondacks are generally rounded mountains rather than angular.

The northern Appalachians, though not as old geologically as the Adirondacks, have experienced many millions of years of erosion. As a result, elevations throughout New England seldom top 1500 meters (4600 feet). Most hills and mountains across the plateau were rounded when scoured by continental glaciers during the *Pleistocene*. Only where the mountains were high enough to remain above the moving ice can one find more rugged relief.

The Green Mountains of Vermont and the White Mountains of New Hampshire are northern New England's two major mountain areas. The Green Mountains are lower in elevation, less than 1500 meters (4600 feet), and were completely covered by ice; their highest peaks are well rounded. The White Mountains, by comparison, rise to 1900 meters (6200 feet); their peaks rose like islands above the thick ice sheets. Thus, the White Mountains' upper slopes are rugged and steep, appearing far more like younger mountains of the western United States.

Although northern New England and upland New York draw character from the region's mountains, its people's homes and livelihoods are in the valleys and lowlands. The region's largest lowland areas are the Connecticut River valley between New Hampshire and Vermont, the Aroostook Valley in northern Maine, and the Lake Champlain Lowland along the northern Vermont-New York border. This last-named area is an extension of the southern Appalachians' Ridge and Valley area and, therefore, an exception to the rule that New England's valleys were formed by erosion. Smaller lowlands border the seacoast, and innumerable streams slice the plateau throughout the area. A detailed map of the distribution of population would look like an inverted map of elevation, with most people living in the lowlands (Figure 7.2).

The Atlantic Provinces' mountains are even lower and more rounded than northern New England's, with few elevations rising above 700 meters (2200 feet). The basic topographic grain of the Atlantic Provinces follows the same northeast-southwest trend that is evident elsewhere in the Appalachians, but further differentiation is less clear. In the southern Atlantic Provinces, uplands in

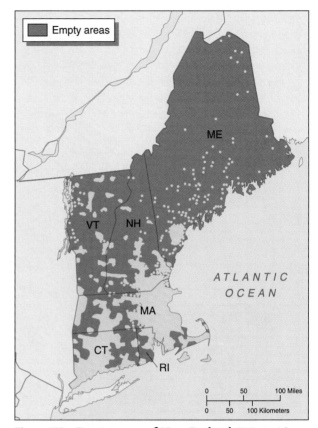

Figure 7.2 **Empty areas of New England** (Adapted from Geography Division, U.S. Department of Commerce, Economics and Statistics Administration, U.S. Census Bureau). The forest lands of northern Maine and mountainous sections of the area mapped remained beyond the margins of most settlement into the twenty-first century.

peninsular Nova Scotia and interior New Brunswick are separated by a wide lowland that extends from the Bay of Fundy to Prince Edward Island (PEI to Canadians).

As in northern New England, the Atlantic Provinces' lowlands are where most of their people live. The central lowland, including Prince Edward Island, is underlain with soft rocks, mostly limestones and sandstones. The Annapolis Valley, in southern Nova Scotia, is part of the same structure that underlies the Connecticut River Valley. These Triassic lowlands are scattered along North America's interior east coast as far south as central North Carolina. Outside the Triassic lowlands, the Atlantic Provinces' larger cities and nearly all of the population of Newfoundland are clustered around the area's harbors.

The rugged and dramatic beauty of this stretch of the Cabot Trail on Cape Breton Island is a delight for tourists but typifies how difficult much of the Bypassed East is for agriculture or urban growth.

Climate

The Bypassed East is no place for people who want a dry, warm climate. It is a place where polar, continental, and maritime weather systems meet. The resulting weather regime is seldom hot, often cold, and usually damp. Because of its location on the continent's eastern side, the Bypassed East is bombarded by wind systems pushing continental conditions into it and limiting any maritime impact away from the immediate coastal fringe. Higher inland elevations further increase the substantial climatic difference between the region's coastal and interior lands.

The cold Labrador Current flows south along the Bypassed East's oceanic margin. Only the most intrepid swimmers are willing to spend time in the chilly coastal waters, even in late summer. Still, coastal climatic conditions are moderated substantially by the water's proximity. Average midwinter temperatures at coastal sites are often 3° to 6° C (5° to 10° F) higher than at nearby interior locations. In contrast, midsummer temperatures are slightly higher in the interior than along the coast, and the coast's growing season is as much as 70 days longer than the interior's.

Overcast skies and fog occur frequently owing to the maritime influence, particularly along the Bypassed East's southern coastline. Southeast winds crossing the Gulf Stream's warmer waters cool as they pass over the Labrador Current's colder waters close to shore, creating dense fog. The fog and heavy cloud cover cool temperatures further during the summer, aggravating agricultural problems in a region with a moderate growing season. Growing crops requiring summer heat and sunlight is difficult in the Bypassed East where the interior's growing season is only 120 days.

Almost all of the region receives substantial and evenly distributed precipitation, usually between 100 and 150 centimeters (40 to 60 inches) annually and spread throughout the year. Snowfall can be substantial, with most places receiving between 25 and 50 percent of their total moisture as snow. Most interior locations average at least 250 centimeters (100 inches) of snow each year, blanketing the ground solidly for 3 to 5 winter months. Near the coast, snow cover is sporadic, with frequent thaws and wind-blown, bare ground.

POPULATION AND INDUSTRY IN A RUGGED LAND

The Bypassed East is not an easy place in which to live and work. Harsh climate, hilly terrain and thin, rocky soils limit agricultural possibilities except in a few particularly well-blessed locations. Substantial mineral deposits are limited. Relative isolation and a small local market also restrict manufacturing growth. Given these characteristics, what is this region's future? However limited, certain geographic advantages remain key to providing the answer.

Early Settlement

This portion of North America was not always bypassed. Its foreland location juts far out into the Atlantic Ocean so that its shores were probably among the first parts of the New World encountered by Europeans centuries before Columbus reached the Caribbean. The first European settlement in Canada was opened in 1605 by the French at Port Royal. Less than a decade later, the English established a series of fishing villages in Newfoundland. By the mid-seventeenth century, many small harbors in central and southern Maine housed British villages.

Two regional resources—fish and trees—attracted early European settlers. Rich fishing banks off the Atlantic Provinces were immediately important. The banks are shallow areas, 30 to 60 meters (100 to 200 feet) deep, at the continental shelf's outer margins (Figure 7.3). Their shallowness allows the sun's rays to penetrate easily through much of the water's depth, encouraging plankton growth. Plankton is a basic food source for many fish. As a consequence, cold-water fish, such as cod and haddock, are abundant. By

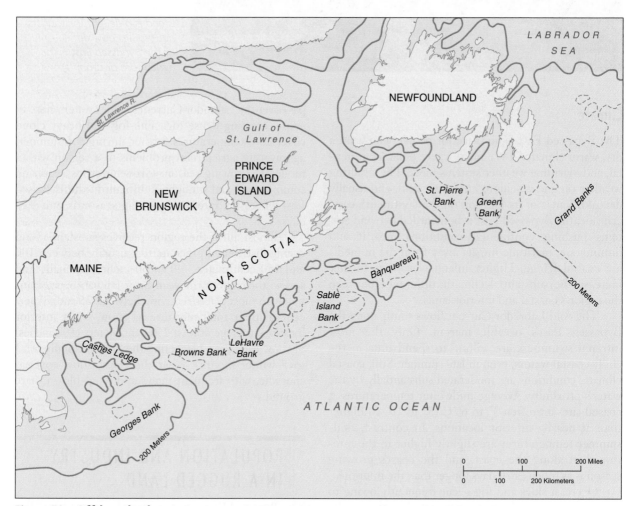

Figure 7.3 Offshore banks. The large areas of shallow, cold ocean water off much of the Bypassed East's coast have been recognized for 500 years as among the richest ocean fishing waters in the world. Sharply lower fish catches after 1990, however, led to concerns about overfishing.

exploiting this resource, European settlers soon established substantial export trade in salted cod.

Trees were the region's second resource lure. By the 1700s, England had already sacrificed much of its own forest cover to agricultural expansion. As England's navy and merchant marine fleets developed, the country was forced to turn to Scandinavia's forests for ship masts and lumber. This arrangement was unsatisfactory, for it made England dependent on foreign powers for a vital military and economic resource. Consequently, the English sought alternatives within territories unclaimed by other European states.

White pine dominated New England's forests. A magnificent tree, it stood straight and reached heights in excess of 60 meters (190 feet). Its wood was clear, light, strong, and easily cut. Maine became a center for ship construction because of these vast forest resources. The best of the white pines were reserved for the Royal Navy's ship masts, but there seemed to be more than enough to satisfy other needs for lumber as well.

Almost all of New England's virgin forests are now gone. Remaining second- and third-growth forests have shorter trees and are insignificant in comparison with the original stands. A few lone giants remain in isolated groves, scant reminders of the forests that once covered the region.

Agriculture was the early settlers' third major occupation, but farms were small and production was limited. Early on, people farmed primarily for subsistence with little surplus produced for market. By the end of the 1770s, most lowlands south of Newfoundland were covered by farms as was much of the sloped land.

Northern New England's agricultural development peaked just after the start of the nineteenth century, but two developments elsewhere in North America began pulling New England's farmers off their land, first as a trickle, then as a flood.

The first development that drew New Englanders away was the opening of western lands. In the United States following the American Revolution, settlement moved beyond the Appalachians onto rich farmlands south of the Great Lakes. East Coast markets were made accessible to midwestern farmers when the Erie Canal was built in the 1820s. Farmers of upper New England's poor agricultural land lost what little market advantage they possessed to crops imported from Ohio and Indiana farms.

During the nineteenth century, about 1 million New Englanders moved off the region's farms and headed west, joining the westward migration to exchange bad land for good. An old saying was that the only crop those New England fields could produce was rocks and their only export was people.

The agricultural settlement history in the Atlantic Provinces parallels New England's experience. The middle St. Lawrence River Valley in southern Quebec and especially in southern Ontario provided agricultural land superior to that of the Atlantic Provinces. Rapid settlement expansion in southern Ontario in the decades after the American Revolution drew farmers from the Atlantic Provinces westward. Construction of the Welland Canal in 1829 to bypass Niagara Falls increased southern Ontario's accessibility.

A second blow to the region's agricultural fortunes occurred during this period with the development of manufacturing in southern New England. This is where the Industrial Revolution began in the United States. As elsewhere, industrial growth created a great demand for labor, and that demand was met first with New England farmers seeking higher wages and steady income. Extensive use of child and female labor, particularly in the textile mills, further enhanced the attraction of manufacturing work over farming by making all family members wage-earners.

Agriculture Today

Since the farm family exodus during the nineteenth century, agriculture's regional importance has continued to decline across most of the Bypassed East. Today less than 15 percent of the land in northern New England (Vermont, New Hampshire, and Maine) is in farms; 100 years ago, this number was closer to 50 percent. Many northern New England towns lost population steadily for a century or more. Forests returned to the sloped land as farming retreated. Even in the valleys, infertile soils, a cold climate, and small farm size made regional agriculture increasingly uncompetitive. As a result, the landscape became one of abandoned farms, scrubland, and woods.

Farming's decline in the Atlantic Provinces was not as great as in northern New England, but it was still substantial. After peaking in the late nineteenth century, the total land in farms declined by about two-thirds. Agricultural productivity remains poor compared to that of the rest of Canada. Regional declines in acreage farmed and amounts produced remain far larger than elsewhere in Canada.

Still, farming continues to be important in a few Bypassed East locations. Concentrated where

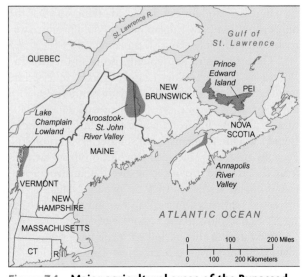

Figure 7.4   Major agricultural areas of the Bypassed East. Most of the region's farming is concentrated in its few lowlands of substantial size.

conditions are most favorable, agricultural efforts specialize in single-crop production. The acidic soils of Washington County in northeastern Maine, for example, support one of the United States' major centers of wild blueberry production. Only Maine, New Hampshire, and Nova Scotia produce wild blueberries for market.

Four of the Bypassed East's areas of agricultural production remain significant and deserve special attention (Figure 7.4).

The Annapolis Valley

Nestled in southwestern Nova Scotia and protected from cold northwest winds by a low mountain ridge, the Annapolis Valley may be the oldest of the region's attractive agricultural areas. The valley, about 130 kilometers by 15 kilometers (80 miles by 9 miles), has long been one of Canada's major apple-producing areas. Apple farmers have suffered recently owing to competition from growers closer to the valley's traditional markets of Great Britain and the urban centers of Quebec and Ontario. Now processing most of the valley's apples rather than marketing them fresh, valley farmers hope to sustain demand by reaching a different market. Farmers have also diversified in recent years, shifting from apples to produce and dairy farming to meet demand in nearby Halifax, the Atlantic Provinces' largest urban center.

Prince Edward Island (PEI)

Prince Edward Island's economy is overwhelmingly agricultural. More than 46 percent of the province's 5660 square kilometer (2038 square mile) land area was in farms in 2001, with over 67 percent of the farmland in crops. The nearly 31 percent of farmland in PEI surpasses the comparable share for every other province and state in North America. PEI's rolling, sloping land of small, green farms is among the most beautiful agricultural landscapes in the United States and Canada. Most PEI crops cannot bear the high cost of shipping to mainland markets and are consumed locally by the island's 135,851 inhabitants (2006).

Tourists visiting PEI are often surprised by and enjoy the picturesque, faintly worn appearance of the farmsteads. Declining demand for PEI's principal seed potato crop due to increased competition from mainland sources, and too many farms too small to be profitable in a modern farm economy, lie behind PEI's landscape appearance. Its comfortable, lived-in aura reflects a stagnant agricultural economy that no longer generates sales sufficient for investments beyond simple maintenance.

The Lake Champlain Lowland

Megalopolis represents the United States' largest concentration of demand for agricultural produce. Much of the demand is for foodstuffs that can be shipped from anywhere in the world. Low-bulk, high-value crops like tea, for example, can be shipped halfway around the globe from distant tea-growing areas with only a small percentage increase in purchase price. Other crops, like wheat or corn, can be handled roughly without damage or spoilage. For these latter products, long-distance shipping in bulk containers is efficient. What's more, frozen and canned vegetables and fruit are processed near where they are produced, so shipping and storage costs are low.

Fresh milk fits none of these requirements. A relatively high-bulk, low-cost product, milk spoils easily and cannot be stored for an extended period. As a result, urban areas usually rely on nearby sources for fresh milk.

Most of the United States and Canada can be subdivided into milksheds, or nodal regions from which a city's population is supplied the fresh milk it consumes. The Lake Champlain Lowland's proximity to Megalopolis allows the Lowland to be part of both New York City's and Boston's milksheds, and a major milk source for their urban populations. As such, the

Prince Edward Island is a land of well-tended farms, rolling hills, and a long, indented coastline. Its agricultural economy suffers because of its distance from the major markets of eastern Canada.

Lowland has a substantial market advantage for milk sales over more distant sources.

An additional advantage for dairy farming lies in the Lowland's mild and moist summers, a climatic condition that encourages fodder crop growth. Cool summers also suit dairy cows well. Vermont has led the United States for some time in per capita production of dairy products, and most of Vermont's production comes from the Champlain Lowland. Dairy farming currently accounts for more than 62 percent of Vermont's crop land. According to the Vermont Dairy Promotion Council, sales from Vermont milk and dairy products totaled $342 billion in 2006. The state currently has about 140,000 dairy cows.

St. John-Aroostook Valley

This area in northeastern Maine and western New Brunswick is the Bypassed East's newest major agricultural area. Commercial farming became important here only late in the nineteenth century, but this was soon one of the leading potato-producing areas in both the United States and Canada. The valley's silty loam soils are ideal for potato growth, and the short growing season encourages a superior crop used widely elsewhere as seed potatoes. Large-scale, mechanized farming predominates. The demand for seasonal labor leads to heavy but only periodic employment of French Canadian farm workers.

As elsewhere in the Bypassed East, the St. John-Aroostook valley's potato growers have experienced economic difficulties in the past several decades. Recent increases in potato consumption primarily benefited Idaho's producers of baking potatoes and frozen and processed potato products. Western potato growers, primarily in Idaho and Oregon, created well-organized merchandising operations, developed many new kinds of manufactured potato products, advertised effectively, and coordinated their marketing. St. John-Aroostook Valley farmers lost many traditional markets to these other potato-producing areas. Consequently, the valley's current agricultural economy is depressed. Potato production is down, and many farmers are out of the business. Still, some of the changes benefiting Idaho producers are being adopted by the St. John-Aroostook Valley producers. For example, one of the world's largest producers of frozen French fries, McCain Foods, Ltd., is located in New Brunswick's St. John Valley and consumes many tons of locally grown potatoes each year.

Forestry

Northern New England offers a prime example of uncontrolled logging's destructive potential. After the original trees were cut, organized reforestation was limited; therefore, much of the current forest of balsam fir, white pine, jack pine, and spruce is lower quality for lumber and pulp production. The present landscape is also far less handsome than the virgin forest it replaced.

Today, none of northern New England's three states is an important lumber supplier, and long-term production is unlikely to increase significantly. Indeed, it is ironic that this area, where the great majority of the land is in forest, lacks a large wood products industry. Trees grow slowly in the cool climate, however, and the lack of thought originally given to reforestation makes lumbering in much of the region unattractive to large-scale corporations.

Northern Maine's pulpwood production is an exception. Here, on some of the eastern United States' least accessible land, forest industries remain important. A small number of private owners control most of the land. Many of the roads and recreation facilities in these Maine woods are also privately owned. The area contains some of the finest white-water canoeing streams in the country as well as large numbers of lakes and marshes. Disagreements over forest utilization versus forest preservation have led to conflicts between the pulpwood producers and canoeists and conservationists.

Forestry is much more important to the economy of the Atlantic Provinces than that of northern New England. About 85 percent of New Brunswick's land area, 74 percent of Nova Scotia's, and 61 percent of Newfoundland's is in forest. Most forestland in the first two areas is privately held, unlike the usual forestland ownership pattern in Canada. Only a small percentage of the Atlantic Provinces' labor force actually works in forestry, but the manipulation of cut wood accounts for nearly half of the region's manufacturing employment. Pulp and paper are its key products, with northern New Brunswick and Newfoundland the most important logging areas. The region's largest paper mills are located on New Brunswick's shore of the Gulf of St. Lawrence. With the mills located close to river mouths, logs are floated to the manufacturing plants, and finished products are loaded directly on large ships for transport to market.

Fishing

Fishing is still important to the Bypassed East's economy, but fishing's future contribution is far from clear. Canada is one of the world's leading exporters of fish products. Most Canadian fish exports come from the Atlantic Provinces, with Nova Scotia leading all provinces in total fish catch each year, followed in order by British Columbia, Newfoundland, New Brunswick, and PEI. In the United States, Maine lobstermen provide 70 percent of New England's total lobster catch, with 72.7 million pounds caught in 2006.

There are two kinds of ocean fishing, inshore and deep-sea. In the Bypassed East, inshore fishing is more important. Small boats and a relatively small capital investment are required, with lobsters and cod being the most valuable catch. Deep-sea fishing on the offshore banks where bottom feeders such as flounder and halibut, as well as cod, are caught requires far larger boats and greater capital investment than inshore fishing.

The inshore fishing industry faces substantial challenges. The challenges are a combination of pollution effects, economies of scale, and overfishing. Inshore pollution is a growing problem, forcing fishermen to move to offshore locations. Inshore fishermen often live in one of the many small villages scattered along the coastline of Nova Scotia and Newfoundland. With small investments and modest incomes, inshore fishermen often cannot obtain funds needed to buy and outfit large, modern oceangoing vessels. When they move onto the banks with their smaller vessels, they compete directly with large fleets of modern ships, including factory ships that process the fish at sea.

Newfoundland's provincial government, concerned about the inshore cod-fishing industry's uncompetitive position, initiated a program encouraging fishermen and their families to move, at government expense, to larger communities. The reasoning went that larger, more effective cooperative fishing operations would be possible in more centralized communities. At the same time, the government's high cost of providing services to the many small fishing communities were cut. So far, the program has been successful in one respect. Of the 1200 small communities targeted (called outports in Newfoundland), more than 500 have been abandoned.

Offshore fishermen in both the United States and Canada face heavy competition from European and Japanese ships. Americans and Canadians lagged far behind their foreign competitors in turning to larger fishing vessels. Even so, North America's offshore industry expanded somewhat, especially in Newfoundland. In 1977 Canada established a 200-mile (320-kilometer) offshore management zone closed to unrestricted fishing by foreign vessels. The United States had established a similar management zone a year earlier. Conflicting claims along their mutual Atlantic Coast led to arbitration between the two countries.

Overfishing may be the most serious challenge for Atlantic Province fishermen. Stocks of cod, pollock, and haddock declined sharply after 1991. The 1994 fish catch was only half that of 1990, and by 1995, the number of Atlantic Province fishermen plying the Atlantic's waters declined to 33,000 from 41,000 in 1990. By 2006, Atlantic fish catches had increased to 881,000 metric tons (live weight) per year, a 240,000 metric ton increase over 1995, but far below the 1.3 million metric tons caught in 1990. Newfoundland, a relative newcomer to Canada, has been particularly hurt by the declining fish stocks.

Newfoundland did not join the rest of Canada until 1949. Following several years when disastrously low European fish prices wreaked havoc with Newfoundland's economy, the province's residents voted by a narrow margin to join Canada. The former name for Canada's eastern provinces, the Maritimes, was changed to the Atlantic Provinces to reflect Newfoundland's incorporation into the country. Although changes in the fishing economy had this direct political consequence, Newfoundland's late decision to join the country hints at the province's isolation from mainstream Canada, an unsurprising feature in the Bypassed East.

Minerals Production

Canadian and American demand for domestic petroleum threatens these countries' offshore fishing industries. Pollution fears about drilling for offshore oil in the rich fishing banks were overruled in the United States in 1979 when the Department of the Interior granted several oil companies exploration leases. Major petroleum and natural gas reserves were discovered in the Hibernia field off Newfoundland's Grand Banks. Production of the world's largest oil drilling platform was begun in Hibernia in the early 1990s and production from test wells commenced in 1997.

In addition to potential environmental impacts from developing the Bypassed East's offshore fields, funding and political issues are also problematic. Development costs will exceed $8.5 billion (Canadian). Ownership conflicts between the provinces and the Canadian central government can be difficult to resolve, although the Canadian and Newfoundland governments reached agreement in 1985 regulating extraction and sharing revenue.

Other than offshore petroleum and natural gas, local minerals are not currently important in the Bypassed East's economy. Iron ore has been mined in the Adirondacks for more than 100 years and the reserves there remain substantial, though total output is relatively small. New York is usually among the top five states in iron-ore production, but only the top two, Minnesota and Michigan, can realistically be called major producing states.

Nova Scotia has significant coal reserves, but deposits are not easily mined. Coal production totaled about 250 thousand metric tons annually in the mid-1990s, and dropped to just 141,000 metric tons in

Halifax, Nova Scotia, is the Bypassed East's largest city. Halifax's business waterfront has benefited from the recent success of the city's modernized port facilities.

2003. The 1999 closure of a large underground mine in Cape Breton accounts for much of the decline in the past decade. Additional mine closures are ongoing. The Atlantic Provinces' coal does not compete well with lower cost coal from more distant sites; for example, American coal undersells it in Ontario's markets.

Coal mining in the Atlantic Provinces was subsidized when its coal was used at a small steel plant in the Nova Scotia town of Sydney. The plant, now closed, was too small and outmoded to be efficient and was kept open for years as a provincial government corporation. Steel produced elsewhere in larger quantity with much better economies of scale was shipped to the Atlantic Provinces and undersold the local product.

Northern New England's igneous and metamorphic rocks make the area an important producer of building stone. Numerous granite quarries are located in central Vermont and along Maine's central coast. Vermont is also the United States' leading marble-producing state. The economic value of these rocks is small compared to minerals industries in other parts of North America, but they constitute an important part of the two states' economies, especially Vermont's.

The Atlantic Provinces' largest mining operation is based on development of iron-ore deposits in Labrador near the Quebec border. Although these deposits are in Newfoundland, they are not in the Bypassed East and have little impact on the region (see Figure 2.9 and Chapter 17). Ore is shipped south by rail to a railhead at Sept Isles, Quebec. There it is loaded on lake vessels and carried up the St. Lawrence River. This mining's only benefit to Newfoundland is from the tax collected.

Cities and Urban Activities

Half of the Bypassed East's population lives in rural areas and half in urban. Few of the region's urban areas are substantial. Northern New England's two largest cities are Burlington, Vermont, with 38,358 residents in 2006, and Bangor, Maine, with only about 31,008 residents that year. In Canada, Halifax, Nova Scotia, is the largest city in the Bypassed East with 359,183 residents in 2001, while St. John's, Newfoundland, and Saint John, New Brunswick, had 172,918 and 122,678 metropolitan residents, respectively. Although these are the largest cities in the region, they are small by North American standards.

Countervailing trends in the region reflect two distinct influences on urban growth and the well-being of the Bypassed East's urban residents. Traditional economic factors, such as regional resource availability and costs and the region's employment structure, do not encourage optimism. Geographic factors, such as location, suggest a tentatively positive view.

Relatively low average per capita incomes are typical in the Bypassed East, and these income levels are partly explained by the small size of major regional centers. Most higher-income occupations in the United States and Canada are located in urban sites. The Bypassed East, therefore, lacks large numbers of urban occupations.

Much of the regional workforce is engaged in primary occupations, such as farming, fishing, mining, and logging, which are traditionally among the continent's lowest paying. A large local market is absent, hampering local manufacturing. Access to major urban areas is poor, so transportation costs are high. As a consequence of this mix, a broad-based economy did not develop from the region's primary industrial base as it did elsewhere in Canada and the United States.

The region's manufacturing economy remains wedded to nonlocal investment or to processing regional raw materials. An example of nonlocal investment is the single IBM plant in Essex Junction, Vermont, that provides 5,000 jobs, nearly 3 percent of all nonfarm jobs in the state.

In a more positive light, the locational advantages that once attracted European settlement and stimulated resource exploitation appear to be assisting the Atlantic Provinces again. Following more than a century of relative isolation, several of the region's cities are growing, benefiting from their location strengths.

Halifax, at the Canadian National Railroad's eastern terminus, is Canada's second leading Atlantic Ocean port, following Montreal. Montreal's St. Lawrence River traffic is greatly reduced each winter, even though a channel is cleared through the ice. As a result, Halifax's ice-free harbor is busiest when Montreal's harbor is restricted by the cold. Halifax developed a diversified economic base as a governmental, financial, and commercial center and is the Atlantic Provinces' most prosperous city. Many Canadians once called Halifax the Gray City because of its drab climate and architecture. New construction and renovation since 1960 greatly changed the city's atmosphere and appearance, and the former nickname no longer applies.

St. John's, on Newfoundland's eastern Avalon Peninsula, is too far east to serve as a major Canadian port of entry. St. John's harbor is excellent, however. As eastern Canada's principal deep-sea fishing port, St. John's should continue to grow. Usually ice free, the harbor serves as an important winter facility for fishing fleets from many countries as they fish the offshore banks.

New Brunswick's Saint John, the oldest incorporated city in Canada, boomed after 1890 when the Canadian Pacific Railway extension across Maine was completed. St. John is more remote and possesses a poorer quality harbor than Halifax, so growth is not as strong or certain. Like Halifax, the port in St. John is most active in winter.

A PERPLEXING FUTURE

Information technology and innovative approaches to education; multinational corporations and entrepreneurial successes; greater personal mobility and leisure time—these and related changes have geographic implications that suggest dizzying possibilities for the future. How the Bypassed East's residents will fare in this future is not easy to predict.

Presently, the region's economic situation reflects its bypassed location. Collecting and processing raw materials, especially fish and wood products in eastern Canada, continues to be the economic emphasis. The region's population remains substantially more rural than Canada and the United States as a whole. The Bypassed East also remains in a transportation shadow, near some of the world's busiest routeways but bypassed in favor of alternative paths providing more effective penetration to North America's interior.

Poverty is greater where the economy depends on primary products and where transport isolation is pronounced. In 2002 average personal disposal income in the Atlantic Provinces ranged from $18,264 (Canadian) in Newfoundland, the country's lowest, to $20,115 in Nova Scotia. Canada's overall average was $22,272. Every province in the region is below the national average. In the United States, Northern New England's 2006 average annual per capita incomes ranged from a low of $30,808 in Maine to $37,835 in New Hampshire (Table 7.1). The U.S. national average was $34,495. Although the statistics are not directly comparable between countries, the general picture of low Bypassed East incomes is clear.

TABLE 7.1

Average per Capita Income for Selected States, 1960–2005, Relative to U.S. Average[a]				
	Percentage of U.S. Average			
State	**1960**	**1970**	**1980**	**2005**
Maine	83	83	82	89
New Hampshire	99	96	98	110
Vermont	85	89	86	95
South Carolina	63	80	76	82
Mississippi	54	65	69	72

[a]U.S. average in 2005 was $34,495.

Source: U.S. Statistical Abstract, 1979, 1990; Department of Commerce, 2005.

Northern New England

The 1980 U.S. census indicated the end of northern New England's long-term demographic stagnation. Maine, New Hampshire, and Vermont in the 1970s were the only states outside the South and West that were growing faster than the national average. New Hampshire, with a population increase of 27.4 percent from 1980 to 1997, continued to grow at a rate well above the national average of 20.7 percent for the period, although Vermont, with a 15.2 percent increase, and Maine, with a 10.4 percent increase, grew at rates somewhat below the average. The 2000 census revealed that these states may be entering into another period of stagnation. While the U.S. growth rate was 13.1 percent for the 1990–2000 timeframe, Vermont's was 8.2 percent, New Hampshire's slipped to 11.4 percent, and Maine's dropped to 3.8 percent.

What was behind the earlier shift in population? Has it come to an end? Most importantly, Megalopolis grew northward. As Megalopolis's cities expanded, new peripheral areas became part of urban America. As people sought residences farther away from immense core cities, Megalopolis's periphery moved out, including a steady northward push into New England. Among northern New England states, this push benefited southern New Hampshire especially.

New manufacturing facilities accompanied or preceded urban expansion into northern New England. Light industry with medium-sized labor force requirements began to select the region for new and expanded plant facilities. The region's small-town and rural environments appealed to employers and their workers, and interstate highway construction during the 1960s provided good accessibility south and west.

But manufacturers are not the only group locating in northern New England. The region's natural beauty and rural isolation have drawn three other groups, which are changing the region's landscape even faster than industrial relocation.

The tourist industry boomed in northern New England after the end of World War II. A rugged coast, abundant opportunities for fishing, skiing, canoeing, and magnificent scenery have all contributed to the growth of tourism in this region.

Fortunately, the region has several tourist seasons, spreading economic benefits through most of the calendar year. Summer is the traditional vacation time for most American families. Then the natural beauty of the area's fall foliage draws thousands who fill the roads to glimpse the forests' tapestry of blazing orange, yellow, and red leaves. Skiing and other winter sports attract large numbers of enthusiasts to snowy slopes during the coldest season. Only spring, a season of mud and biting flies, does not fit the travel habits of touring Americans.

The Adirondack's economy is also heavily dependent on tourism. Lake Placid, home of the 1932 and 1980 Winter Olympics, may be the best known ski area, but it is only one of many. The Adirondack State Park is the United States' largest state park. New York state policy encourages outdoor recreation on parklands, and development pressures are growing on nearby private lands.

Another group that is changing the region's landscape are second-home owners. Owning another home and time-share ownership are more prevalent among Americans than ever before. Vacation homes are being built along northern New England's seashores, around its lakes, and across its mountains. Typically, owners occupy these second homes a few weeks or months each year, after which the dwellings are rented for as much of the year's balance as possible. In some northern New England counties, there are more part-time dwellings than homes occupied year-round.

At present, the region's local construction industry is flourishing as it meets continuing demand for second homes. The longer-term impacts of this trend are less clear but are probably positive in economic terms. As with tourism, second-home development arises from the region's proximity to Megalopolis. Nearly a quarter of the United States' population lives within a day's drive of northern New England. Residents of Quebec and Ontario also find it a nearby and attractive place.

Retirees are a third group that is helping to change northern New England. Many of Maine's coastal communities, small college towns in Vermont and New Hampshire, and old villages throughout the region have become popular retirement centers. In both the United States and Canada, the number of retirees will

Tourism is a cornerstone of northern New England's economy. Located near Megalopolis, attractions such as skiing and snowboarding resorts are within a day's drive of 50 million Americans.

increase steadily for at least the next three decades. As they spend their retirement incomes, retirees make significant contributions to the local economy. They also change their communities' political and cultural landscapes.

Eastern Canada's Prospects

The Atlantic Provinces' economic prospects are cloudier than those of northern New England. The Provinces' natural beauty is more distant from Canada's large metropolitan region than northern New England's is from Megalopolis. New Brunswick and Nova Scotia are unlikely to become part of Canada's economic core whereas the United States' manufacturing core already reaches into New Hampshire. If Quebec were to secede from Canada, given Quebec's location between the Atlantic Provinces and the rest of Canada, the economic and political impacts of secession on the Atlantic Provinces are unpredictable.

The Canadian government, recognizing the relative poverty of the Atlantic Provinces, moved to improve the region economically. Canada established a Department of Regional Economic Expansion (DREE) in 1969 to encourage employment growth in the country's slow-growth areas. Although

eliminated in 1982, about half of all DREE funds distributed during its existence went to the Atlantic Provinces, with Quebec receiving most of the rest. Direct payments were made to private industry, encouraging entrepreneurs to locate and helping them succeed in less developed areas. Grants were also made to improve the region's transport facilities and water and sewer systems and to fund worker development programs.

The impact of federal efforts in the Atlantic Provinces is difficult to assess. On the positive side, personal incomes began to converge with national average incomes after 1950. The convergence process may have intensified in the late 1960s, but recession in the late 1970s and early 1980s was especially hard on the Atlantic Provinces (Figure 7.5). Probably as a reflection of improving regional incomes, the long-term pattern of outmigration from the Provinces slowed and may have turned around in New Brunswick and PEI. The larger urban areas in the Atlantic Provinces are also growing. Despite these trends, the area's economic growth continues to lag behind the rest of Canada's.

Apparent advantages explaining recent growth in northern New England are slower to develop in the Atlantic Provinces. The Atlantic Provinces are too far from Canada's large urban regions to benefit from

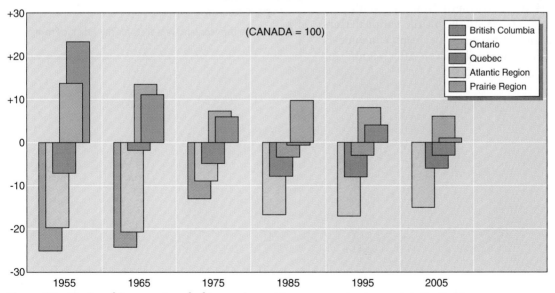

Figure 7.5 Regional income trends, by province. New resource developments in the Prairie Provinces and federal policies encouraging growth in low income areas, particularly Quebec and the Atlantic Provinces, led to a lessening of regional income disparities by the mid 1970s. By 1995, the Atlantic Provinces and Quebec had fallen significantly behind other parts of Canada once again.

urban overflow. Although the interstate highway system improved northern New England's accessibility, distances are greater in Atlantic Canada. The Provinces have high-quality, modern highways, but the roads carry little traffic. Even so, governmental and private investments aimed at attracting tourists are apparent on the landscape.

Still the Bypassed East

The Bypassed East may be a region in early transformation as its extended period of isolation ends. Northern New England is close to Megalopolis, and millions live next door. City dwellers visit the region in droves, and some move there permanently. One testimony to change is the fact that northern New England's population is increasing far more rapidly than that of the United States as a whole. Even the long-term outmigration pattern in the geographically isolated Atlantic Provinces is changing. Jobs outside the primary economic sector are increasing substantially. Overland transportation routes are much improved. When northern New England's choicest second-home and retirement sites are taken, the Atlantic Provinces' attractions may seem less distant and more appealing than they do now.

How all this will play out, and how quickly, remains to be seen. Northern New England's personal income levels are still below the national average. The Atlantic Provinces' per capita income is increasing but only slowly. Even after significant investment by Canada's federal government, these provinces remain Canada's poorest. Significant offshore petroleum and natural gas development may yield regional revenue, but this development poses a threat to the region's traditional fishing economy. Petroleum development may also come into conflict with tourist, second-home, and retirement industries, all three of which depend on the Provinces' natural beauty.

The regional economy of eastern Canada and northern New England is expected to improve. Real economic growth, however, as well as the landscape changes that accompany this growth must first overcome the region's decades of isolation. The pace of change in the Bypassed East will remain slow.

Review Questions

1. What are the major factors that have limited manufacturing in New England?

2. What is the significance of the term "transportation shadow" as it applies to the Bypassed East?

3. Discuss the major benefits of the region's physical environment as they relate to economic opportunities and development.

4. How can tourism be considered a "mixed-blessing" to the region's economy?

5. What are the economic prospects for the Atlantic Provinces?

CHAPTER 8

APPALACHIA AND THE OZARKS

Preview

1. Although separated by more than 300 miles, Appalachia and the Ozarks constitute a single region based on their similar topography and settlement history.

2. European settlement in the Appalachians, mostly by the Scots-Irish, occurred late in the colonial period, but tapered off after western areas better suited to farming became available.

3. There are differences between northern and southern Appalachian cultures even though both sections share a relative lack of economic opportunities.

4. The region is relatively poor and its economy has been dominated by coal mining and farming, but tourism is beginning to be significant in places.

5. The Tennessee Valley Authority, Appalachian Regional Commission, and Arkansas River Navigation System continue to be vital players in the economic development of Appalachia and the Ozarks.

The Appalachian Mountains' rugged landscape played a major role in the settlement history of the United States. The mountains were a barrier to early overland travel from the Atlantic Coast. Without them, European settlement may have spread inland more thinly and more quickly. Because few good land passages cut through the Appalachians, East Coast cities sprang up at certain locations and grew at different rates. Indeed, the Appalachians help define North America's resource geography and chart an economic history and population distribution that would have been very different without the mountains.

The Appalachian region's human geography remains intertwined with its topography. Local relief is greater than 500 meters (1600 feet) in many areas, and it is sometimes greater than 1000 meters (3200 feet). Many slopes are steep, limiting how the land can be used. The mountains continue to create problems for transportation, and this affects the pattern, pace, and forms of regional economic development. Even future impacts due to the region's proximity to Megalopolis and Southern growth centers are connected with Appalachia's mountainous character.

Far to the west, beyond the Mississippi River valley, the smaller Ozark-Ouachita mountainous region echoes Appalachia's history and character. These two mountainous territories are separated by 475 kilometers (300 miles) but form a single region because their topography and settlement history are similar. Both demonstrate an unusually close association between topography and human settlement issues (Figure 8.1).

A VARIED TOPOGRAPHY

The Appalachians' characteristic ruggedness is far from uniform. The uplands are composed of three physiographic provinces, each having a distinctive

Appalachia's hilly and irregular topography makes farming difficult or small-scale, or both. This farmer might grow a little corn or vegetables, but most of the cleared land is used as pasture.

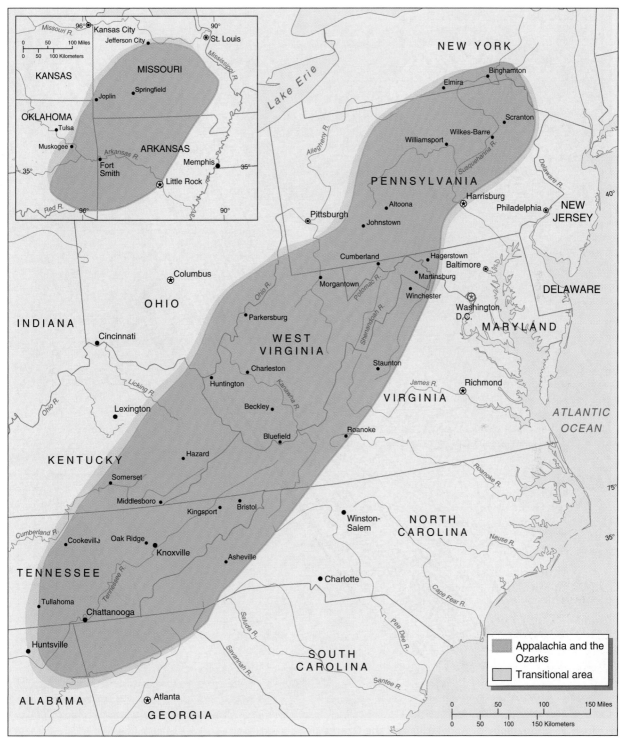

Figure 8.1   Appalachia and the Ozark.

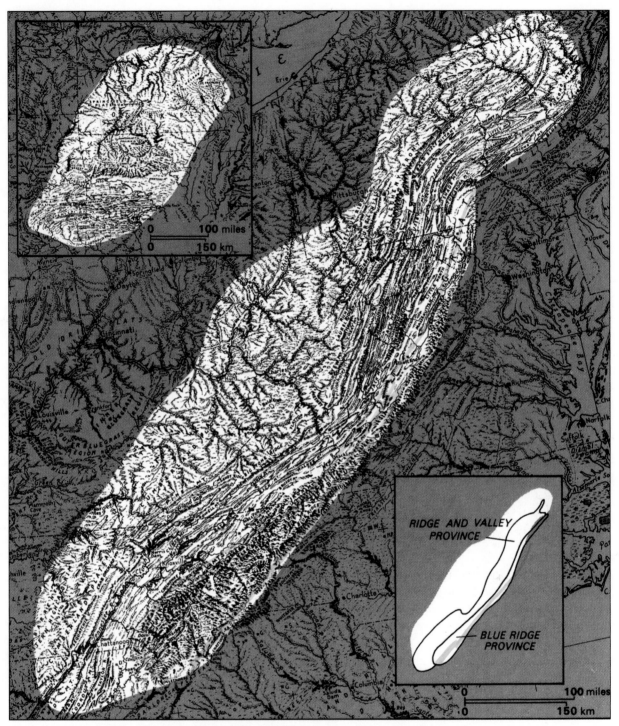

Figure 8.2   Topography of Appalachia and the Ozarks (reproduced from Raisz, *Physiographic Diagram* of the United States). The linear topography of the eastern Appalachians and southern Ozarks contrasts sharply with the jumbled surface to the west and north. This difference emphasizes the importance of underlying rock structure on topography.

topography (Figure 8.2). Substantial differences in the pattern of human occupation also exist in each section. These provinces lie in parallel belts with northeast-southwest axes.

The Blue Ridge is the Appalachians' easternmost topographic belt. Composed of ancient Precambrian rocks, the Blue Ridge has been severely eroded since tectonic plate collisions in the Atlantic crumpled North America's eastern coast beginning 460 million years ago. The Blue Ridge's current elevations are only a fraction of their former levels, some of which geologists estimate may have topped Mt. Everest's present 29,028 feet (8848 meters) above sea level.

Erosive activity has been greatest along the Blue Ridge's eastern edge. Over time, the immense mountain chain eroded to the rolling, remnant upland plain of the adjacent Piedmont physiographic region (see Figure 2.1) and the present Blue Ridge.

Changes in elevation between the Piedmont and the Blue Ridge are usually abrupt and substantial. Similarly, the two provinces' human geography since the arrival of Europeans has been quite different. For these and related reasons, the Piedmont and Blue Ridge physiographic provinces are placed in different regions, even though both share a common geological origin.

The Blue Ridge's elevation and width increase from north to south. South of Roanoke, Virginia, the Blue Ridge is the most mountainous part of the Appalachians. In Pennsylvania and Virginia, the Blue Ridge is a thin ridge between the Piedmont and the Great Valley to the west. Along the North Carolina-Tennessee border, the Blue Ridge broadens to a width of nearly 150 kilometers (95 miles).

For decades after Europeans settled the East Coast, movement west beyond the Blue Ridge funneled through the few natural gaps. South of Pennsylvania, the Potomac River, the James River at Roanoke, and the New River in northwestern North Carolina flow through the best of these low openings in the Blue Ridge. As late as the middle of the nineteenth century, the Blue Ridge was considered a nearly impenetrable barrier. Stonewall Jackson repeatedly surprised Union forces during the Civil War with his ability to charge across the Blue Ridge quickly with unexpectedly large numbers of men.

West of the Blue Ridge lies the Appalachians' Ridge and Valley section. The Ridge and Valley is the easternmost portion of a great expanse of sedimentary rock beds lying between the Blue Ridge and the Rocky Mountains. The eastern sediments were severely folded and faulted to form the Ridge and Valley's characteristically linear topography.

The Ridge and Valley is about 80 kilometers (50 miles) wide on the average. Ridges composed of relatively resistant shale and sandstone follow the Appalachians' northeast-southwest trend and rise 100 to 200 meters (300 to 600 feet) above intervening valleys. Through the ridges, rivers cut a few gaps across the area's grain.

Intervening valleys several miles wide lie between the ridges. The valleys, many floored with a limestone sedimentary layer, contain some of Appalachia's best farmland. When limestone is broken down by erosion, minerals—like lime—needed for productive farming in the eastern United States become available in the soil.

Between the Blue Ridge and the Ridge and Valley sections is the Great Valley. Typically hilly rather than flat and extending virtually the entire length of the Appalachians, the Great Valley is an historically important routeway. In Pennsylvania, it is called the Lebanon or Cumberland Valley; in Virginia, it is the Shenandoah Valley; and in Tennessee, it is the Tennessee Valley. Whatever its name, the Great Valley has tied Appalachia's people together.

The Appalachian Plateau is the region's westernmost province. Known as the Cumberland Plateau in Tennessee and the Allegheny Plateau from Kentucky northward, the Appalachian Plateau is bounded on the east by a steep scarp called the Allegheny Front. This scarp was the United States' most significant barrier to early westward settlement expansion. Only the Rocky Mountains equaled it in blocking the easy spread of American settlers in the efforts to move west.

The Appalachian Plateau's topography was formed as streams eroded the interior lowland's uplifted horizontal beds. Erosion created a rugged, jumbled topography, with steep, sharp ridges bordering narrow, curving stream valleys. A more gentle and rounded landscape appearance resulted in the north where the Plateau's northern portion in New York and Pennsylvania—called the Allegheny Plateau—was smoothed by Pleistocene glaciation.

Except for limited areas, such as portions of the Cumberland Plateau, level land is scarce on the Appalachian Plateau. Most communities are squeezed into small spaces in the stream valleys (Figure 8.3). Physical differences between the Ridge and Valley and the

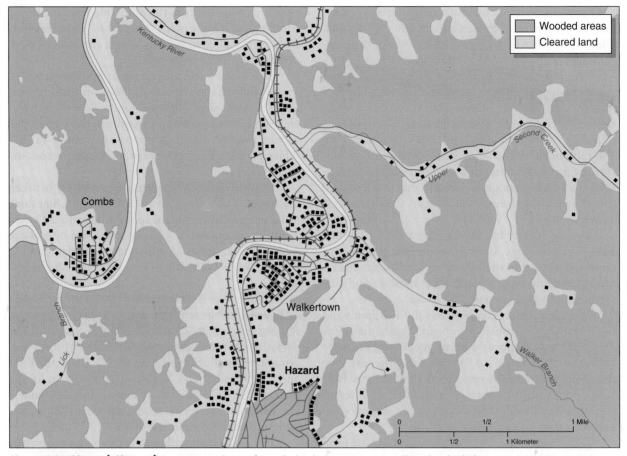

Figure 8.3   Hazard, Kentucky. Many residents of Appalachia live in its narrow valley. The shaded wooded areas corresponded closely with uplands or areas of substantial local relief; cleared land most often marks the valleys of this rugged region.

Plateau regions, and their different human geographies, are clear to anyone flying overhead. The symmetrical, parallel ridges and valleys of the former suddenly give way to the Plateau's rumpled, apparently patternless landscape.

The Ozarks-Ouachita uplands follow a topographic regionalization roughly similar to the Appalachians, but the grain is east-west instead of northeast-southwest. The more southerly Ouachita Mountains exhibit a series of folded parallel ridges and valleys similar to the Appalachians' Ridge and Valley. The Ozarks section is separated from the Ouachitas by the structural trough of the Arkansas River valley just as the Appalachians' Great Valley separates the Blue Ridge and the Ridge and Valley. Similar in appearance to the Appalachian Plateau, the Ozarks are a group of hilly, irregularly eroded plateaus.

APPALACHIA'S EUROPEAN PEOPLE

Early Settlement

Mountains and forests immediately catch the attention of Appalachia's casual visitors, but its people give the region much of its special character. The term "hillbilly" and its associated images grew out of generations-long experiences rooted in Appalachia. Far more mythical than real, this characterization of Appalachia's people nevertheless guarantees the region a distinctive spot in America's folk landscape.

European settlers did not push past the Blue Ridge into the Appalachian Highlands until late in the colonial period, 150 years after they first occupied land along the

East Coast. The easiest and first-used passageway into the Great Valley beyond was located in southeastern Pennsylvania, where the Blue Ridge is little more than a range of hills. Pennsylvanians found the mountain lands to the north and west inhospitable. Consequently, European settlers gradually spread down the Great Valley into Virginia. Most were either German or Scots-Irish. They were soon joined by members of other ethnic groups moving inland from the southern lowlands.

By the late eighteenth century, Europeans began settling in surrounding highland valleys and coves. For reasons that are still in historical dispute, the Scots-Irish played a major role in this movement. Some assign cultural motivations, arguing that these people were particularly footloose and independent, constantly searching for freedom on the frontier.

Others maintain that the group represented the main immigrant segment at the time the coves were settled; hence, they were most likely to contribute to frontier settlement. Whatever the case, the Scots-Irish, joined by many English and Germans, dominated early European settlement in the highlands.

American settlers moving into the Appalachians after the United States won political independence from Britain found adequate agricultural potential for small farms. About 10 to 20 hectares (25 to 50 acres) of cleared land in a stream valley was about all a single farmer could handle. The forests teemed with game, timber was plentiful, and domestic animals could graze in the woods and on mountain pastures. By the period's standards, this was reasonably good land. Thus, although the more rugged areas were not occupied by

Narrow mountain valleys and coves, poorly connected to outside highways by ill-kept roads, continue to be home to many of Appalachia's poorest people.

settlers, a large farming population soon scattered throughout the region.

As good as the land was for small-scale farming, the land was poorly suited to large-scale agriculture compared to land farther west in the lowland South. Rugged topography, coupled with a cool upland climate, rendered most of Appalachia unacceptable for a plantation economy. Only in some broader lowlands, such as the Shenandoah and Tennessee valleys, did a few large plantations develop. In consequence, many southern Appalachian counties have small black populations, and southern Appalachia's politics has always differed from the lowland South's.

Appalachia gradually grew more isolated as flatter, richer western agricultural lands were cultivated and grain production and farming subsequently mechanized. Small Appalachian farms became economically marginal as farming costs rose, and economies of scale were possible only for larger operations.

Famous pathways through the region, such as the Cumberland Gap at the western tip of Virginia and the Wilderness Road from there to the Bluegrass Basin of Kentucky, were winding and difficult. As a consequence, travel was avoided whenever possible. East-west routes between the northeastern seaboard and the Great Lakes followed the Mohawk Corridor and the flat lake shore of Lake Ontario, thereby avoiding the northern Appalachian Uplands.

Easy routes did not exist across the southern Appalachians. Major railroad lines skirted the highlands, inordinately benefiting those places through which the rail lines passed (see Figure 5.2). Atlanta's early growth, for example, resulted largely from its role as a rail junction at the mountains' southern edge. Appalachia's rural residents, living in the rugged reaches beyond Atlanta's immediate hinterland, were unaffected by such urban transportation benefits. Limited accessibility, or geographic isolation, was a major element in the poor economic opportunities available to Appalachia's residents.

Urbanization developed slowly in Appalachia, particularly southern Appalachia. In part, the region shared the South's emphasis on agriculture, extending this emphasis long after manufacturing and urban living developed in the rest of the country. In addition, Appalachia's people had few products to sell, the demand for urban goods and services was limited, and the sparse transportation network did not encourage urban development. As a whole, the region existed largely outside the pattern of economic development common elsewhere where transportation improvements fostered urban economic growth, and urban economic growth encouraged transportation improvements.

With little urban development, no antebellum plantation economy, and limited agricultural land, few new immigrants were added to the early European mix of Scots-Irish, English, and German settlers. The area's Anglo population remained overwhelmingly northwestern European in ethnic background, as well as Protestant and increasingly conservative religiously. Appalachia's people tended to remain where they were born. Over time, attachments to family, land, and the local community grew, providing the basis for the region's distinctive culture.

Southern Appalachian Cultures

Behind Appalachia's cultural distinctiveness is the regional population's relative immobility. In a nation where immigration, education, and commerce added more and more disparate elements to the total population, Appalachia became increasingly unusual because it simply remained unchanged until well past the middle of the twentieth century. This, in turn, may have led to even greater regional isolation. As Jack Weller argued:

> The fierce loyalty of mountain people to home is mostly a loyalty to the only culture in which they feel secure and which operates in ways they know and appreciate. . . . For mountaineers, moving is a kind of death to his way of life. It cuts him off from his sustaining roots.[1]

Henry Caudill, in *Night Comes to the Cumberlands*,[2] noted that few native West Virginians left the state during the Great Depression, whereas more than three quarters of all of West Virginia's foreign-born coal miners moved away to metropolitan areas outside the state.

In addition to being almost exclusively white, Anglo-Saxon, and Protestant, many in Appalachia are also poor. Appalachia is easily the United States' largest predominantly white, low-income region. Large swatches of the region—most notably a band from West Virginia through eastern Kentucky and Tennessee—are occupied by families living at or below the federally defined poverty level (Figure 8.4). The government's estimate of the income a family of four needs to meet a minimally acceptable standard of living was $18,400 in 2003.

In portions of Appalachia, poverty is largely attributable to the region's economic base. Appalachia is a major coal-producing area. Labor demand there declined steadily following the outbreak of World War II as the coal mining process was mechanized. Indeed, places such as eastern Kentucky where coal mining is concentrated were especially hard hit by high unemployment and continue to show high levels of poverty.

Appalachia's population is conservative, both politically and religiously. Many conservative Protestant churches in the United States trace their roots to Appalachia, and the region is part of what is called the Bible Belt. Politically, most elected officials are also conservative, although strands of rural populism are found.

One feature demonstrating how the region's roots differ from the South's is the Republican Party's traditional strength in Appalachia. Although Republicans are now well-represented across the South, the Appalachian sections of North Carolina, Georgia, Tennessee, and Alabama supplied many of the South's small trickle of successful Republican politicians beginning shortly after the Civil War when the rest of the South was solidly Democratic. The Missouri Ozarks have also often sent Republicans to Congress.

Today, Appalachia's legendary isolation is gone. No longer realistic is the old image of mountain people living in mountain coves and far up steep valleys and having almost no contact with the outside world. The reasons for this change will be discussed shortly. But a people's values change much more slowly than their technology and their social institutions. As Thomas Ford observed:

> So it is not implausible to suppose that the value heritage of the Appalachian people may still be rooted in the frontier, even though the base of their economy has shifted from subsistence agriculture to industry and commerce and the people themselves have increasingly concentrated in towns and cities.[3]

---

[1] Jack E. Weller, *Yesterday's People: Life in Contemporary Appalachia* (Lexington: University Press of Kentucky, 1965), p. 86.

[2] Henry M. Caudill, *Night Comes to the Cumberlands* (Boston: Little Brown, 1962), p. 179.

[3] Thomas R. Ford, *The Southern Appalachian Region: A Survey* (Lexington: University Press of Kentucky, 1962), p. 29.

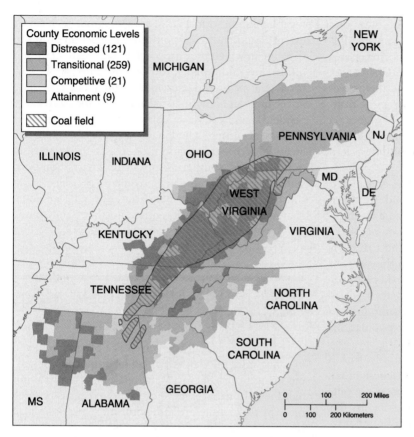

County Economic Levels
- ■ Distressed (121)
- ■ Transitional (259)
- □ Competitive (21)
- ▨ Attainment (9)
- ▨ Coal field

MICHIGAN

NEW YORK

ILLINOIS   INDIANA   OHIO

PENNSYLVANIA   NJ

WEST VIRGINIA   MD   DE

KENTUCKY   VIRGINIA

TENNESSEE   NORTH CAROLINA

SOUTH CAROLINA

MS   ALABAMA   GEORGIA

0   100   200 Miles
0   100   200 Kilometers

Figure 8.4 Poverty counties, 2002. Appalachian counties experiencing distressed economic levels are located mostly in the south or are coal producers.

These observations remain apt. The region's economy has changed greatly, but its people's beliefs and values have changed more slowly.

Much of what is true about southern Appalachia fits the Ozarks as well. Take the stereotyped Ozark hillbillies; they are as much a part of Ozark legends as their brethren in Appalachia. And, as in southern Appalachia, the legends are inappropriate. The ethnic background of settlers in both regions and the subsequent experiences of their descendants were nearly the same, and so people in both places have similar attitudes. Indeed, except for the absence of coal mining in the Ozarks and the physical distance separating these areas, southern Appalachia and the Ozarks have much in common.

Northern Appalachia

Northern Appalachia's land and people share only a few of southern Appalachia's regional characteristics. Certainly, the Appalachians from Pennsylvania north have mountainous topography, and settlers there experienced the larger region's early economic development

problems. Beyond that, the similarities between southern and northern Appalachia grow more tenuous.

Although people in the northern Appalachians are by no means wealthy, they endure less poverty than their southern neighbors (see Figure 1.5). Also, in contrast to southern Appalachia, northern Appalachia's early northwestern European settlers were followed by immigrants from other parts of Europe. This is especially true in Pennsylvania and northern West Virginia where coal mining attracted East European immigrants in the late nineteenth and early twentieth centuries. New York's highlands, on the other hand, were settled by other New Yorkers and farmers from New England. Finally, many cultural practices in the northern Appalachians are not at all the same as those of the southern highlands. Religion provides a notable example of this cultural difference. Fundamentalist churches are less common in the north; Catholics and members of Eastern Orthodox churches are in the majority in many northern Appalachian counties, especially in Pennsylvania (see Figure 3.6).

Transportation within the northern Appalachians also became far better than in the southern Appalachians.

In part, the mountains are less continuous, lower, and therefore, more easily breached. But as the upper Midwest's economy blossomed and the continent's major area of commerce and manufacturing formed (Chapter 5), people in the northern Appalachians found themselves fortunately located. Transport lines connecting the manufacturing core's eastern and western portions were soon built across the region. As a consequence, far more development occurred there, especially in central and western Pennsylvania and New York.

Given these distinctions, geographers considered the question: "Should the northern and southern Appalachians be considered part of the same region?" What they concluded was that in spite of northern and southern Appalachia's differences, the two parts have more in common than they do with the states of which Appalachia is a part. Certainly, northern Appalachia is distinct from its surrounding areas. It is also linked to southern Appalachia by the mountains and valleys and by the relative poverty their populations share.

West Virginia is the only state entirely within Appalachia. Every other state's capital, largest urban area, and major economic development areas are outside Appalachia's regional margins. Tennessee fits this general pattern, for example, even though both Knoxville and Chattanooga are located within Appalachia. Nashville's 2006 population of 552,120 is larger than both Knoxville's (182,337) and Chattanooga's (155,190) combined.

Appalachia's separation from the states in which it is located was demonstrated tangibly during the Civil War. In the South at that time, people in the mountainous areas saw little reason to join their states in secession. The plantation economy and its way of life were foreign to those in the mountains. Opposition to secession was strong and widespread throughout the Appalachian South. Opposition was strong enough among Virginia's mountain people that they voted to withdraw from the rest of the state, forming West Virginia in 1863. But even so, other states' ties remained strong. In spite of considerable dissension, southern Appalachian populations accompanied their states out of the Union.

The portions of states outside Appalachia's margins nevertheless have substantial influence within Appalachia. As elsewhere in the United States, states fund public education, and part of that education is exposure to the state's history and politics. State aid and state governmental programs are important to local economic efforts. Newspapers and, to a lesser extent, radio and television often restrict attention to the state in which they are located, with state news and information dominant. And because most states levy taxes and all provide services to their citizens, state governmental activities are important for every state resident. In short, a state's boundaries function as a container uniting what may otherwise be disparate areas. This orientation toward one's state also helps set northern and southern Appalachia apart. Southern states share many similarities (Chapter 9), just as do those in the Northeast, and these broader regional characteristics help differentiate northern and southern Appalachia.

ECONOMIC AND SETTLEMENT PATTERNS

Appalachia is generally perceived as a rural region, and in some ways this view is valid. The proportion of people living in urban places in Appalachia is only about half the national average. According to census reports, most of the people are either rural or rural nonfarm residents. Rural nonfarm residents live in rural areas but work in urban places. The highest rural population densities in the United States are in eastern Kentucky and other parts of Appalachia, with many counties having rural densities of 60 or more per square kilometer (150 or more per square mile). Appalachia's high rural density is supported by an economy dependent on small farms and mining.

Agriculture

Owner-operated farms are more prevalent in Appalachia than in any other region in the United States. Kentucky and West Virginia lead the country in that category. Farms with more than 6 hectares (15 acres) of cropland in one North Carolina Appalachian county were recently found to be owner-occupied in an astounding 99.3 percent of cases. In addition, farm tenancy did not develop in Appalachia. This is a region of independent small farmers, however modest the farm income may be.

The best description of Appalachian agriculture may be too few resources divided among too many people. Throughout Appalachia, the average farm is small, containing only about 40 hectares (100 acres). The region's rugged topography, poor soil, and short growing season limit available cropland and require greater emphasis on pasturage and livestock. Because fields are small and scattered among valleys, large farm machinery cannot be used efficiently.

The net result of all this is that average farm incomes are low. To make ends meet, many Appalachian farmers operate their farms on a part-time basis. By supplementing their incomes with other jobs, they can retain their land. Aggravating the situation, average farm size is declining in parts of the region as farms are divided through sale or inheritance. Smaller farms, like part-time farming, reinforce the growing importance of off-farm income for many Appalachian farmers.

The typical agriculture in the region is called general farming, and Appalachia is the United States' major general-farming region. This means that no crop, farm product, or combination of products dominates the local farm economy. Extensive animal husbandry is the most common and probably best agricultural use of the region's steep slopes. Specialty crops, such as tobacco, apples, tomatoes, and cabbage, are locally important in some valley areas, with small burley tobacco plots being most common in the southern Appalachians.

Corn, another traditionally important cash crop, was illegally distilled into the region's famous moonshine and rushed to lowland markets by legendary drivers who anticipated today's sport of stock car racing. High production costs, alternative job opportunities, improved detection, and expanded sale of legal alcoholic beverages combined to cut moonshine production drastically in recent years. Marijuana may have replaced moonshine as the region's primary contraband export. Corn, normally used on farms for animal fodder, remains the region's leading row crop.

Exceptions to the pattern of semimarginal agriculture exist within Appalachia. Virginia's Shenandoah Valley, for example, was for decades the state's breadbasket. Competition from wheat grown on fertile grasslands in the South and Great Plains forced Shenandoah farmers out of the national wheat market by the late nineteenth century. Although winter wheat is still grown, hay, corn for fodder, and apples are now the Shenandoah valley's major crops, with turkey raising important in some locales. It remains a fruitful and beautiful farming region.

Dairying and apple production are key in central Pennsylvania's many valleys. The Tennessee valley is a substantial agricultural district, with emphasis on fodder crops and livestock. Even so, these relatively large productive areas are exceptions.

Coal

Although farming is part of the economic and cultural tapestry in Appalachia, it is inadequate in explaining Appalachia's large rural population. Farming's chief partner across much of the region is coal. Almost all of the Allegheny Plateau is underlain with a vast series of bituminous coal beds, which collectively comprise the world's largest coal district (see Figure 2.8). Coal seams as thick as 3 meters (10 feet) interbedded the plateau's flat-lying sedimentary rocks. Over time, the coal seams have been exposed by the same streams that created the plateau's rugged topography. The horizontal character of the coal beds is clearly evident on topographic maps where symbols for mines are marked for miles along the same elevation contour line.

Appalachian coal became significant shortly after the American Civil War. New types of coke-burning iron and steel furnaces were invented, creating a demand for coke processed from bituminous coal. The thick coal seams of southwestern Pennsylvania and northern West Virginia fueled Pittsburgh's rise, garnering it the nickname "Steel City." When the nation turned to electrical power in the twentieth century, Appalachian coal stoked electric-generating facilities along much of the East Coast and across the interior manufacturing core.

After almost a century of growth, the coal industry fell into decline following World War II. Production dropped as petroleum and natural gas replaced coal as major fuels. Coupled with mine mechanization required to keep mining costs competitive, this switch in energy sources led to drastically high unemployment in many coal mining areas. Out-migration followed. Many coal counties lost a full quarter of their populations between 1950 and 1960. The resulting economic depression, blending with Appalachia's general poverty, created areas—like eastern Kentucky—with particularly severe problems.

Coal production increased after 1960 to record levels in the 1980s (Table 8.1). But Appalachian coal, heavily used by the United States' steel industry, again faced lower demand when national economic recession in the early 1980s lowered steel production. The steel industry used approximately 75 million tons of coal annually in the mid-1970s, but by the late 1980s only

TABLE 8.1

Bituminous and Lignite Coal Production, Major Mining States, 1947–2002 (in thousands of short tons)				
State	**1947**[a]	**1970**	**1995**	**2005**
Eastern States				
Alabama	19,048	20,560	24,640	21,339
Illinois	67,680	65,119	48,180	32,014
Indiana	25,449	22,263	26,007	34,457
Kentucky	84,241	125,305	153,739	119,734
Ohio	37,548	55,531	26,118	24,718
Pennsylvania	147,079	80,491	58,893	67,494
Virginia	20,171	35,016	34,099	27,743
West Virginia	176,157	144,072	162,199	153,650
Western States				
Colorado	6,358	6,025	25,710	38,510
Montana	3,178	3,477	39,451	40,354
North Dakota	2,760	5,693	30,112	29,956
Texas	—	—	52,648	45,939
Wyoming	8,051	7,222	263,822	404,319
Total United States	630,624	602,932	1,032,974	1,040,227

[a]1947 was year of maximum production prior to the 1970s.

Sources: U.S. Bureau of Mines, *Minerals Yearbook* (Washington, D.C.: U.S. Government Printing Office, 1948, 1970); Energy Information Administration, *Quarterly Coal Report* (Washington, D.C.: U.S. Department of Energy, 1996); *Annual Coal Report*, 2005 (Washington, D.C.: U.S. Department of Energy, 2006).

half as much was needed. This fluctuating demand affected the coal industry in Appalachia and, therefore, its workers.

Demand for Appalachian coal was also affected in the 1980s by the shift of domestic U.S. coal mining to other parts of the country. While discovery and mining of new deposits means that the United States produces more coal than ever before, Appalachia sees modest benefit. An increasing share of U.S. production is low-sulfur coal from the west (Chapter 13). Cushioning the decline in Appalachian coal output, nearly 45 million tons of Appalachian coal is exported annually, almost all of it shipped through the port of Norfolk, Virginia. The thick seams and mechanized mining in the Appalachian fields make this coal cheap on the world market, and it finds ready buyers abroad. The primary destination for U.S. coal is Canada, followed by Japan, Brazil, and France.

Another set of changes also added to fluctuations in coal demand. Concern about air pollution in the 1960s led the federal government to discourage using coal in power plants (Figure 8.5). However, as electrical power demand grew and concerns over nuclear power safety and the cost and availability of petroleum increased, the government changed its stance about coal. The government actively encourages coal use in electric power generation, although recent, growing concerns over global climate change may once again temper policy toward coal use.

Newly built generating plants use huge quantities of locally mined coal to produce electricity. Much of the electricity generated in Appalachia is transmitted to consumers outside the region from generating plants that are increasingly visible on the central and northern Appalachian Plateau landscape.

Proven U.S. coal reserves are huge. Experts project reserves will last at least three more centuries at current use rates. Domestic liquid petroleum reserves, on the other hand, have declined steadily for decades. These two facts emphasize the future importance of coal.

Pilot projects to extract natural gas from coal have been successful, although the process is expensive. But natural gas extraction costs will become economically

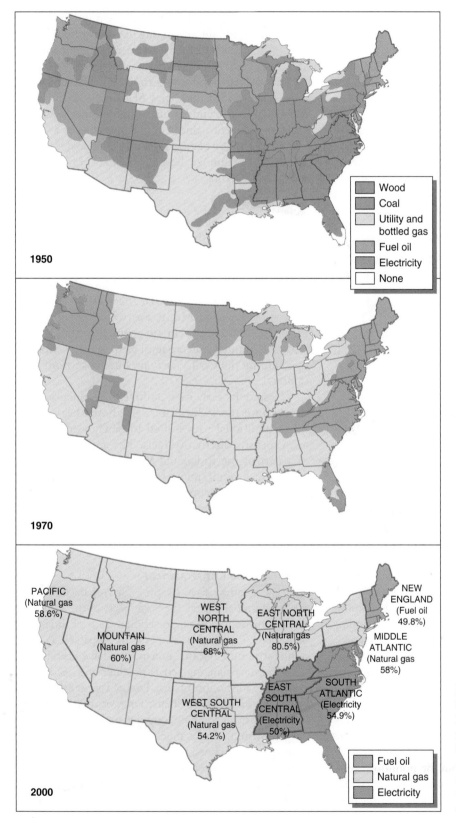

Figure 8.5 Primary home heating fuel, 1950, 1970, and 2000. Coal was replaced by natural gas and electricity as a home heating fuel between 1950 and 1970. Aside from the Northeast, most regions of the country now rely almost exclusively on natural gas or electricity.

competitive as technology improves. Extracting petroleum from coal is also technologically possible, yet costly. When and if either competitive extraction goal is achieved, U.S. coal reserves will be consumed more quickly than is presently projected.

Mining Methods

Appalachian coal is mined in several ways. The earliest method, underground (or shaft) mining, is still the chief extraction technique in northern Appalachia. The shaft, often times more than 500 feet in length, is dug perpendicular or at a slope in order to reach the coal seam and serves to connect the underground workings of the mine with the tipple and processing plant above ground. Modern techniques using huge mobile drills and continuous mining machines rip tons of coal per minute from the seams and deposit it on conveyor belts for the trip to the surface.

Surface mining is more prevalent in Appalachia's central region, primarily eastern Kentucky, western Virginia, and southern West Virginia. Three principal surface mining techniques are used in the region: contour mining, area mining, and mountaintop removal. Surface (or strip) mining is far less expensive than underground mining as long as the coal seams are relatively shallow. With this method, large machines remove rocks above a horizontal coal seam and then lift out the uncovered coal. In the case of contour stripping, the waste rock, called *overburden*, is simply dumped down the slope. Over time, strip mine extraction along seams on a hillside creates a peculiar, stepped appearance that looks from a distance as though a series of boxes were piled on top of one another, each smaller than the one below.

Area mining is used in locations with flatter terrain, such as in southeastern Ohio. Area stripping is still done on the surface, but the scale of the operation and contiguous area is much greater where the land is flat. Given a gently rolling parcel of land, the stripping proceeds via making a series of parallel cuts across a wide area. With each cut made to access the coal seam, the overburden is cast behind to fill in the previous cut. Massive scoop shovels, such as the Gem of Egypt, or tremendous draglines, like the Big Muskie, each as large as multistoried buildings, remove the overburden and dig out the coal seam. These immense machines are also used in western coal mines.

Mountain-top removal is used on a much smaller scale. The technique is generally best suited for extremely hilly areas in West Virginia and parts of eastern Kentucky, where the coal seam might be situated close to the surface among dome-like hills. In this case, virtually all of the overburden is cast downslope and the hill is significantly flattened.

About half of the coal extracted in Kentucky and most taken out in Ohio and Alabama is strip mined. Most coal from Pennsylvania, Virginia, and West Virginia, or about two-thirds from Appalachia as a whole, is shaft mined. By comparison, nearly all coal is surface mined in Wyoming, Texas, Montana, New Mexico, and North Dakota.

Anthracite

Although bituminous fields are abundant on the Appalachian Plateau, anthracite from the Ridge and Valley province's northern tip in Pennsylvania was the earliest Appalachian coal mined.

Anthracite is much harder and lower in moisture content than bituminous coal. Because anthracite is smokeless when burned, it was widely used to heat homes well into the twentieth century. Anthracite was also used to smelt iron ore until techniques were developed during the Civil War to produce coke from bituminous coal.

Despite its early uses, anthracite production grew slowly until the 1920s. Afterward, home heating and other traditional uses shifted to newly available and less expensive alternative fuels. The anthracite seams, as a part of the Ridge and Valley, are fractured and folded. As a result, mining costs are high, which puts anthracite at an economic disadvantage. As a consequence, current production is minimal. Anthracite mining is unlikely to experience a resurgence even though substantial anthracite reserves remain.

Weighing the Consequences

Coal has been a mixed blessing for Appalachia. Long the economic mainstay for large parts of the region, it directly or indirectly supported hundreds of thousands of people. Coal seems integral to Appalachia's economic structure, but the human costs have been high.

Counterbalancing coal production's economic benefits is a list of problems. Tens of thousands died in mine-related accidents. Black lung, caused by breathing coal dust for too long, crippled and killed countless others. Strip mining is much less hazardous than shaft mining to the miners' health. Accident rates per ton of coal mined are lower, and the risk of black lung from breathing coal dust is greatly reduced. Even so, shaft mining is a risky and physically demanding enterprise.

Evidence of coal mining is often visible on the surface. Large structures, such as this coal breaker in Ashley, PA, still stand in areas where mining took place.

Regional unemployment is still widespread. Recent increases in market demand were met by applying greater mechanization, not by hiring more workers. New manufacturing plants, attracted by available labor, have located in the region, but most pay low wages and the labor force is largely comprised of women.

Most corporate mineral rights were obtained early and cheaply. The original agreements often circumscribed the taxes that can be charged. As a result, although some Appalachian states initiated or increased surcharges on locally mined coal, coal taxes remain low. Benefits for source-area economies are reduced, because most mining profits leave the region.

Additional negative consequences of coal mining arise from mining's impact on Appalachia's physical environment. Strip mining creates unsightly gashes across the landscape. Every coal mining state has passed laws requiring that strip-mined lands be reclaimed. The federal government entered the picture by passing a Surface Mining Control and Reclamation Act in 1977. The Act requires mining companies to prove they can reclaim the land before a mining permit is issued. The Act also mandates that, after mining is completed, the land is restored to its premining purposes.

Reclamation legislation does not address the current environmental impacts of earlier mining activity.

Acid mine drainage, the acidic groundwater seeping into streams from mines and mining waste, kills fish and plant life along thousands of miles of Appalachian creeks and streams. Narrow, winding valleys that would have been sparsely settled without mining employment are overcrowded and subject to repeated flooding. Even Appalachia's rich summer vegetation often cannot hide vast, gray slag heaps and coal dust. A beautiful part of America has been severely damaged. Full environmental recovery will take centuries.

Disputes over whether or not mined land should be reclaimed are an excellent example of competing values and perspectives. On one hand, some might take the position that restoration hurts mining companies and therefore damages the regional economy. It is extremely difficult to return rugged topography to its former shape, and land reclamation costs are very high, estimated at $7500 to $25,000 per hectare ($3000 to $10,000 per acre) in the late 1980s, the most recent figures available. Pennsylvania reclamation efforts in the 1990s cost about $5500 per acre. On the other side, some might argue that without reclamation, the benefits from coal extraction are too heavily weighted in the companies' favor. Reclamation is an investment in the future, the period after the mining companies depart. Not degrading the lives of Appalachia's future residents should be considered part of doing business today.

Couched as economics versus the environment, or as immediate gain versus long-term well-being, arguments over strip mining and reclamation requirements are likely to continue.

Other Mining Activities

Regional mining activities are not restricted to coal. The Ozarks' tristate district, where Oklahoma, Kansas, and Missouri meet, has long been a major lead mining area. Southeastern Missouri, just outside Appalachia, has produced lead from surface mines for more than 250 years. Missouri has supplied most of the lead mined in the United States and currently produces more than three quarters of the country's total production.

Drilling for oil is another important extraction process. The first oil well in the United States was drilled in northern Pennsylvania in 1859. That state led the country in petroleum production through most of the nineteenth century. Today the area supplies only a small part of the nation's crude oil needs but remains a producer of high-quality oils and lubricants.

Finally, southeastern Tennessee is the United States' most significant area of zinc production. In addition, several mines around Ducktown, near the North Carolina and Georgia borders, yield the only significant copper produced east of the Mississippi River.

REGIONAL DEVELOPMENT PROGRAMS

States usually pay greater governmental attention to places in the state where economic and political influence is strongest. For states containing sections of Appalachia, this means governmental attention usually has focused on issues in non-Appalachian parts of their state. Similarly, political currents that affected federal policy kept national attention on issues that did not directly address Appalachia's problems. This inattention—at both state and federal levels—changed during the 1930s Depression with the first of three intervention programs.

The Tennessee Valley Authority

Appalachia's rivers, like its coal, are a mixed blessing to the region. Some streams are long and deep enough to be transportation routeways. And water power offered early settlers the means to establish gristmills and sawmills.

The rivers had a dark and temperamental side, however. The southern Appalachian highlands are North America's moistest area east of the Pacific Coast. Heavy rain is frequent and is a source of runoff that can flood the region's narrow valleys. The desire to control one of Appalachia's major rivers, the Tennessee, led to the largest and arguably the most successful regional development plan implemented in American history.

The Tennessee River's headwaters lie in Appalachia's mountains between Virginia and Georgia, and it flooded as often as any river in the South. The Tennessee also flowed through some of the inland South's poorest areas. Early in his first term, President Franklin D. Roosevelt conceived a plan to harness the Tennessee River and its tributaries, thereby improving economic conditions throughout the entire Tennessee valley.

Improving navigation on the river was seen as part of the federal government's constitutionally designated control over interstate commerce. The Tennessee Valley Authority's (TVA) first charge, therefore, was to develop the Tennessee River for navigation. Under the TVA's direction, a 3-meter (9-foot) barge channel was built in the river as far upstream as Knoxville. Other navigation improvements included constructing or purchasing dams to guarantee stream flow.

Since the 1930s, the Tennessee valley's economic development has been a logical extension of TVA's initial commitment. The TVA currently controls 45 dams, including 29 hydroelectric dams. Some of these dams were developed privately, and TVA control was subsequently acquired. Flooding was reduced, and electric-power generating facilities have been built. Most dams on the Tennessee and Kentucky rivers have power-generating facilities (Figure 8.6).

The TVA supplies electricity across the region. As power needs grew, demand exceeded the TVA dams' capacities to generate electricity. So the TVA then constructed other kinds of power sites. Today, over 85 percent of the electricity produced by TVA comes from thermal plants, including 13 coal-burning and three nuclear-powered facilities. As a result of these thermal power plants, the TVA uses tens of millions of tons of coal annually and is Appalachia's largest coal user. The TVA has recently added facilities that utilize wind, solar, and methane gas resources. In total, the TVA provides electricity to over 8 million people.

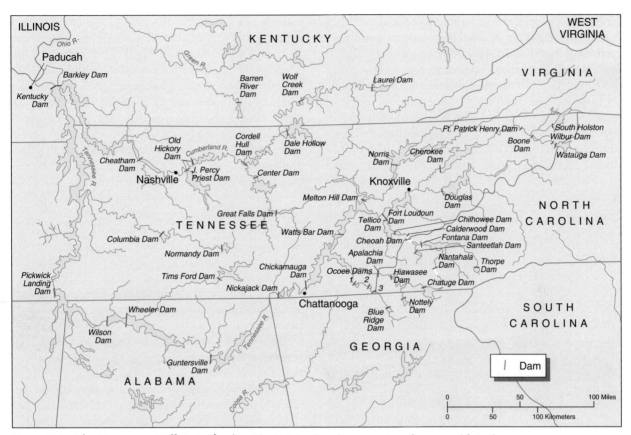

Figure 8.6   The Tennessee Valley Authority. The TVA has long been a major influence on life in the Tennessee Valley, with the more than 40 dams on the area's streams the most obvious aspect of that influence.

TVA power was once cheaper than most alternative sources in the United States. Inexpensive electricity attracted several heavy power-using industries to the Tennessee valley. Most notable was the large aluminum processing facility at Alcoa, south of Knoxville. The country's first atomic research facility was placed at Oak Ridge, west of Knoxville, partly because large amounts of power were available. Manufacturing expanded in Knoxville, Chattanooga, and the tricities of Bristol, Johnson City, and Kingsport, Tennessee. The TVA also developed artificial fertilizer industries, another heavy power-consuming activity.

Although the TVA's dams helped control floods and produce electric power, the TVA instituted additional beneficial programs as well. Above the dams, the TVA initiated a program to help valley farmers control erosion. Holding precipitation runoff at the farms would reduce flooding and slow the rate at which reservoirs filled with silt. To accomplish these ends, the Authority oversaw a careful inventory to assess the region's agricultural potential, constructed thousands of farm ponds, and offered advice on erosion control.

In addition to the water itself, the TVA manages 293,000 acres of public land and 11,000 miles of shoreline in the Tennessee Valley. Major public recreation areas were developed on some of this land, and recreational boating was encouraged. The Tennessee valley has some of the United States' finest inland water-recreation spots.

TVA proponents argue that much of the area's significant manufacturing growth since World War II resulted directly from the Authority's work. This position is weakened to an unknown degree because major manufacturing growth occurred in other Southern areas without the TVA, such as the Piedmont between North Carolina and Alabama and the lower Mississippi River valley. Although TVA contributions to flood control, cheap water transport, and power production undoubtedly had a major impact on regional manufacturing growth, the roots of this growth are not attributable solely to the TVA.

Nonetheless, the TVA clearly changed the area's economy. Criticized from time to time as an example of inappropriate governmental intervention, it is generally well regarded within the Tennessee Valley. With over 13,000 workers, the TVA is one of the region's more important employers. Tennessee, once one of the South's poorest and least industrial states, is now a regional leader in manufacturing.

Appalachian Regional Commission

The United States Congress passed the Appalachian Redevelopment Act in 1965, extending the 1961 Area Redevelopment Act. The 1965 Act created the Appalachian Regional Commission (ARC), making this agency responsible for an area extending from New York to Alabama. Since its inception—and like the TVA—the ARC has spent billions of dollars to improve the region's economy. But unlike the TVA, the ARC's primary thrust is to improve Appalachia's highways, hoping to decrease local isolation and encourage the location of new industries.

Since government investing in transportation to encourage regional economic growth is a well-established practice in the United States, the ARC's commitment is not unusual. The Erie Canal, opened in 1826, was funded by the state of New York. Other canals were built by the federal government in Ohio, Illinois, and Indiana during the 1830s and 1840s. More than 40 million hectares (100 million acres) of federal land were given to western railroads to help finance rail construction.

Virtually all of twentieth-century America's roads and highways were built and are maintained by local, state, and federal agencies. Because construction costs are large and economic returns usually long term, private investment in transport infrastructure is unusual. It has been more usual to use private funds for other forms of communication—notably, telephones—but only because the government initially guaranteed a monopoly to investors in exchange for extensive regulation.

The ARC is different from the TVA in other important ways as well. First, the ARC requires federal and state cooperation, with each governmental level paying part of the total costs. Second, the ARC worked to improve public and vocational education. Third, regional economic planning was required to make maintenance of physical environmental quality a priority.

Requiring state governmental cooperation partially explains the ARC's extensive influence. States, not the federal government, defined the Commission's target area. Because no state wanted to be excluded, eventually the entire physiographic region became eligible for federal matching funds through the ARC.

Flooding is an ever-present problem on the Appalachian plateau, where the towns and streams crowd together in narrow valleys. Harlan County Kentucky faced the effects of such crowding when muddy waters inundated the area in March 2002.

The Arkansas River Navigation System

The Arkansas River Navigation System is the third governmental intervention program to address Appalachia's problems. Constructed during the 1960s and 1970s and dedicated in 1971, the Arkansas River Navigation System established a 3-meter (9-foot) navigation canal up the Arkansas River from its confluence with the Mississippi River to Catoosa, Oklahoma, just downstream from Tulsa. Barge traffic increased as a result, and hydroelectric power is produced at dams constructed to stabilize the river's flow.

APPALACHIA'S FUTURE

Appalachia's problems are well known. The term *Appalachia* widely connotes isolation and poverty. Poverty will probably persist for some time, especially in mining areas. The natural environment is beautiful but, in some areas, abused and degraded.

Nonetheless, portions of Appalachia have experienced real change. Parts of the southern highlands in Georgia, the Carolinas, and Tennessee are witnessing a boom in recreational and second-home construction. Land values in these places are increasing rapidly, attesting to this development. Communities such as Highlands and Cashiers in North Carolina are summer, and quite often winter, havens for lowlanders. Manicured, well-cared-for golf courses and communities stand in striking contrast to the old image of a slightly ramshackle Appalachia.

The Blue Ridge and Virginia's Shenandoah Valley are experiencing a similar boom, fueled mostly by people from southern Megalopolis. Ski resorts attract vacationers as far south as North Carolina's southwestern mountains. The area around Branson, Missouri, has benefited economically as this small town on the Ozark Plateau became a national performance center and destination for legions of country music fans. The Ozarks and Ouachita Mountains are also dotted with new second-home and retirement communities. The long-term pattern of outmigration from the region, though not entirely over, is diminishing. The gap between per capita income levels in Appalachia and the rest of the United States is still very real but has narrowed. The economic future is positive across much of Appalachia and the Ozarks.

Review Questions

1. What are the three major physiographic provinces in Appalachia? What are the criteria for this regional scheme?

2. How does Northern Appalachia differ from Southern Appalachia?

3. What have been the major social, economic, and environmental impacts of coal mining in Appalachia?

4. What were the initial goals of the Tennessee Valley Authority? Discuss the major contributions of the TVA.

5. What were the original objectives of the Appalachian Regional Commission? Discuss the major impacts of the ARC.

6. Has the Arkansas River Navigation System met regional development expectations?

CHAPTER 9

THE CHANGING SOUTH

New England is provincial and doesn't know it, the Middle West is provincial, and knows it, and is ashamed of it, but God help us, the South is provincial, knows it, and doesn't care. **Thomas Wolfe**

Preview

1. Although there is great diversity within the region, the South remains distinctive culturally and economically within North America.

2. Because of its settlement history and the role played by slavery, the South today continues to have a large African-American population.

3. Racial discrimination and segregation was a reality in the region's past but has much less effect on the region's politics and geography today.

4. The South's economy has changed dramatically, increasingly shifting from agrarian to industrial, although poverty persists in many areas.

5. In recent decades, Southern cities have grown dramatically and changed the rural character of areas within the region.

The South is singular in the depth and persistence of its regionality. Its people have a deeper sense of where they live as a culture region than do residents of any other part of the United States. Certainly, people elsewhere, such as in northern New England's "Down East," the Plains and Prairies, the Southwest, and the Pacific Northwest, possess cultural characteristics and views of where they live that contribute to their senses of their own territory's distinctiveness. But the South's sense of itself is unusually strong, forged by attitudes, habits, and behavior rooted in the region's turbulent past.

Most Southerners are proud to identify themselves as Southern, gaining sustenance from the region's identity as a unique part of the United States. The South's regionality is reinforced by the nearly consistent image people living elsewhere have about it. Slower to be acknowledged are the significant changes well underway in the region.

A REGION APART

Southern culture is usually viewed as uniform, but this view is far from accurate because strong differences exist within the region. The Gulf Coast, the Southern highlands, the Georgia-Carolinas' Piedmont, and many portions of the northern interior South, among other sections—each possesses its own version of Southern culture, and each version is at variance with others. Native Virginians, for example, know they differ from Georgians, and both groups recognize their differences with Southerners from Tennessee or Mississippi.

These variations within the South complicate efforts to formally define Southern culture, but Southerners find comfort in a shared sense of being Southern. Newcomers may not recognize the basis for the comfort, but those reared in the South are aware of it

The South contains clear contrasts between its past and its future. Within this region's modern, thriving cities, such as Charlotte, North Carolina, live people who self-consciously maintain attitudes and traditions not too different from those of their great-grandparents.

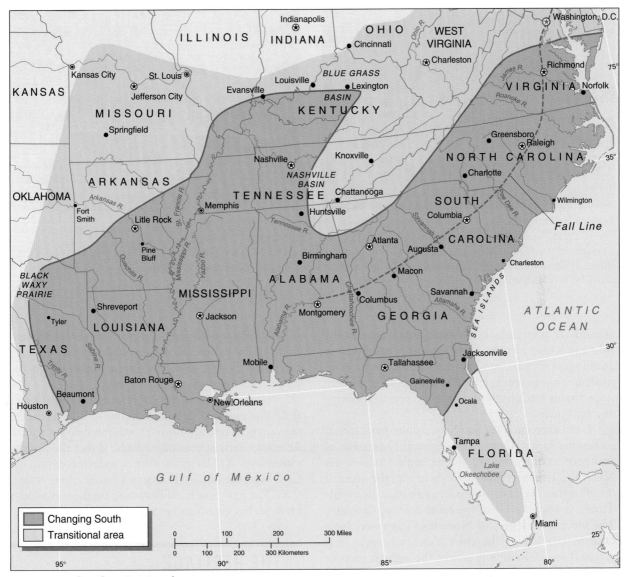

Figure 9.1 **The Changing South.**

as well as the broader region in which they live, regardless of other differences.

The South can be viewed as a geographic composite of beliefs, attitudes, patterns, habits, and institutions associated with life in the entire group of southeastern states (Figure 9.1). The extent of Southern culture, however it is defined, is not coincidental with the census South. Peninsular Florida and coastal Texas, locationally the most southern territories of the continental United States are the least imbued with the South's culture.

The personality of a place can be gleaned by passing through the landscape and living among its people with one's senses open, and the characteristics giving

the South its personality are numerous, as for any region. Throughout much of the rural South, one can still stroll down a dusty, red dirt road in the summer with an open pasture or cropped field on one side and a mixed, but mostly pine woods on the other. A green wave of heavy *kudzu* vine cascades across the woods' edge and extends tentative ribbons of vegetation out into the road. As recently as a generation ago, a single-story, broad-porched frame house with a painted metal roof might have contained a family preparing a customary evening meal of ham hocks and snap beans, sweet potatoes, and early turnip greens, followed by cornbread and molasses. Many houses in the rural

Regionally distinctive house-types may suggest aspects of the region's cultural underpinnings. The shotgun house (left), shown here in New Orleans, is a modest form of housing found throughout the region, but is less prevalent in other areas of the United States. The Leverett House in Beaufort, SC (right) represents early upperclass housing in the South. Large porches provided residents a cool place to escape the heat.

South have pretty much the same appearance today, although higher incomes allow replacing the old frame dwellings with neat, modular homes, and most rural roads are now paved. In addition, the spread of grocery store chains has greatly reduced the region's dietary distinctiveness.

Local expressions of Southern culture have spread across the larger region and beyond. One style of Southern country music, for example, focuses on Nashville, Tennessee; another style carries the name of the Bluegrass Basin in Kentucky; a third, called Old-Timey, is associated with the southern Appalachians. The blues are a legacy of Memphis, Tennessee. New Orleans jazz carries the city's name, wherever it is played. Gospel music is prevalent throughout most of the South, and several styles of gospel exist.

Stock-car racing, a highly popular "good-old-boy" tradition, had a Southern concentration until late in the 1990s. Indeed, this sport is still more intensely associated with the South than with any other region in the country, though its popularity has spread across the United States. This expression of Southern culture originated almost entirely in North and South Carolina, ostensibly born from moonshine runners who once raced their cargoes along back roads at night from Appalachia's foothills to urban centers on the Piedmont.[1]

Too often, a region's culture is perceived superficially and in caricature. Expressions of Southern culture, such as its music and sports, only scratch the surface of a rich and complex subject.

Today, Southern culture is in rapid transformation. The old South's rigid attitudes and practices, the region's heritage, are deeply rooted in colonial North America's earliest economic goals. When the South's expressions of those goals were overthrown during the Civil War, the old South began a century-long transition. The new South, still drawing on the old South's heritage, has grown and changed from that transition.

THE HERITAGE

Europe's earliest colonial intentions in North America were commercial and exploitative. The South's location and the economy that arose there satisfied these intentions.

North America's coastal plain south of Delaware Bay, especially south of Chesapeake Bay, contained many fertile areas suitable for agriculture. Good soils, combined with long, hot summers, regular rainfall, and mild winters permitted settlers to raise crops not typically grown in Great Britain and northern Europe, crops with ready buyers in European markets.

Rivers navigable to small boats meander across the South's Atlantic coastal plain, that low, level swath of land between the sea and the higher, hillier Piedmont.

---

[1] Richard Pillsbury, "Carolina Thunder: A Geography of Stock Car Racing," *Journal of Geography* 73, no. 1 (January 1974): 39–47.

Because the navigable rivers permit crops to be carried to market ports easily, settlement expanded freely, particularly between the James River in Virginia and the Altamaha River in Georgia.

Agrarian Settlement Patterns

Compared to the North, European population densities remained low throughout most of the South until well into the nineteenth century. Early on, urban concentrations larger than village size either clustered around port cities like Norfolk, Wilmington, Charleston, and Savannah, or at the heads of navigable rivers, where cities like Richmond, Columbia, and Augusta took root.

Aside from such urban clusters, the South stayed strongly rural and agrarian as colonial settlers spread small farmsteads into the hinterlands. From seventeenth-century beginnings, they established a way of life that did not weaken until late in the nineteenth century and remained significant until after World War II.

Of course, more is behind early Southern agrarian culture than a sparse population, few large cities, and a supportive physical environment. In timeless fashion, economic goals motivated European exploration, not human curiosity and a search for knowledge. The greatest long-term return was generated by establishing highly structured cash crop farming. In the Atlantic Southern lowlands and on the Caribbean islands, this goal fed the development of highly structured plantations devoted to cash crop production.

Plantations came to dominate the early Southern colonial economy. As early as 1612, tobacco was grown along Virginia's James River and in northeastern North Carolina. By 1695, plantation owners raised rice and indigo in and around coastal swamps in the Carolinas and Georgia. Cotton production, initially concentrated on the Sea Islands between Charleston and Spanish-held Florida, slowly gained importance until about 1800. Then, its production spread inland with enough speed and dominance to become "King Cotton."

Besides plantations, privately held small farms dotted the South. Mostly family-operated, they sustained owners' families with food and sometimes meager cash incomes to buy what could not be grown or cottage-made. Though numerous and viable contributors to the South's agrarian aura, those small farmers were only an echo of the plantation owners' economic dominance. Plantations were the high-volume business enterprises whose laborers sketched the outline for the South's distinct agrarian culture and sculpted its spatial organization.

Fueled by the emphasis on the plantations' cash crop exports, small market centers like Camden and Orangeburg, South Carolina, emerged to become collection and transshipment points for products within their geographic reach. Thus, the South's transportation network evolved to move goods from the interior directly to coastal export centers like Charleston, South Carolina.

This arrangement had key cultural effects. Although scattered cities in the South like Atlanta, Richmond, and Charleston maintained social and economic links as they bustled with banking, mercantile, and social events, smaller marketplaces stayed relatively isolated from each other. Because their focus was on their hinterlands, few interconnections existed (see Figure 5.2). As a consequence, distinctively local allegiances became the norm, and most of the South's rural population stayed relatively isolated. Isolation is not uncommon among rural people, but it was especially pervasive throughout the South, contributing to the region's admitted and unashamed provincialism.

The South's Peoples

In addition to reinforcing strong rural provincialism, the South's plantation system had other effects on the region's culture. Large-scale plantation agriculture required a sizable annual investment, and much of that investment was labor. Low population densities and cheap land in the South led to the heavy use of slave labor. In the late 1700s, Native Americans captured in intertribal warfare by tribes friendly to colonial forces were added to the slave ranks which were fast growing with Africans brought manacled in ships' cargo holds. But the Indian slaves tended to slip away if they survived. Plantation owners, therefore, focused on Africans who lacked familiarity with the land and did not have tribal groups that could shelter them.

As slavery became entrenched, it retarded voluntary immigration to the South. Potential but less affluent European settlers found more work and livelihood opportunities in the North and settled there. Since early in the nineteenth century, therefore, fewer of the South's white people were foreign born than in any region of the country.

Although the picture is now rapidly changing, even today the overwhelming majority of Southern whites are of English and Scots-Irish descent. Many can trace ancestors to early European colonists of the late 1700s through mid-1800s. Indeed, significant immigration to the United States and Canada from countries outside the British Isles did not occur until the 1840s. So while immigrants from non-British cultures brought new skills and knowledge to alter local habits and attitudes elsewhere in the country throughout the nineteenth century, their contributions were lacking in the South. This sort of isolation secured and colored the threads of the South's particular cultural tapestry, whose scenes depict a blend of Scots-Irish, English, African, and Native American influences.

The Impact of Slavery

The Southern colonies' heavy use of African slaves lies at the crux of a fundamental aspect of Southern culture. The first Africans arrived in Virginia in 1619, ten years after the English established the initial James River settlement and a year ahead of the May-flower's arrival at Plymouth, Massachusetts. Thus, while slaves were not imported in large numbers until the early eighteenth century, blacks were integral to the South's organization and social environment from the beginning of European settlement. By the first U.S. census in 1790, blacks comprised 39.1 percent of Virginia's population, 43 percent of South Carolina's, 25.5 percent of North Carolina's, and 35.4 percent of Georgia's. Even in Maryland, whose northern border edged the Mason-Dixon line and whose people were divided in their North-South ties, the population in 1790 was 33 percent black.

The slaves' presence meant that elements of African cultures became key parts of Southern life. Socially and psychologically isolated and disoriented, imported Africans retained and transmitted many of their original mannerisms to Southern whites. The slaves' impact on speech patterns, diet, and music in the South is undisputed. The cadence of conversation, a table laden with fried okra or gumbo, and a gospel choir are characteristically Southern; they are also characteristically African. Undoubtedly, more subtle contributions enrich the dominant Southern white culture as well.

Other indisputable cultural consequences of black slavery in the region are less positive. For slaveholding to have existed effectively, there could be no empathetic identification of the slave owner with the slave. If the owner put himself in the slave's position, he would recognize the moral and emotional implications of the slave's situation, and this recognition would threaten to destroy the institution.

Southern whites got around any moral dilemma posed by slavery by adopting a view of blacks that was similar to the dominant view held in Europe until the late eighteenth century. Namely, drawing on religious and scientific views of the time, they believed non-whites were evolutionarily inferior and, thus, socially and culturally inferior. Such justifications kept moral qualms about slavery at bay.

By the turn of the nineteenth century, however, opposition to slavery was gaining strength outside the South. As pressure to eliminate slavery arose from outside, justifications for it became more intense and self-righteous within the region.

Despite the moral implications of slavery, Southern whites' attitudes toward blacks were ambiguous, with bigotry and affinity tied to proximity, the relative sizes of both groups, and the economic aspects of the master-slave relationship. More than 90 percent of all U.S. blacks lived in the South until the beginning of the twentieth century. Blacks were present throughout the region in large numbers except in the Appalachians and peninsular Florida.

As overt as racial bigotry was and, unfortunately, still is in some places, the long and intimate relationship between Southern whites and blacks produced an affinity between them that was often not apparent to or comprehended by non-Southerners. Before and after abolition, day-to-day living brought blacks and whites together in many circumstances. Based on more than 200 years of lopsided interaction, both sides knew what to expect from the other, so individuals usually conducted themselves without apparent tension and even a kind of easy affection. That racial tensions existed and frequently flared into violence cannot be denied and underlines the historically rooted inequality of the old South's black-white relationships. Nevertheless, intimacy between the races was and is as much a part of Southern culture as the region's long history of violence between blacks and whites.

In tracing the South's cultural threads, two other populations, neither Anglo nor African in ancestry, had key impacts. These are Cajuns in southern Louisiana and Indian (Native American) tribes. The name Cajun derives from Acadian, for these Catholic, French-speaking people descend from French exiles who left Acadia (now Nova Scotia and New Brunswick) when the non-Catholic British took the region from the French in 1763. The Cajuns stayed in Louisiana after France sold the Louisiana Territory to the United States in 1803. Today known for their French dialect, spicy food, and music, Cajuns are a distinctive part of the South's culture.

Many American Indians were forcibly removed from the South by the 1830s. Governmentally sanctioned purge campaigns led by Andrew Jackson and massive relocations like the tragic Trail of Tears, where Cherokee Indians were rounded-up and walked from their Smoky Mountains homes to Oklahoma, had devastating consequences for the South's indigenous peoples. Nonetheless, many families escaped and several tribal groups clung near ancestral lands. The largest historic tribes now in the South are the Eastern Band of the Cherokee along the Blue Ridge in southwestern North Carolina and the Choctaw in central Mississippi.

Besides those known from earliest colonial records, several other large but "newer" tribes exist today, born of coalitions of small, nearly extinct tribes who banded together for strength, material help, and comfort. The largest of these include the Lumbee in southeast North Carolina, the Seminoles in southern Florida, and the Catawba in northwest South Carolina.

Religion is important in the South. In many churches, congregation members openly express their deep and fully felt personal religious experiences.

These and a great many smaller clusters of Sioux and Iroquois managed to elude removal by publicly denying their Indian heritage while privately clinging to it.

Religious Patterns

Though not directly tied to the plantation system, another facet of the South's culture is Southerners' adherence to evangelical Protestant religions. Small, white clapboard churches lit by ordinary glass windows and decorated with straight-back chairs flanking a plain altar still dot the countryside. Sometimes led by preachers whose sole authority lies in being "called," they consistently draw congregations every Sunday from the scattered rural and small-town populations.

Although various Protestant congregations like the Pentecostals, along with Methodist and Episcopal congregations, are numerous, the Baptists have numerical dominion in the South (see Figure 3.6). Freewheeling, energetic Baptist evangelists entered the Southern Piedmont and hill country hard on the heels of mid-eighteenth-century settlers and found them eager listeners. Frontier hardship, isolation, and individualism were compatible with a religion adopted through intensely emotional, but personal, conversion experiences.

As a group, Baptists were resistant to the rigidities of formal institutional organization, and this permitted the religion's expansion across the sparsely settled South. Any individual who was spiritually born again and felt the call to preach could do so without either the formal approval of the church hierarchy or denominational constraint. Because most nineteenth-century immigrants to the United States settled elsewhere in the country, the religious pattern established in the South by the early 1800s was not affected by the religious affiliations of later immigrants.

Prelude to Change

The South's settlement pattern and economic organization changed dramatically by the time the Civil War broke out. Generally, slave densities were lower in the South's border states. But the commercial promise of agricultural land in the Louisville and Bluegrass Basin sections of Kentucky, the Nashville Basin and the loess plains of western Tennessee, and the tobacco and hemp plantation country along the Missouri River westward from St. Louis lured slaveholding plantation owners. Plantations also brought blacks to northern

Florida's panhandle and the state's interior highlands almost as far south as present-day Ocala. By the 1850s, they reached Louisiana and the east Texas plains.

Quite likely, blacks lived in almost every county outside the southern Appalachian highlands, but they comprised a large proportion of the local population only in original plantation areas like south-central and tidewater Virginia, northeastern North Carolina, and tidewater Georgia and South Carolina. Their numbers were high, too, in the new lands most suited to large-scale cotton production, such as Piedmont South Carolina, the inner Coastal Plain of Georgia, Alabama's Black Belt, the Mississippi valley, the loess plains south of Memphis, and the black-soil prairies of eastern Texas.

By the end of the Civil War, the South's economic underpinnings were ravaged, crippling it both politically and financially. Railroads were torn up and equipment confiscated; shipping terminals were demolished; and most of the industrial base was destroyed. Confederate currency and bonds were worthless. Cotton stocks awaiting postwar sale were confiscated by Union forces. Farms and fields were in disrepair, and implements and livestock were lost. Large landholdings were broken up or heavily taxed. Investment funds had either been used up during the war or were drawn off during the decade after the war. With slavery eliminated, the region's labor supply was uncertain.

The South recovered from the war slowly and developed in ways that drew largely on underlying cultural patterns that neither war nor economic turmoil weakened. These echoes of cultural patterns reverberated across the South for almost a century after the Civil War and prepared the region for its current geography.

CONSEQUENCES AND TRANSITION

The South's culture and economy underwent changes wrought by the Civil War. Southern whites reacted to black emancipation by settling into institutionalized segregation. Even though free from institutionalized slavery, Southern blacks had few opportunities until World War I. What's more, Southern attitudes and feelings of isolation from the rest of the country became more, not less, entrenched than in the antebellum period. Each consequence interacted with the others and affected the South's evolving economic, social, demographic, and political patterns.

Persistent Poverty

Disintegration of the South's antebellum economy led to difficult times during the 12-year Reconstruction period (1865–1877). There had been great loss of life and destruction of transportation and manufacturing capacity, and the rigid slave-labor-dependent plantation economy was demolished. Although the prewar North had begun to industrialize and accept the accompanying change as something to be expected, the South, having lost so much else, clung to tradition.

For Southerners, the heavy taxes and rebuilding costs imposed during Reconstruction required intense exploitation of the one key resource they still had—land. Thus, cotton production continued to dominate the South's economy.

Other factors needed for economic development, however, were less available. Local capital was scarce; most of it was consumed by the war effort or drawn off after the war by the North's taxation. Credit was in short supply. Farmers and small entrepreneurs lucky enough to get loans faced sharply increased interest rates, and they found themselves continually in debt. The obligations this scenario generated perpetuated Southern dependence on agriculture, maintaining a high demand for the region's historically important cash crops, cotton and tobacco.

The bustle of commercial activity, fundamental to New Orleans, is caught vividly in this 1872 lithograph of the city. Riverboats and ocean schooners vie with smaller vessels and rafts on the Mississippi before competing with others for space along the riverbank to load or unload their goods.

Emancipation complicated the picture. No longer could black people be forced to work as slaves. The white labor pool had been decimated by war casualties. For hopefuls of either race, the dream of Yankee-supplied economic independence never materialized and gradually faded. Because few jobs were available in the South's small towns, rural residents had to make whatever arrangements they could with the remaining white landowners. *Sharecropping* became the means to survival and the way of life for most blacks, just as it was for many poor whites who had lost their land.

Sharecroppers were frequently in debt to the landowner (just as the landowner was in debt to bankers and merchants), and they were not permitted to leave the sharecropping arrangement until debts were paid. Sharecropping's bonds were reinforced for the black population with "Black Codes." Black movement outside agricultural areas was restricted, and educational opportunities were limited. For most rural blacks, life as a plantation slave was exchanged for a constricted agrarian existence and deep indebtedness to white landowners or local merchants. Some blacks owned the land they worked, but only 2.5 percent of black farmers had this benefit in 1900.[2]

Southern black farmers were hampered by poor access to credit, farms too small to be highly productive, and the racism of the region's culture. Such conditions kept their annual incomes very low. Therefore, geographers explained much of the South's distribution of poverty by identifying where rural blacks lived (compare Figures 9.2 and 1.5). For a full century following the Civil War, the pattern of Southern poverty remained strongly associated with the distribution of the region's black population.

Rural Southern poverty was not exclusively a black experience. Even during antebellum times, most Southern whites were small-scale farmers. Relatively few had the money or experience to operate a large landholding successfully. Because cotton plantations were established early and took up the best agricultural land, poor whites lived in three zones. They occupied lower-quality agricultural areas north and south of predominantly black areas of Georgia, Alabama, and Mississippi, the peripheral southern Appalachians, and land tucked among the large plantations on locally poor land. After the Civil War, hard times pushed many into sharecropping arrangements little different from those affecting rural blacks.

Resident Native Americans not living on reservations found themselves in the same situation as rural blacks and poor whites. But because census and other official documents tended to classify them as black, white or other, depending on skin color, many of their precise settlement areas and conditions are difficult to pinpoint.

The Seeds of Economic Change

About 1880, the South's economy entered a new phase. Manufacturing developed rapidly, led by growth in the cotton textile industry. This promised changes in employment opportunities. For much of the nineteenth century, U.S. cotton textile manufacturing was based largely in New England. But in the years prior to the turn of the twentieth century, it started shifting south. The industrial migration was so steady that by 1929, 57 percent of the nation's cotton textile spindles were in the South. This figure is more than two and a half times the share existing in 1890 and chronicles the expansion of textiles and apparel industries in pockets of the region.

High levels of *underemployment* in the Piedmont South lured New England's textile and apparel manufacturers. Textile and related industries are labor intensive, and the South had extensive pools of people wanting work. Another attraction was the opportunity to modernize factories. New England plants were increasingly obsolescent as technological improvements occurred. Unlike the situation in the South, higher-value industries in New England drove up local taxes and land and power costs. As a result, New England textile manufacturers looked elsewhere. Once relocated, textile and apparel manufacturing plants spread quickly across the Carolina Piedmont and northern Georgia. They subsequently drew natural and synthetic fiber industries producing the material used in textile manufacturing. Both groups benefited by locating near each other.

The impact of this budding development on general income levels in the South was scant. Because Southern blacks lived in the region's best agricultural areas—where most cotton was grown but new industries were resisted—they could not take advantage of jobs associated with the region's early industrialization.

[2] For a clear and thorough discussion of this arrangement, see James S. Fisher, "Negro Farm Ownership in the South," *Annals*, the Association of American Geographers, 63, no. 4 (December 1973): 478–489.

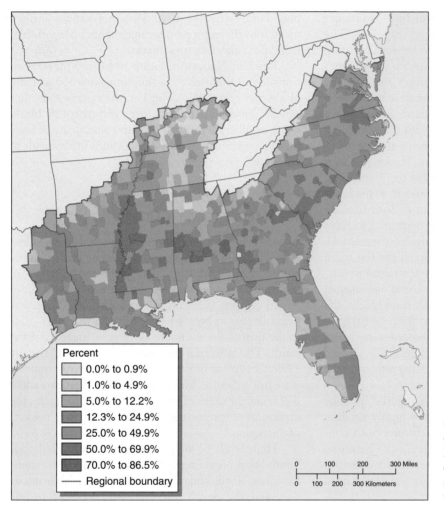

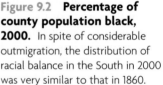

Figure 9.2 **Percentage of county population black, 2000.** In spite of considerable outmigration, the distribution of racial balance in the South in 2000 was very similar to that in 1860.

In any case, textiles are a low-wage industry looking for cheap labor, so even among workers who got industry jobs incomes rose slowly.

Textiles and related manufacturing were not the only major sources of industrial opportunities during the 1880s and 1890s. Railroad reconstruction and other public improvements increased money flow and stimulated town growth along railway lines. Cigarette manufacturing sprang up in the tobacco-growing regions of North Carolina and Virginia. As the railroad network strengthened, timber resources were exploited. Much of the timber left the South as a raw material, but furniture manufacturing arose in North Carolina and Virginia. After 1936, pulp and paper manufacturing emerged throughout the South. These industries are still regionally important (Figure 9.3).

The South's iron and steel industry was also stimulated during the last quarter of the nineteenth century. Although a small and scattered iron industry had existed for decades, iron-making technology improved enough by the 1870s to excite the rise of regionally important iron production centers around Chattanooga, Tennessee, and Birmingham, Alabama. Local capital and entrepreneurial initiative help explain Chattanooga's rise, but Birmingham also benefited from a large, nearby deposit of high-quality coking coal. These two cities, together with Atlanta, formed a vital and linked industrial triangle in the South by 1900.

The Atlanta-Birmingham-Chattanooga triangle was centrally located in the South and could have been a more important industrial focus than it was for the region, broadly improving the South's labor skills,

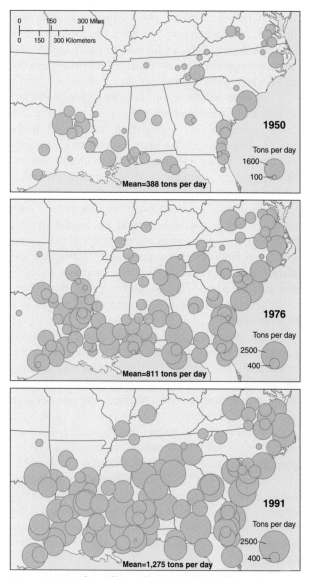

Textile and apparel industries have a high machinery-to-labor ratio, require a large amount of floor space, and demand relatively low skills from their labor force. The women tending these machines represent the largely female workforce in such low-wage industries.

Figure 9.3   Pulp mill production, 1950–1991. (1950 and 1976 maps courtesy of J. F. Hart.) The tremendous growth in the exploitation of Southern forests during the second half of the twentieth century is indicated by changes in pulp production. Most increases in production occurred initially on the Atlantic and Gulf coastal plains.

income levels, and general economic welfare. But the triangle's potential was dampened by corporate manipulation of the market. Because labor costs in Birmingham were low on a national scale and raw materials were near, Birmingham manufacturers could produce steel at much lower costs than could those in Pittsburgh, the North's primary center of steel production. But instead of boosting Birmingham's

competitive position, these conditions were neutralized when discriminatory shipping rates were imposed on Birmingham-manufactured products.

With no interference from the federal government, Pittsburgh's United States Steel Company imposed on all users of Alabama steel what came to be called the Birmingham differential, or a pricing policy of "Pittsburgh plus." Under this company policy, consumers of Alabama steel paid the price of steel at Pittsburgh plus $3.00 per ton ($2.72 per metric ton) plus regular freight costs from Birmingham. Thus, Atlanta firms found it cheaper to order steel from Pittsburgh than from Birmingham, even though a Pittsburgh-Atlanta shipment traveled more than four times the distance than a shipment from Alabama.

The Pittsburgh-plus pricing practice was eventually ruled illegal and stopped, but not before the policy crippled the competitive cost advantage of Alabama steel at a time when the country's economy was expanding rapidly. While in place, the practice contributed to Southern industry's slow growth, and it helped reinforce the region's sense of isolation.

Thus, changes in Southern economic development were irregular and slow for more than half a

The abundant pine forests of the South produce about two-thirds of U.S. pulpwood. Pulp mills such as this one at Plymouth, North Carolina, transform wood waste to pulp through a process that requires wastewater aeration, as can be seen in the background. Note the heavily wooded, flat landscape.

century after the Civil War. Certainly, the South was not resource poor, but the average annual incomes of its people were far below the national level. While industrial growth helped raise some incomes, the industries were chiefly the kind that paid workers low wages. Most Southerners continued to depend on agriculture, and many of these people remained poor.

Geographically, the greatest manufacturing employment growth occurred in the Piedmont sections of Virginia and the Carolinas and in the Atlanta-Birmingham-Chattanooga triangle. The Gulf Coast also experienced growth as local resources were developed (Chapter 10). The number of poor white families in the South was large, but an even larger proportion of black families lived in poverty. Most blacks lived outside areas experiencing industrial growth and found themselves locked into restrictive agricultural arrangements. They also faced increasing racial repression just as economic opportunities increased for whites.

Institutionalizing Racial Segregation

Even before the Civil War, relations between whites and blacks were not uniform in the South. Most whites were not like Simon Legree, just as most blacks did not fit the Uncle Tom image. During the initial years after the war, antiblack action by whites was localized. This let some former slaves gain title to good agricultural land, enter professions as educators, scientists, lawyers, and doctors, or be elected to political office.

But after federal troops left the region in the late 1870s, an institutionalized alternative to slavery suffused Southern life; segregation laws began to be passed. These restrictive laws, called Jim Crow laws, required racial separation in Southern life. By the end of the century, virtually total segregation existed in every Southern state and was sanctioned legislatively.

Although a nineteenth-century Supreme Court ruling concluded racial separation was legal if equality was ensured, *de jure segregation* did not lead to separate but equal facilities and opportunities for whites and blacks. Whites presumed themselves superior to blacks and used the force of law to institutionalize this belief.

Formal segregation, the partition of people into separate societies according to criteria such as religion, race, or ethnic background, has many geographic expressions. Blacks and whites in the South met, communicated, and worked together or near one another. In many ways, they lived interconnected lives. But segregation institutionalized a physical distance between the groups wherever possible. Two sets of schools operated. Two sets of restaurants, recreation facilities, park benches, drinking fountains, restrooms, and other points of potential public contact existed. Housing had its white areas and black areas. Blacks found occupational access restricted, with only the less desirable employment available to them. They also found overt and covert restrictions on their efforts to vote, almost totally disenfranchising them in the South before World War I.

The South's post-Civil War geography blended two very different human landscapes in the same place at the same time—one black, one white. In some places, including the residential areas of Mississippi, Louisiana, and eastern Texas, the two worlds hardly overlapped. Yet, there was a great deal of geographic overlap in workplaces and retail shopping, even if blacks and whites did not interact socially. When individual blacks resisted discrimination, they were targets of violent repression. Lynching became an ugly aspect of post-Civil War white Southern culture.

Even the Supreme Court's separate but equal doctrine was not followed. When the South's scarce financial resources were allocated for public services, black facilities received less than half of the amount

TABLE 9.1

Selected Characteristics of Racial Inequality in Southern States, 1930							
		Annual school expenditures per pupil ($)		Number of one-teacher schools per 1000 pupils		% illiterate	
State	% Black	White	Black	White	Black	White	Black
Alabama	35.7	27.43	8.31	2.03	5.04	4.8	26.2
Arkansas	25.8	22.75	10.42	4.89	4.95	3.5	16.1
Florida	29.4	34.59	10.56	n.a.	n.a.	1.9	18.8
Georgia	36.8	24.64	6.53	1.48	6.49	3.3	19.9
Kentucky	8.6	22.22[a]	21.90[a]	7.17	8.99	5.7	15.4
Louisiana	36.9	36.41	8.78	0.79	4.02	7.3	23.3
Maryland	16.9	9.41	33.89	1.74	4.52	1.3	11.4
Mississippi	50.2	30.60	4.79	1.51	6.72	2.7	23.2
North Carolina	29.0	28.76	12.70	1.22	3.13	5.6	20.6
South Carolina	45.6	37.53	6.65	1.07	4.27	5.1	26.9
Tennessee	18.3	28.97	15.92	3.42	5.02	5.4	14.9
Texas	14.7	38.76[b]	16.02	1.75	4.19	1.4	13.4
Virginia	26.8	29.44	13.06	2.89	4.59	4.8	19.2

[a]Does not include Louisville.

[b]Includes salaries of superintendents.

Source: Charles S. Johnson, *Statistical Atlas of Southern Counties* (Chapel Hill: University of North Carolina Press, 1941).

budgeted for white facilities. The difference between white and black schools by 1930, for example, was striking (Table 9.1). Every Southern state except Tennessee, Kentucky, and Maryland spent more than twice the amount per pupil on white schools than black schools. White schools in South Carolina and Mississippi received well over *five* times as much per pupil as black schools. These figures suggest the pressures that eventually led to the 1954 U.S. Supreme Court decision, *Brown v. Board of Education of Topeka*, reversing the existing separate but equal doctrine and ending de jure segregation.

Outmigration

The South's persistent poverty and racial repression led to one of the greatest population redistributions in U.S. history. For the first 50 years after the Civil War, black migrants trickled away from the South. Then, suddenly, what had been a trickle of black emigrants became a flood that continued across the next half-century.

By the time World War I started, black poverty and the near-subsistence economy of sharecropping had merged with Jim Crow laws and violence to push blacks (and poor whites) out of the South. To make matters worse, a severe boll weevil infestation affected cotton production catastrophically as the war broke out in Europe, further reducing the South's economic underpinnings.

Jobs pulled people north. World War I shut off the supply of European immigrants, but industrialists continued to need more workers. They found an alternative source in the large unemployed and underemployed Southern labor pool. Labor recruiters and press campaigns reached into the South to glorify the virtues and opportunities of Northern manufacturing employment and Northern life in general. With hope and seeking opportunity, people trekked north. Once the outflow began in earnest, relatives, friends, and former neighbors sent information and financial support home, encouraging others to undertake the long move north.

Although almost all Southerners leaving the region moved to cities, people were selective about their regional destination. For instance, those leaving the South's Atlantic states traveled mostly to Northern cities east of the Appalachians, such as Washington, D.C., Baltimore, Philadelphia, New York, Hartford, Providence, and Boston. Emigrants from the south-central states of Alabama, Mississippi, Kentucky, Tennessee, Arkansas, and Louisiana either moved along major transportation lines to St. Louis, Indianapolis, Gary, and Chicago, or went to manufacturing cities such as Cincinnati, Detroit, Cleveland, Pittsburgh, and Buffalo. People migrating westward from the South moved primarily from Louisiana, Texas, Arkansas, and Mississippi to western cities, such as Denver, Portland, Seattle, and the California cities. Regardless of where outbound Southerners migrated, distance and transportation affected their choices and explains, geographically speaking, the pattern of their distinctive migration streams.

This 1923 editorial cartoon from the New Orleans Times-Picayune *reflects a Southern viewpoint of Northern labor recruiters as sharp-eyed outsiders attempting to sway contented black labor with false stories. But in fact, black labor was not contented in the South and wages were higher and hours shorter in the North.*

The South's economy suffered from this exodus. The people who left were self-selected. Most black outmigrants, for example, were between the ages of 18 and 35 and they sought factory or service jobs. Thus, their economically productive years were spent outside the region. Of those who stayed behind, many were in their later productive years, retired, or not yet in the labor force (Figure 9.4). Racial limitations on opportunities in the South's professional occupations meant that many of the region's highly trained young blacks were gone by midcentury.

Sectionalism

The South's sectionalism after the Civil War is easily visible in its political behavior. The Solid South was a label used for decades to indicate that the entire region voted as a bloc, often in direct contradiction to national trends. Southern voters associated the Civil War and Reconstruction with the North and the Republican Party, so Southern whites became stubborn opposition Democrats. Long after blacks were disenfranchised and the Northern tenor of both parties changed, the South continued to vote Democratic, at least at the local level.

Manifestations of the South's sectionalism can be extremely subtle, but they nevertheless reflect regional culture. Interwoven with the other major strands of widespread poverty, explicit racism, and a continued agrarian and small-town orientation was Southerners' sense of their region as having always been different from the rest of the country. This feeling created an isolation and regional consciousness that reinforced existing differences.

As Thomas Wolfe pointed out, Southerners' sense of region is strong and conscious. Perhaps because of this regional identification and the disdainful attention it received from non-Southerners, the slow erosion of sectionalism's potency was little noticed for many years. Even after it was clear that the South was changing, observers continued to refer in sectional terms to the rise of the New South.

THE NEW SOUTH

The New South is unfolding from the old South. Its present spatial and regional characteristics are built on patterns that evolved over decades and, in some ways, over centuries.

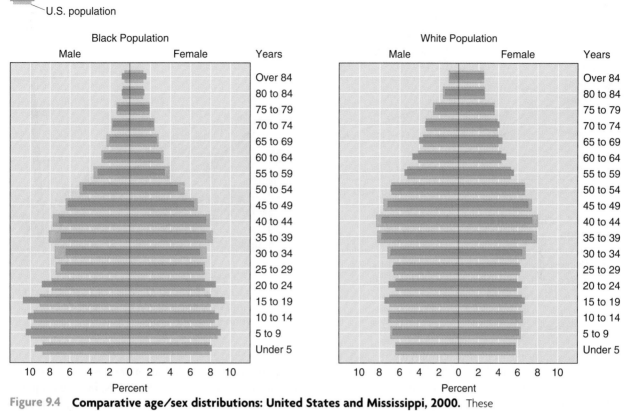

Figure 9.4 **Comparative age/sex distributions: United States and Mississippi, 2000.** These graphs, called age-sex pyramids, suggest that Mississippi blacks left the state in significant numbers after 1920 and that there was some departure of whites after World War II, as well.

The key to recent changes lies in the South's gradually fading regional isolation. Prior to World War II, most of the South's people and certainly their leaders appeared to believe the region had successfully seceded from the United States decades earlier.

After the later 1930s, however, and especially since the end of World War II, trends and pressures external to the South infiltrated the region and broke down its isolation. Some changes were caused by events affecting the entire nation's economy and are not particular to the South. Other changes followed from purposeful federal intervention in the South's affairs. And some aspects of the New South have come about as the South's distinctive culture matured.

Economic and Urban Reorganization

The South's economic structure changed after the mid-1930s, departing dramatically from a mid-Depression economy that was little different from the South of 1870.

In the 1930s, the South remained dominantly agrarian, with agriculture supported by heavy use of animal power (usually mules) and hand labor. Sharecropping and tenant-farming arrangements were widespread. Agricultural products were not processed in the region, ensuring that most sale value was added outside the South. The South remained capital deficient, with few investment funds available from regional sources. Southern industry was low-wage or oriented toward narrow local markets. The region's urban structure reflected the South's economic orientation with small market centers, railroad towns, textile mill towns, and county seats being the South's most pervasive urban form.

All this changed quickly during the half century after World War II. As early as the 1950s, more than half of the region's labor force was engaged in

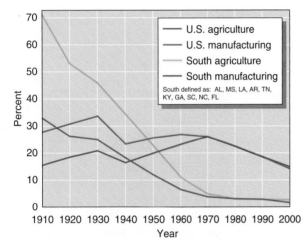

Figure 9.5 **Changes in the South's economy, 1910–2000.** Southerners were much more heavily engaged in agriculture than Americans as a whole and much less heavily engaged in manufacturing until after World War II. By 1970, there was no significant difference between the South's economy and the rest of the country's.

Opportunities were few and life was hard across much of the South until the region began its economic transformation following World War II.

urban-based, nonagricultural employment, and the proportion in agriculture continued to decline (Figure 9.5). Agriculture's decline paralleled a sharp increase in manufacturing employment and employment in service activities. Furthermore, the South's industrial mix diversified; no longer is Southern manufacturing limited to the early stages of raw materials processing.

Agriculture also diversified. Among the region's traditional cash crops, cotton is still the most important. Other significant, traditionally grown crops include tobacco, sugarcane, peanuts, and rice. But all of these crops, except rice, were under strict federal acreage controls since the 1930s and 1940s, with rice under controls since about 1951. As a result, growers reserved their best land for cash crop production and, total output remained high.

Cotton is no longer exclusively a Southern product. Irrigated cotton is grown in the Texas Panhandle, along sections of the Rio Grande, and in Arizona and California. Approximately 4400 acres of upland cotton were harvested in Texas in 2002, nearly as much as the 5435 acres harvested in all of the traditional Southern states. Within the South, soybeans have replaced cotton as the most valuable cash crop produced. In 2002, the U.S. soybean crop was worth $12.4 billion compared to a $3.4 billion cotton crop. New storage and

shipping technology led to a modest comeback by Southern-grown cotton in the 1990s, but the crop will never again dominate the region's economy as it once did. Price drops have also hurt cotton growers. The average price for a pound of cotton reached a high of $0.86 per pound in 1995, but dropped all the way to $0.34 in 2001. By mid-2007, however, cotton had rebounded to an average high of $0.57 per pound.

Although cotton production declined, livestock industries and other crops increased sharply. Beef production improved greatly after World War II as farmers enhanced pastures with better grasses and fodder crops and applied more fertilizer. New cattle strains were developed to thrive in the hot, humid Southern summer (see Chapter 10).

Within the last 40 years, national broiler and chicken production concentrated in the South. Now clearly a regional industry, approximately 75 percent of all U.S. commercial broilers are Southern in origin (see Map Appendix). Chicken prices remain relatively low as growers apply modern farming and management techniques. Highly automated production in large volume along with new technology and a well-developed, spatially concentrated marketing network combine to keep this agricultural industry concentrated in the South.

Even more dramatic than the South's food product diversification are transformations in the means of farm production. Southern agriculture is no longer an animal-powered, hand-labor operation. Wherever possible, machinery is used in manufacturing and agriculture as it is elsewhere in the country.

When Southern farming mechanized, the traditional sharecropping system virtually disappeared. Accompanying this change, average farm size in the South increased sharply. Much of the rural population that wished to leave the countryside for urban employment has now done so. Those who remain are either successful full-time farmers, part-time farmers who work urban jobs to supplement farm income, or individuals with few formal skills eking out a living at very low incomes.

Of those who left the rural South in the last 50 years, virtually all migrated to urban centers, many to cities located within the South. Southern rural-to-urban migration increased rapidly after World War II as the region's economy participated in the United States' post-Depression expansion. Both blacks and whites made up this city-bound movement, pulled by greater opportunity. In 1940, there were only 35 cities with populations greater than 50,000 in the South. By 1950, the number had increased to 42 cities, a jump of 20 percent, and by 1996 it had reached 110, an increase of more than 300 percent over 1940. Although the South is still

strongly rural in character, the regional trend is toward a level of urbanization typical of the United States as a whole.

Atlanta's transformation in recent decades symbolizes the South's participation in modern America's urbanization. Atlanta is centrally located in the region, a vital transportation hub, and represents almost literally the South's phoenix-like rise from the ashes of the Civil War.

Today's Atlanta is a business, financial, and commercial center without peer in the South. Enterprises from Coca-Cola to CNN originated in Atlanta and continue to be based there. During the first two-thirds of the 1990s, the Atlanta metropolitan area took in more new land area than any city in North America. One result is that commuters there average more miles traveled per day than residents of any U.S. metropolis, including Los Angeles.

When this metropolitan area of 3,541,230 people (in 1996) represented the United States by hosting the 1996 World Olympics, Atlantans were proud. They knew their city had been acknowledged internationally as equivalent to other Olympic host cities, including Los Angeles, Moscow, Madrid, Tokyo, and Athens. The city's metropolitan area has since grown to 5,478,667, based on 2006 U.S. Census data.

The pull to the South's cities was stimulated by industrial growth matching Southern agriculture's

Centennial Olympic Park. Located near the center of the South, Atlanta's energy and magnetism attracted the 1996 Olympic Games.

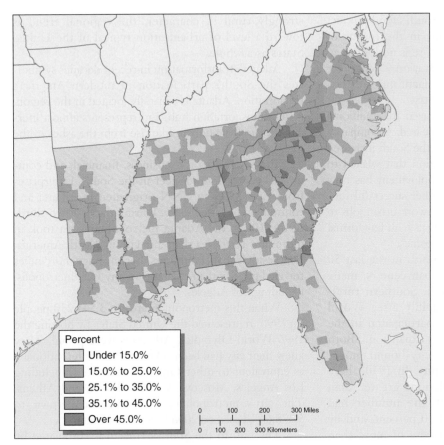

Figure 9.6   Percentage of nonagricultural labor force employed in manufacturing, 1950. Except for portions of the Piedmont and other scattered pockets in the South, manufacturing employment remained low across most of the region in 1950.

changes and producing an industrial mix typical of the United States as a whole. The nonagricultural labor force in manufacturing increased greatly in almost every part of the region (Figures 9.6 and 9.7).

The South's traditional industries, such as steel (Alabama), tobacco products (North Carolina and Virginia), and textiles (northern Georgia, the Carolinas, and southern Virginia), remained regionally important for a while but lost their dominance as other manufacturing such as aircraft production and electronics appeared.

Most significantly, as the average Southern consumer earned higher wages, the regional market drew many consumer-goods manufacturers into the South, and service industries expanded rapidly. Regional banking thrived, and some firms, like Bank of America of Charlotte, North Carolina, competed successfully with the country's largest financial institutions.

The demand for nonagricultural labor increased, improving incomes further and strengthening local markets. As this trend continues, most of the South's remaining low-wage industries (Table 9.2) will be driven out by the competition of high-wage activities for a limited labor supply. As a more skilled labor force is demanded, educational improvements will be essential, or the transition to a higher-wage economy will be slowed.

Overall, economic activity in the New South reflects patterns found in other regions of the Sunbelt. Even though the South continues to contain areas with very low per capita income, the region's persistent poverty is fast disappearing.

Government Intervention

The South's most significant changes were triggered by direct governmental involvement, first from the federal level until the 1970s, then from state initiatives. Governmental intervention occurred in economic and social matters. Furthermore, economic intervention has most often been a stimulus to development and not merely a regional welfare payment meant to

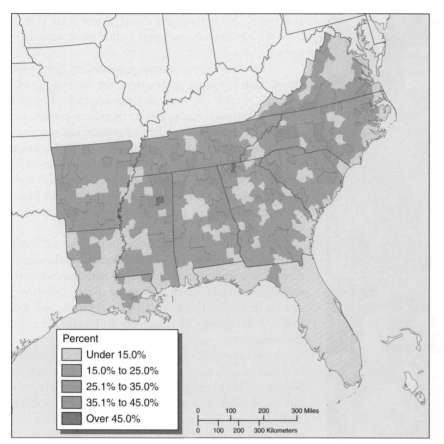

Figure 9.7 Percentage of nonagricultural labor force employed in manufacturing, 2000. Within one generation, the South experienced a tremendous growth in manufacturing employment (see Figure 9.6). By 1990, few parts of the region were without a significant manufacturing work force.

Percent
- Under 15.0%
- 15.0% to 25.0%
- 25.1% to 35.0%
- 35.1% to 45.0%
- Over 45.0%

TABLE 9.2

Index of Average Hourly Wage in Manufacturing Industries, 1957–2006 (U.S. average = 100)						
State	**1957**	**1967**	**1977**	**1989**	**1998**	**2006**
Alabama	86	85	87	90	90	92
Arkansas	71	71	76	82	82	79
Florida	79	84	82	86	84	87
Georgia	75	78	79	89	88	87
Kentucky	96	96	101	103	102	100
Louisiana	94	96	102	109	110	106
Mississippi	68	72	74	79	79	81
North Carolina	70	72	73	84	87	86
South Carolina	70	73	76	85	78	89
Tennessee	81	81	83	92	88	83
Virginia	79	81	83	96	95	99

Sources: Calculated from U.S. Bureau of Labor Statistics, *Employment and Earnings, 1939–1974, 1991, 1998, & 2006,* and *U.S. Statistical Abstracts,* 1978.

Chicken production requires heavy investment and large-scale operations. This facility is typical of those found in all major production regions across the South.

balance income deficiencies. The regional development characteristics of the Tennessee Valley Authority (TVA) are an excellent example for a portion of the South (see Chapter 8).

The federal Agricultural Adjustment Acts (1935 and later) provided the main stimulus to the market growth that transfigured the South's economy.[3] Southern per capita incomes in the early 1930s were about one-third the national average, and the region's economic structure had changed very little since Reconstruction, a full 60 years earlier. The Acts were aimed

[3] Merle C. Prunty, "The Agrarian Contribution to Recent Southern Industrialization," in *The South: A Vade Mecum*, James R. Heyl, ed. (Washington, D.C.: Association of American Geographers, 1973), pp. 108–118.

at farmers throughout the country, but their consequences for the largely rural South were especially strong.

Before the Agricultural Adjustment Acts took effect, farm product prices were set by supply and demand in the international marketplace. For the South, this meant that prices for Southern cotton, for example, fluctuated according to production successes or failures in the world's cotton-growing areas. More importantly, farm labor in the cotton South (basically tenant and sharecrop farmers) was in competition with hard-pressed cotton producers along the Nile or Indus rivers, or elsewhere in what was still a colonialized world economy.

Agricultural wages and prices in the United States were adjusted upward under the Agricultural Adjustment Acts to reflect more accurately national industrial wage differentials. The result was a sharply improved market in the South for manufactured goods, and this initiated the upward development spiral still affecting the region.

In an act of federal intervention widely recognized as significant to the South's social structure, the U.S. Supreme Court struck down in 1954 the racial separate but equal doctrine permitted almost 70 years earlier. This decision initiated changes in the South's social geography, changes that are still reverberating in every part of the country where race affects opportunity.

Racial segregation's geographic and institutional consequences were not quickly affected simply by declaring the practice illegal. The antiblack component of white Southern culture had been nurtured for too many years to change rapidly, but even in this aspect the New South began a process of accommodation and reconciliation that made it considerably different from the old South.

During the South's economic revitalization, many blacks migrated from rural areas but stayed within the region. By 1980, more than two-thirds of the South's black population lived in the region's major metropolitan areas, a proportion that remained constant into the 1990s. Black outmigration from the region continued at a high rate, as discussed earlier, but by the early 1970s the numbers began to change. Better economic opportunities in the South's urban areas began to be more available to blacks than previously. By 1980, the Census of Population confirmed that more blacks were entering the South than were leaving it. In 1998, the U.S. Census Bureau issued a report indicating that more blacks were moving to the South than to any

Throughout the upland South, traditional barns are still used to air-dry burley tobacco. Across the piedmont and lowland South, flue-cured tobacco barns have been replaced by more efficient sheds that look like abandoned truck trailers. The adoption of new technology is changing the cultural landscape.

other part of the country (Figure 9.8). The 2000 census reconfirmed this continuing trend.

All this suggests that regionally based racial push factors are no longer significant in national population redistribution. In other words, the South is no more threatening racially than any other region of the country. It also confirms the South's growing economic maturity, for migration within the United States has followed economic opportunity. Finally, it suggests that Southern culture, already strongly influenced by its black population, will continue to benefit from greater than average black cultural contributions.

Also new to the South is an influx of people with Asian or Latino heritage. Especially noteworthy is the Latino group. Although Mexican agricultural migrant workers have been used in the region for decades, Latino men and women from various Middle American countries now live in most small towns across the South and work in a wide range of service jobs. The numbers remain small, but in a region whose people's ancestry was almost entirely northwestern European or African, the change is striking.

Cultural Integration

Common to much of what has happened in the South since the mid-1930s is a gradual decline in the region's distinctiveness. What is occurring in the New South makes it more and more like the rest of the eastern half of the country, but with a definite Sunbelt vitality. A significant infusion of Northern migrants, especially

Percent of population

	1900	1910	1920	1930	1940	1950	1960	1970	1980	1990	2000
Northeast	4.4	4.9	6.5	9.6	10.6	13.4	16.0	19.2	18.3	18.7	17.6
Midwest	5.6	5.5	7.6	10.6	11.0	14.8	18.3	20.2	20.1	19.1	18.8
South	89.7	89.0	85.2	78.7	77.0	68.0	59.9	53.0	53.0	52.8	54.8
West					3.8	5.8	7.5	8.5	9.4	8.9

Figure 9.8 Changes in regional distribution of the U.S. black population, 1870–2000. Following World War I, the South's black population left the region in very large numbers, outmigration that continued until the 1970s when the departures declined sharply and reversed by the 1990s.

to Southern metropolitan growth centers, has made some of these cities less distinctively Southern in culture and more just urban. The issue would be argued vigorously and at length by many native Southerners, but the region's leading city, Atlanta, is losing the informality and old-boy network that once made it distinctively Southern. Atlanta is now much less a Southern city than a fast-paced, overcrowded American city on the rise. If this transformation is happening to Atlanta, are Charlotte, Richmond, Birmingham, and Nashville far behind? Racial discrimination was never peculiar to the South, although it was often more blatantly expressed and formally institutionalized there than elsewhere. With the end of overt, legalized forms of racism, interracial attitudes in the South are no different than in the rest of the country. In many ways, the easy familiarity coming from generations of living together and the stark attention to racial matters required during Southern desegregation mean that interracial interaction in the New South is often less self-conscious than elsewhere in the United States.

As the South experienced these changes, the region's sense of isolation and sectionalism declined.

Southerners have adopted national trends and habits, and Northerners and Westerners have demonstrated an affinity for what Southerners recognize as their own. Improved communications are broadening the national exposure of Southern writers, poets, artists, and others who draw deeply on the region's cultural heritage.

Politically, Southern elections and administrations are no more tumultuous than elsewhere and usually no more race-oriented. Since the Civil Rights Act of 1964, blacks have had less difficulty registering to vote and exercising that right. The full political spectrum is represented among Southern elected officials, although the majority tend to be traditional conservatives. At the national level, for example, the large, conservative wing of the Republican Party is led by Southerners.

All of this is reducing Southerners' defensiveness and therefore reducing the region's sectionalism. Southerners generally remain aware of and take pleasure in their cultural distinctiveness, but the grounds for this self-awareness are disappearing. Before too many decades pass, the region's cultural provincialism will be no stronger than that found in other parts of the United States and Canada.

OLD SOUTH AGRICULTURAL CASE STUDY

The Yazoo-Mississippi Delta

The Yazoo-Mississippi Delta is a microcosm of the Old South even though its history is modified by its physical geography. The Delta, as it is known in Mississippi, occupies an eastern portion of the central Mississippi River's alluvial floodplain (Figure 9.9). Small local variations in relief occur, but the Delta's overwhelming visual impression is of a perfectly flat landscape. Filled in since the last Ice Age by glacial outwash and flood-deposited alluvium, the Delta drops only about one-third of a meter (1 foot) of elevation for every 3.2 kilometers (2 miles) of the 320 kilometers (200 miles) between Memphis and Vicksburg.

Poor internal drainage, frequent flooding by the Mississippi, and the presence of Choctaw and Chickasaw Indians kept European settlement tentative in the Delta for many years. However, the lure of fertile soil, proximity to river trade, and the profits to be made from cotton gradually drew plantation agriculture into this hazardous region after the 1830s.

As elsewhere in the South, cotton production claimed large numbers of slaves. The Delta was so exclusively plantation

country that by 1860 blacks outnumbered whites by more than 3 to 1 with some areas exceeding a 10:1 ratio.

Land development was slow, however. Floods inundated the Delta in 1858, 1862, 1865, 1867, 1868, 1871, and 1874. Their frequency, combined with the Civil War's ravages, kept the area largely wilderness. The tremendous pressure on the South's productive land after the war and political pressure by railroads and land speculators during the period from 1875–1925 finally attracted federal efforts to control flooding. These efforts were only gradually successful, but the Delta attracted both whites and blacks as land was cleared, swamps drained, and new areas planted in cotton.

The Delta changed but mostly in Old South ways. Cotton acreage increased almost 500 percent between 1880 and 1920. Towns within the Delta stayed small; Greenville, at a population of 11,560 in 1920, was the only town with more than 8000 people. While no longer slaves, the Delta's majority black population continued to labor in the fields as sharecroppers.

An especially severe flood in 1927, followed by the national distress of the Depression, led to more decisive governmental

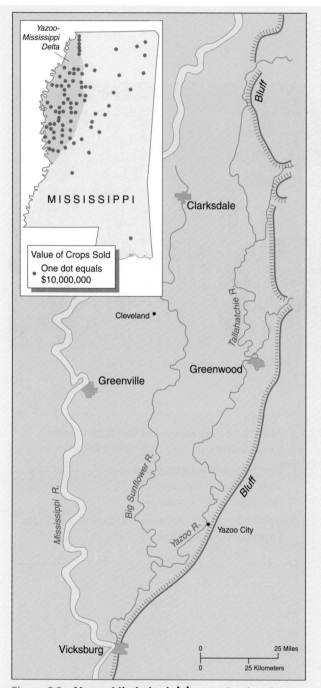

intervention in the Delta's economy. Large landowners recovered financially with the Agricultural Adjustment Acts and cultivated more acres. Although the amount of land planted in cotton was controlled from Washington, D.C., rice and soybeans soon proved profitable additions, especially since both crops did well on land too wet for cotton.

Agricultural expansion during the 1930s and 1940s occurred just as blacks left the Delta region in increasing numbers. During the half-century after 1920, the Delta's black population declined 33 percent. Landowners responded to the labor shortage by increasing farm mechanization. Because cotton had always been especially labor-intensive, this transformation of agricultural operations was more extreme among cotton-growers than among farmers of other crops elsewhere in the country. And because mechanized cultivating, harvesting, and applying of fertilizers and pesticides requires considerable capital and large-scale operations, small farmers in the Delta lost out to the most successful landowners.

In addition to these features—dependence on agriculture; a large, poor, black sharecropper population; an absence of major urban centers; a decline in cotton's relative importance; and changes stimulated by federal intervention and mechanization—the Delta also highlights Old South patterns in its grudging industrial growth. Later to appear and less intense than elsewhere in the South, a few new manufacturing activities began to surface in the Delta after World War II. The success of agriculture (as well as the valid fear of floods) probably retarded industrial entry for many years. Even with heavy outmigration, successful farm mechanization eventually generated a surplus labor force. As Delta manufacturing grew modestly, so did employment, rising from 6491 in 1955 to 15,079 in 1970. By 1997, more than 20,000 people worked in manufacturing.

The Yazoo-Mississippi Delta is very rich agriculturally, increasingly diverse economically and possesses significant residual poverty (especially among its majority black population) in spite of the presence of both old and new wealth. As such, it is also representative of the larger and older South.

Figure 9.9   Yazoo-Mississippi delta. With rich soil, flatland subject to flooding and used intensively for agriculture, the delta continues to have very few urban places.

Land in the Mississippi Delta is flat, and most is intensively used. Here, a farmer tends to a cotton field.

Review Questions

1. How did the early plantation economy impact the spatial organization of the South?

2. Explain the rise and decline of the South's once dominant textile industry.

3. Discuss the push and pull factors that led to the tremendous out-migration of African-Americans from the South.

4. What are the major components of "Southern culture"?

5. What are the major changes that have occurred in the South's economic structure since the mid-1930s?

6. How has government intervention served to stimulate the South's economy? Explain the initiatives and their impacts.

7. Explain how the Yazoo-Mississippi Delta can be described as a microcosm of several aspects of the South.

CHAPTER 10

THE SOUTHERN COASTLANDS: ON THE SUBTROPICAL MARGIN

Preview

1. The climate of the Southern Coastlands supports a large production of subtropical crops such as citrus fruits and rice.

2. Natural hazards, especially hurricanes, are a threat to the region every summer.

3. Because of its coastal location, the region is an important part of the U.S. trade network.

4. Prosperity in the Southern Coastlands is divided between areas that benefit from tourism and recreation and areas based on industrial production and trade.

5. Florida is an important culture hearth for the Latino population in the United States.

The United States' Southern margin is divisible into two sections that are almost equal in length but very different in character. The Southwest Border Area (Chapter 14) spreads across the western part and shares a land boundary and cultural affinity with Mexico. The Southern Coastlands takes up the eastern half, picking up from the mouth of the Rio Grande and ranging through Florida's peninsula to coastal North Carolina (Figure 10.1).

Uncommon in the United States, the Southern Coastlands is a humid, subtropical place that appeals to visitors, attracts new residents, and has farms that grow crops distinctive to its environment. The Gulf of Mexico's warm waters give the region's climate a balmy, maritime influence. The region's southernmost point, Key West in the Florida Keys, lies only 95 kilometers (60 miles) from the Tropic of Cancer.

All in all, the Southern Coastlands derives its character from its location at the continent's southern margin. The humid subtropical environment is one consequence of this location. But the region's role in generating U.S. trade with the rest of the world, and the coastal region's singular resource pattern, also help make up the Southern Coastlands' complex identity.

Many Southern Coastlands activities depend directly on the unblemished character of the region's natural environment. Other activities treat the environment merely as a setting. When these alternative approaches take place near each other, conflict occurs.

Development in the Southern Coastlands' eastern half is largely recreational and includes abundant retirement communities. Development in the region's western half is more heavily dependent on resource extraction. These two uses, each dependent on a different component of the Southern Coastlands' regional geography—subtropical environment and location on the continent's southern margin—contain the seeds of conflict and are the two primary themes for this region.

The region's advantages for international trade, especially with Latin America, indicate the Southern Coastlands' third theme. Lying between the U.S. South and Latin America, the Southern Coastlands has aspects of old South culture; but as part of the continental margin, the region is economically different from the rest of the South, which contributes to its special character.

The Southern Coastlands' exposure to Latin American cultural influences also makes it different from the South. Initially buffered by the Gulf of Mexico and the Caribbean Sea, the Southern Coastlands felt a muted Latin influence until well after World War II. Then, the growth of the Cuban-heritage population in southern Florida and the intensification of trade between Latin America and the United States, much of it flowing

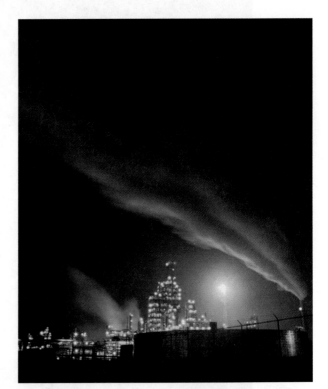

Oil refineries, such as this one in New Sarpy, Louisiana, punctuate the landscape of the Southern Coastlands. Economics and environmental integrity often confront each other across the region.

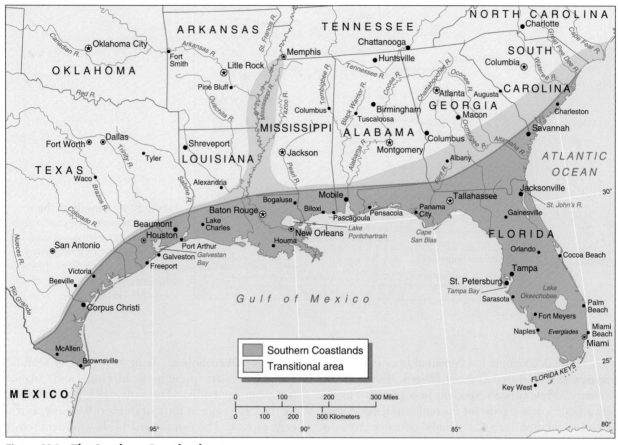

Figure 10.1 The Southern Coastlands.

through southern Florida, deepened the Southern Coastlands' distinctive character.

THE SUBTROPICAL ENVIRONMENT

The Southern Coastlands' climate has a great impact on the region's human geography. Humid and subtropical, the climate gives the region mild winter temperatures, hot summers, and a long growing season.

The growing season, or the potential growth period for agricultural crops, is the average number of days between killing frost in the spring and fall. Across almost all of the Southern Coastlands, the growing season lasts at least 9 months (Figure 10.2). But in much of the region, its average exceeds 10 months. The southern half of Florida, for example, is unlikely to experience frosts every year. Only in Hawaii, southern California, and southwestern Arizona do farmers benefit from growing seasons equivalent to those of the Southern Coastlands.

In addition, farmers throughout most of the Southern Coastlands also benefit from the region's abundant precipitation. Using Houston, Texas, as a reference point, places east of Houston generally receive more than 125 centimeters (50 inches) of rainfall per year. With most rain falling between April and October, the moisture joins plentiful sunlight and warm temperatures to support vigorous plant growth.

Agriculture

Southern Coastlands farmers produce crops that are grown in few other parts of the United States and Canada. Citrus fruits, rice, and sugarcane are the region's most important specialty crops.

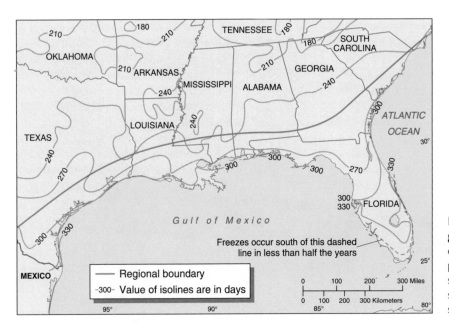

Figure 10.2 Length of the growing season. The combination of a southerly location and maritime proximity leads to a long growing season and, across most of the region, sufficient precipitation for reliable specialty crop farming.

Florida's economy has benefited from citrus since Spaniards first introduced the fruits in the sixteenth century. While citrus is especially important in Florida, its production is significant in southern Texas as well. Outside the Southern Coastlands, citrus is grown in large quantities only in Arizona and California. Indeed, only California exceeds Florida in the production of lemons and navel oranges (Table 10.1).

Primary commercial citrus farming areas in Florida have gradually migrated south along the peninsula's interior. Today, most citrus is grown south of 29 degrees north latitude, between Ocala and Lake Okeechobee. About 40 percent of Florida's citrus groves are found between Tampa and Orlando (see Map Appendix). The shift south was prompted by the yearly threat of frost in Florida's more northerly groves. The southward push would be more extensive, but swampy soils and standing water in the Everglades prevent citrus growers from moving into Florida's totally frost-free margin.

Of the seven major citrus fruits grown in Florida, oranges and grapefruit are the most important. Orange production increased steadily until 1978, when 7.4 million tons were harvested. While this figure dipped to 6.1 million tons in 1992, production jumped to 8.4 million tons in 1993. Production figures topped 10 million tons for the first time in 1997 but have declined since 2000 (Table 10.2). More than 80 percent of Florida's orange crop is processed rather than sold as fresh oranges. By processing oranges into frozen concentrate, Florida has developed a sizable food-processing industry. This spreads the crop's economic benefits among more of the state's people than if the fruit were shipped out fresh. Processing also allows year-round sales, so returns are not limited to the harvest period.

Farmers grow grapefruit in the same area as oranges, but the grapefruit has less economic impact; demand is lower, and output is only about one-fourth that of oranges. Elsewhere in the Southern Coastlands,

TABLE 10.1

Production of Citrus Fruits, in Florida and California, 2005–2006 (in boxes)		
Citrus Fruits	**Florida**	**California**
Oranges (all)	147,900,000	57,500,000
Tangerines	5,500,000	3,600,000
Tangelos	1,400,000	0
Grapefruit	19,300,000	6,000,000
Lemons	0	24,800,000

Source: United States Dept. of Agriculture, National Agricultural Statistics Service, "Citrus Fruits 2006 Summary, September 2006."

TABLE 10.2

Orange Production in Florida, California, and the United States (in 1000 tons), 1989–2007			
Season	Florida	California	United States
1989–1990	4,958	2,677	7,745
1990–1991	6,822	961	7,848
1991–1992	6,291	2,528	8,909
1992–1993	8,397	2,505	10,992
1993–1994	7,848	2,385	10,329
1994–1995	9,248	2,100	11,432
1995–1996	9,149	2,175	11,426
1996–1997	10,179	2,400	12,692
1997–1998	10,980	2,588	13,670
1998–1999	8,370	1,350	9,824
1999–2000	10,485	2,400	12,997
2000–2001	10,049	2,213	12,390
2001–2002	10,350	2,100	12,543
2002–2003	8,865	2,363	11,313
2003–2004	10,890	1,952	12,930
2004–2005	7,560	2,235	9,982
2005–2006	6,656	2,156	8,898
2006–2007	6,300	1,726	8,115

orange and grapefruit farming is under irrigation in the extreme southern portion of Texas between Brownsville and McAllen.

Because citrus are tree crops, fruit must be hand-picked, often from the top of a long ladder. Each year during harvest, growers need significant short-term labor. Thousands of migrant laborers come to the dense, parklike groves for annual bouts of intense physical effort. As temporary workers, the laborers are usually housed in cheaply constructed structures. Many bring their families. While their day's work may yield a respectable pay, laborers endure long workdays. Fundamentally, citrus farming is an industry required to devote more than one-third of its total production expense to hired labor; yet its workers cannot earn enough at one location to satisfy annual living requirements.

Another specialty crop, sugarcane, is grown only in the Southern Coastlands and Hawaii. Sugarcane is a tropical plant, requiring more than one year to mature fully. It is not at all frost-tolerant, and its water requirements call for at least 125 centimeters (50 inches) of rain per year. Although both temperature and water requirements limit sugarcane cultivation in the continental United States, cane grows successfully in Louisiana and Florida. In Louisiana, rainfall is sufficient for cane production; in Florida, rainfall is supplemented by irrigation.

To avoid frost damage, farmers harvest immature cane early. Even though the sugar content is low before the cane fully matures, federal import controls and price supports are usually enough to ensure farmers a profit.

Rice is a third Southern Coastlands specialty crop, but one with fewer climatic demands than sugarcane or citrus. Given sufficient water, rice matures within one growing season at a pace roughly proportional to the amount of summer heat received. Within the Southern Coastlands, irrigated rice grows in Louisiana (373,000 acres in 2007) and in Texas (146,000 acres in 2007). In its geographical spread, rice aptly demonstrates the transitional nature of the Southern Coastlands' boundary. Farmers grow rice north of the Gulf Coast in the Mississippi Alluvial valley, which cuts through Arkansas and Mississippi. Even though these parts of Arkansas and Mississippi are dissimilar to the Southern Coastlands in most ways, their warm climate supports rice production to the tune of 1.3 million acres in Arkansas and 189,000 acres in Mississippi in 2007 (see also Map Appendix).

Although less obviously a set of specialty crops, a variety of vegetables are produced by Southern Coastlands' farmers for winter sale. The region's subtropical climate complements the cold winter continental climate of North America's manufacturing core. Each winter, most fresh vegetables reaching northeastern urban markets are grown in Florida or the southern margins of other Gulf Coast states, although some are now shipped from the extreme southwestern United States, South America, and South Africa. In geographers' parlance, the proximity of Southern Coastlands producers to northeastern markets gives them a distinct locational advantage over California (and more distant) vegetable growers.

Generally, climatic conditions are favorable for agriculture, but the Southern Coastlands is a flat, low-lying region. Soil conditions are less uniformly beneficial than climate. Soils range from the fertile but poorly drained muck of the Louisiana coast and Mississippi Delta to the extremely sandy soils in north and central Florida.

Florida is swathed by immense orange groves, some of which thrive through irrigation.

Attempts to drain the region's swamps have frequently failed for both practical and political reasons. Florida's extensive Everglades provide an example. It is both troublesome and expensive to construct and maintain drainage lines in this large, flat area. But beyond the engineering and financial challenges, the Everglades are a unique, vast, and fragile ecosystem dependent on slowly moving freshwater entering the

After Florida's orange crop is picked by hand, oranges are loaded in bulk for shipment.

Growing conditions for sugarcane are below optimum along the Gulf Coast, but cane can be grown. The use of machinery in large-scale operations permits profitable production where smaller scale and hand labor would not.

marshland. When part of the marsh is drained, several detrimental effects result. The *water table* is lowered, permitting extensive encroachment of salt water; grasslands are more exposed to fire; and the delicate life-support balance for wildlife, shellfish, and vegetation is upset. But however pressing agricultural demands are, southern Florida's tremendous urban growth and the consequent demand for land and drinking water is a much more serious and long-term threat to the ecological balance of the Everglades.

Where the Southern Coastlands does not need drainage, farming benefits from heavy irrigation. For example, Florida's central highlands, which lie about 30 to 46 meters (100 to 150 feet) above sea level, are underlain by sandy soils with moderately poor to very poor water-retention capacity. Citrus- and vegetable-growing areas yield annual output as much as ten times more valuable when crops are irrigated than when precipitation is the sole source of moisture. With this degree of productivity improvement possible, the Coastlands' distinctive subtropical environment has developed agriculturally beyond the levels found across the interior South.

Amenities and Age Structure

Personal disposable income and retirement financial security increased in the United States and Canada after World War II. As a result, recreation and retirement are now major industries seeking to attract large numbers of consumers. The Southern Coastlands' subtropical climate entices millions to visit or move south.

Geographer Edward Ullman[1] argued that as economic reasons for migration become less important, people will choose to live where they find it pleasant. Ullman suggested that regions with an outdoor climate similar to a comfortable indoor climate are most preferred. In the United States and Canada, the two places best approximating this comfortable feel during winter months are coastal southern California and the Southern Coastlands. Air-conditioning gives these regions' summers nearly equivalent appeal.

Amenities have been important stimuli in the regional growth of Florida and the Gulf Coast. Census figures leave no doubt that Florida is one of the prime destinations of people following their retirement. In 1950, 12.4 percent of Florida's citizens were over 60 years of age. By 2000, the proportion over 60 was estimated to be more than 22.2 percent. In comparison, continental manufacturing core states like Ohio and Connecticut had populations above age 60 in 1950 of 13.7 and 13.3 percent, respectively. The proportions of people 60 or older in these states were much lower than Florida's in 2000, with 17.4 percent in Ohio and 17.7 percent in Connecticut.

Although mild winters draw large numbers of retirees and their steady incomes to the Southern Coastlands, tourists are the group making the larger economic contribution. New Orleans, for example, attracts vacationers with its unique culture and array of metropolitan leisure opportunities. Along the coast east of New Orleans, Mississippi's short stretch of what locals call the "Mississippi Riviera" booms with construction of hotels, motels, restaurants, and beaches where visitors can bask in subtropical breezes.

Florida is the Southern Coastlands' most magnetic tourist attraction. With long beaches on both the Atlantic Ocean and the Gulf of Mexico sides of the peninsula, it draws winter vacationers from Megalopolis and North America's industrial interior. Luxury hotels lining the Miami Beach oceanfront and luxury estates along the coast at Palm Beach are the cores of tourism-based urban development extending more than 160 kilometers (100 miles) along Florida's southeastern coast. Today, strong competition from other tourist destinations and an aging physical plant have diminished the appeal of many of the region's older hotels, but their earlier magnetism transformed the face of coastal southern Florida. Other seeds of

[1] Edward L. Ullman, "Amenities as a Factor in Regional Growth," *Geographical Review* 55 (1954): 119–132.

The coastal resources of peninsular Florida are well suited to recreation, as illustrated in this section of Fort Lauderdale where yacht canals have been constructed to permit maximum access to the water.

recreational development sprouted along Florida's Atlantic Coast between Miami Beach and Jacksonville. Although Florida's Gulf Coast is less well endowed with attractive beaches, considerable growth exists from Tampa Bay south to Naples.

So strong is the demand for subtropical amenities that recreation spots have now spread north along the Atlantic margin well past coastal Georgia into coastal South and North Carolina and beyond. The Georgia and South Carolina coasts, such as at Georgia's Sea Islands and South Carolina's Hilton Head, Charleston, and Myrtle Beach, are experiencing intense development.

Not all of the tourist attractions in the Southern Coastlands depend on natural beauty. Walt Disney's decision to construct Disneyworld, an eastern equivalent to his highly successful Disneyland in southern California, radically changed the area around Orlando. Disneyworld draws millions of out-of-state visitors to south-central Florida and has attracted many other theme parks by promising regular, high-volume tourist traffic and tourist spending.

The central Florida recreation complex stimulates considerable spillover urban growth. Orlando and nearby cities are an inland link between Florida's east coastal and west coastal urban clusters. A small megalopolis extending from Miami to Tampa-St. Petersburg

via Orlando contained more than 9 million people according to the 2000 census. Although not the only reason for Florida's urban coalescence, the more than $65 billion in taxable sales related to tourism and recreation in Florida during 2006 starkly indicates the economic power of amenities.

Hazards

Winter Freeze

The Southern Coastlands' subtropical environment is not entirely beneficent. Vegetable growers deceived by the typically mild winters, for example, may attempt to produce crops year-round. But when the occasional yet inevitable mid-winter frost reaches southern Florida, considerable crop damage and financial loss are the results.

Even a ripening Florida citrus crop can be seriously harmed by winter freezes. Citrus growers in Lake County, Florida, for example, can expect the first frost about December 15. But they can face a hard freeze as much as 3 weeks earlier. They use smudge pots, oil heaters, and other partially effective devices to protect their investment—sometimes with less than satisfactory results. Not as well publicized as citrus damage is the potential harm from untimely wintery blasts for sugarcane growers in Louisiana.

For decades, general affluence and low transportation costs have allowed more and more people to enjoy environmentally pleasant places. This greater use has often made the places less pleasant, but they nevertheless remain destinations for newcomers.

The short and inconsistent frost period and generally mild freezes also create problems for agriculture because, although controlled by expensive chemical means, insect and parasitic pests thrive during the long growing season and create a constant threat to farm profits.

Hurricanes

More sporadic, more dramatic, and locally more destructive in the Southern Coastlands than elsewhere in North America are hurricanes (Figure 10.3). Large cyclonic tropical storms, hurricanes are generated by intense solar heating over large bodies of warm water. Under proper conditions, a massive low-pressure atmospheric pocket forms, drawing in warm, moist air. The earth's rotation and spheroidal shape mean that air flowing toward the hurricane's center establishes the storm's broad, counterclockwise rotational pattern. The deeper the pressure pocket, the faster the wind flows and the more destructive the storm's potential.

Hurricanes can affect settlements far from the Southern Coastlands, but that region is where hurricanes concentrate. Areas facing the tropical waters of the Caribbean Sea and the southern Atlantic, and the entire Gulf coastal lowland are extremely vulnerable to the hurricanes' fury. The potential for damage was demonstrated dramatically in 1989 when Hurricane Hugo

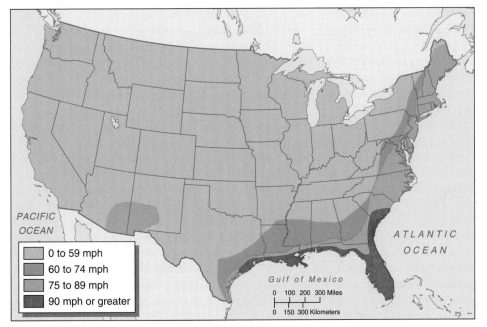

Figure 10.3 Hurricane Hazard Zones. Data from the National Hurricane Center reveal distinct hazard zones based on incidence of hurricane winds over the past 100 years.

Johnny Montgomery, 82, wades through flood waters in the aftermath of Hurricane Katrina, which inundated much of New Orleans, Louisiana and the Gulf Coast in September, 2005.

bowled ashore just north of Charleston, South Carolina, and in 1992 when Hurricane Andrew passed over Miami. More recently, in 2003, Hurricane Isabel caused widespread damage, although it caused greater devastation farther north along the coast of the Carolinas.

On August 28, 2005, the National Hurricane Center focused its attention on Hurricane Katrina, which had continued to evolve in the Gulf of Mexico. The storm had grown to an enormous category 5 hurricane, extending more than 1000 miles in diameter and exhibiting winds of up to 175 mph.

Katrina made landfall in Louisiana on August 29th as a category 4 hurricane, featuring winds of up to 140 mph and generating storm surges from 24–28 feet. By the time the storm moved on, more than 90,000 square miles of the Southern Coastlands had been devastated. The official death toll was listed as 1277 with thousands left homeless; the damage was estimated at over 200 billion dollars. More than one million residents were evacuated from the region and millions were left without power from Louisiana to Florida. In Louisiana, damage to property and infrastructure was exacerbated when levees were breached and widespread flooding ensued around the already devastated city of New Orleans. Along the gulf coast of Mississippi, storm surges laid ruins to coastal areas. The extent of second and third order effects such as oil spills, water pollution, fish and animal kills, disease incidence, and looting were even more difficult to measure and some derivatives of the natural disaster are still unfolding.

The aftermath of Katrina showcased the challenges of coordinating and undertaking a massive humanitarian relief effort by local, state, and federal agencies, and reminded us of the dangers associated with developing and inhabiting marginal lands, the vulnerability of the impoverished, and the risks associated with living in hurricane-prone regions.

Because the heaviest damage from a storm is usually limited to a narrow swath of several dozen miles as the storm moves onshore, much of the Southern Coastlands has not been affected for years. Indeed, hurricanes are variable in occurrence and strength, and the lengthy period between direct coastal strikes by hurricanes at any specific site tends to make newcomers and some long-term residents complacent about the tremendous force and destructive power of major storms. So coastal development intensifies, squeezing into space along the waterfront and increasing the dollar value of structures exposed to hurricane damage.

Water

The Southern Coastlands is gradually succumbing to a slow-acting hazard. Ironically, while most of the Southern Coastlands receives abundant rainfall and no part of the region is far from the sea, water is being drawn from some local sources faster than it can be recharged naturally. As surprising as it may sound, major sections of the region are facing a situation in which water demand exceeds water supply.

The water supply options available to growing municipalities are limited. If rivers are too far away and the land is too flat to construct large reservoirs, water must be withdrawn from the water table by drilling wells and pumping it out. Geographically, the Southern Coastlands fits this bill. Its problem appears where major urban developments lie at a distance from rivers that might otherwise be used as a water supply.

In central Florida, water is pumped from the water table. In the past, rains were usually sufficient to fully recharge it. But now, even the area's normally abundant rainfall is not always enough. Tremendous growth between Cape Canaveral and Tampa, with the greatest population and business concentration in Orange County's Orlando area, has spread roads, parking lots, homes, motels, amusement parks, retail outlets, and all other facets of retirement and tourist development across the land. Rainfall in these built-up areas is carried away through storm drains and does not percolate to the water table. The result is reduced recharge of underground water just as water extraction is increased.

As the water table level drops, the land loses its support from below and the surface collapses, sometimes very suddenly. If unsupported land subsides slowly over a large area, the decline may not be apparent. But when the surface collapses suddenly, a *sinkhole* may be formed. A sinkhole abruptly opened on May 13, 1981, swallowing an automobile dealership in Winter Park, Florida. It brought national attention to this potential hazard, if not to its underlying cause. The state will not disappear in a catastrophic, Atlantis-like manner, but Winter Park's experience and smaller versions in the years since show that the environmental repercussions of development are not always immediate or obvious. Neither are solutions to problems caused by excessive underground water withdrawal. In the meantime, urban settlement continues to spread across the land.

ON THE MARGIN OF THE CONTINENT

The Southern Coastlands' subtropical environment reflects its location along North America's southern margins. The region's climate and vegetation are reflections of its absolute location, or site. Its relative location, or situation, however, is more effectively understood in terms of two other characteristics. The Southern Coastlands is a key exchange area between North America's continental interior and overseas markets, especially markets in Middle America, the Caribbean, and South America. The Southern Coastlands also contain mineral resources that play a role in the region's trade and economic importance.

Trade

The Gulf of Mexico has a shallow, emergent coastline containing many high-action beaches. Much of it is either backed by extensive swamps or partially shielded behind offshore sandbars. Coastwise shipping uses the protection provided by the sandbars in the *intracoastal waterway* system. However, most bays lying behind the bars are too shallow to give good anchorage for large ships engaged in transoceanic trade.

Most of the Southern Coastlands' larger ports developed either on large river estuaries or at anchorages a short distance inland from the region's river mouths (see Chapter 3). Pensacola, Florida; Mobile, Alabama; and Galveston and Corpus Christi, Texas, all developed in the shelter of large bays, each of which bears the port city's name. Jacksonville, Florida, on the St. Johns River; Brownsville, Texas, on the Rio Grande; and New Orleans, Louisiana, on the Mississippi River, are river ports. The Southern Coastlands' largest city, Houston, was not originally a port city but has become one. Construction on the Houston Ship Channel across shallow Galveston Bay began in 1873, giving Houston access to ocean trade.

The ability to reach inland sources of exchange is even more important in generating trade than physical facilities. Each Southern Coastland bay providing good harbor facilities is the outlet for a river draining a part of the interior. Although these rivers are not equally navigable, they assisted early settlement expansion, and some are still used by small barges.

Rail connections strengthened each harbor's access to North America's major inland markets. Trade was also improved by dredging the river flowing into the coastal harbor or straightening it to improve navigation. Jacksonville, for example, was an early terminus for railroads entering Florida from Georgia. Jacksonville's hinterland also extended west into the state's panhandle and south into Florida's agriculturally rich central highlands. Ports that might have competed with Jacksonville were too distant or too low-quality to attract Jacksonville's shippers. Consequently, the city was well established before highway connections reinforced its growth. In Alabama, Mobile's access to the interior now extends to the Ohio River via the Tennessee-Tombigbee waterway.

As for Southern Coastland cities, New Orleans is in a class by itself. Spain founded the city early in European colonial times. France soon succeeded Spain, seeking a southern counterpart along the Mississippi River to Quebec's and Montreal's St. Lawrence River access to the North American interior.

Because of its site, New Orleans became both a control point and a shipping focus for the entire Mississippi River system. With proper caution, the Mississippi is navigable far to the north, allowing early shippers to reach into the continent's agricultural heartland using shallow-draft, paddle-wheel steamers. The river's main tributaries are also navigable, extending the waterway system west into the Great Plains (see Chapter 12) and east into the continent's evolving manufacturing core (see Chapter 5).

This high-altitude photo of New Orleans dramatically demonstrates the city's intense urban development within a large meander of the lower Mississippi River and the trade and transfer point advantages the site affords.

New Orleans was born within a large river meander on the low-lying Mississippi River delta. This site puts the city only a few feet above mean sea level, making flooding a continuing threat. But in geographical terms, the city's situation compensated. It gave merchants such tremendous advantages that the city's population grew quickly in the nineteenth century. New Orleans was by far the largest urban place south of the Ohio River and east of the Rocky Mountains until after World War II.

New Orleans lies midway between the Southern Coastlands' recreation sites along the eastern Gulf and the region's industrial centers to the west. The city's French colonial heritage is consciously maintained in its French Quarter, or the *Vieux Carré*. A strong non-Anglo cultural imprint exists outside the Quarter as well, in a distinctive mix of local Creole, Cajun, and European heritages. The consequent blend of cuisines, along with a wealth of jazz and Dixieland music performances, eighteenth-century architecture, and an anything-goes reputation draws millions of tourists to New Orleans each year. Casual visitors, there for recreation, may be surprised by the heavy barge and ship traffic on the river and by the heavy industry supported by this traffic. New Orleans is one of the busiest ports in North America.

The other major city in the Southern Coastlands' western portion—the region's largest by 1970 and the United States' fourth largest in 2000—sharply contrasts with New Orleans. Houston is new, while New Orleans has a long history of cultural and economic importance. Although Houston is not located on a river or even on Galveston Bay, its port connections became significant after World War II when the local petrochemical industry arose. Houston established urban prominence on the crest of mineral resource exploitation and Gulf Coast access.

When Gulf Coast oil fields west of New Orleans came into production during the early 1900s, Houston was a moderate-size city of fewer than 75,000 people. By 1930, the city had multiplied four times to almost 300,000. Through the Great Depression and World War II, Houston continued to grow, doubling its population again by 1950. By the time the 1980 census was taken, the city itself had grown to 1.6 million people, ranking it fourth in the United States behind New York, Chicago, and Los Angeles. By 2006, Houston's metropolitan population had grown to more than 5,539,949.

Resources

North America's true continental margin does not coincide with the visible, sand-glistening seacoast. Instead, it is defined by a shelf of land at the continent's edge that lies below the sea. In some cases, the shelf juts into the sea only a few miles. But along much of the Atlantic Coast and in the Gulf of

The spires of St. Louis Cathedral overlook Jackson Square in the heart of New Orleans' French Quarter, where the city's French cultural inheritance has been preserved. Vibrant and elegant, and not badly damaged by Hurricane Katrina and its aftermath, the French Quarter draws thousands of visitors to New Orleans each year.

Mexico, the continental shelf spreads more than 80 kilometers (50 miles) from shore before its edge drops off into the deep, abyssal plain. Mineral exploration along the coast from the Rio Grande to the mouth of the Mississippi River led to the discovery of extensive petroleum and natural gas deposits, both onshore and off (Figure 10.4).

Texas and Louisiana are currently two of the three leading petroleum-producing states; Alaska is the third. Their output far exceeds that of the two next most important, California and Oklahoma. While Texas and Louisiana possess large producing fields inland from the Gulf, the coastal fields are major contributors to the totals of both states.

The search by petroleum companies for new deposits along the Gulf Coast extended seaward by mid-century. Drilling for and extracting petroleum from the Gulf's continental shelf far beyond sight of land was expensive and technologically difficult but eventually successful. But economic factors aside, exploration success created new political problems. Conflicting state and federal claims flared for the first time over the issue of which governmental level held jurisdiction over continental shelf resources. Texas and Louisiana had immediate stakes in the argument, and other coastal states, especially those bordering the Gulf, asserted an interest.

The United States' federal organization permits litigation between individual states and the national government. The outcome of a complex set of court cases made jurisdiction over the Gulf's continental shelf resources different for the various Southern Coastland states. Now, Florida and Texas are sanctioned to claim up to 9 miles seaward, whereas Louisiana, Alabama, and Mississippi can claim only 3 miles.

Continuous increases in domestic consumption of petroleum, natural gas, and petroleum products, combined with uncertain foreign supplies, led the national government during the early 1970s to open commercial bidding on offshore tracts between Biloxi, Mississippi, and Tampa Bay, Florida. By the early 1980s, bids opened for tracts off the Atlantic Coast as well. Opposition to exploratory drilling in the eastern Southern Coastlands has been strong from environmental and fishing groups, along with resort and recreation interests, based on the Coast's amenity characteristics. Not surprisingly, Florida led the opposition, a reflection of its long Gulf coastline and its orientation toward recreational resources.

Oil spills or leakage can have disastrous short-term effects on tourist activities, but many conservationists fear the effects can be more serious for the region's ecological health. Extensive mangrove swamps, coastal marshlands, and commercial fishing beds exist in delicate ecological balance. This balance can be badly upset for extensive periods if oil drilling mishaps or shipping accidents spread pollution. The classic disagreement between those who would develop consumable resources and those who want to preserve the natural

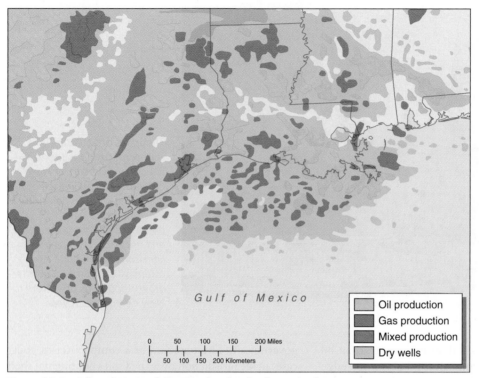

Figure 10.4 **Oil and Natural Gas Production in the United States.** Given the coastal location of these resources, the long arc of petroleum and natural gas deposits paralleling the Louisiana-Texas Gulf Coast has generated a large refining capacity at Texas and Mississippi River ports.

environment's balance will not be resolved easily along the Gulf Coast.

Natural gas, often found in conjunction with petroleum deposits, is another vital Gulf resource. For years, natural gas was burned off as an unwanted byproduct of petroleum production. But its energy potential was eventually realized and generated a search for natural gas independent of petroleum exploration. Now numerous, scattered Gulf Coast natural gas deposits spatially intermingle with the region's long arc of petroleum deposits. Pipelines carrying the gas radiate from main coastal production centers to the primary consumption points across the continent, especially to the manufacturing core.

Two minerals less commonly associated with petroleum and natural gas also are found in the geologic formations of the Texas and Louisiana coastlands. Specifically, subsurface rock contortions trapping petroleum and natural gas were formed in the region by the gradual upthrust of huge *salt domes*. Far less valuable than either mineral fuel, large quantities of rock salt are nonetheless mined in southwestern Louisiana. More valuable than salt,

but not as valuable as petroleum or natural gas, is sulfur, found in the *caprock* of many salt domes. Large sulfur deposits at Beaumont, Texas, and across the state boundary near Lake Charles, Louisiana, supply all U.S. sulfur needs. Additional deposits inland and beneath the continental shelf will ensure this mineral's abundance for many years. Phosphate from substantial deposits in Florida is also of national significance.

Industrial Development

The industrial growth of the Southern Coastlands west of New Orleans was stimulated partly because of coastal location and partly because of the tremendous quantities of capital generated by petroleum and natural gas extraction. Wealth accumulated in Texas and Louisiana brought industries to extraction sites. Petroleum refineries sprang up outside all major ports from Corpus Christi, Texas, to Pascagoula, Mississippi, with intensive concentrations around Houston, Beaumont, and Port Arthur.

As petroleum is refined into fuels and lubricants, a number of petrochemicals become available. Industries using petrochemicals as their raw material were drawn to the western Southern Coastlands.

Highly industrialized countries such as the United States and Canada depend on a strong petrochemical industry. Natural gas and petroleum products are used as chemical components when manufacturing many everyday products. Plastics, paints, antifreeze, fertilizer, insecticide, and prescription drugs are all manufactured in chemical plants located along the western Gulf Coast (Figure 10.4). Other chemical industries not dependent on petroleum and natural gas production, such as those producing sulfuric acid, superphosphate fertilizer, and synthetic rubber, are major consumers of the region's abundant sulfur and salt.

The locational coincidence of oil, natural gas, sulfur, and salt, together with capital investment capabilities, supported rapid economic and population growth, especially in urban centers like Houston. There is a downside to this positive picture, however: The region's economy can plummet when declining oil prices sharply reduce the investment capital available.

Yet there is more to the location of Southern Coastlands industry than capital availability and raw materials proximity. For example, two large oil fields extending from southeastern New Mexico to central and eastern Kansas produce vast quantities of petroleum, natural gas, and capital but generate much less local industrial development than occurs along the Gulf Coast. The difference is largely a matter of accessibility.

Because water transportation is cheaper per ton-mile than land transport, even though slower, Southern Coastlands industry has an advantage. Its finished products are shipped efficiently to Megalopolis's ports by ocean carrier and to the interior manufacturing core by barge using the intracoastal waterway and the Mississippi River system.

Just as coastal location facilitates shipment out of the region, products and raw materials are moved into Gulf Coast industries more efficiently. Bauxite is an example of such a raw material. Bauxite ore, from which aluminum is eventually refined, has a low aluminum content when mined. Ideally, the ore is refined near where it is extracted because moving the ore's waste along with the useful portion is not efficient. Shipping by water, such as to the Gulf Coast, addresses part of the refining demands.

Aluminum refining also requires large amounts of electricity. Since most bauxite the United States imports originates on the energy-poor Caribbean island of

More than 40 miles inland from the Gulf of Mexico on the vast coastal plain of Texas, industry near the head of the Houston Ship Channel is supplied by ocean-going vessels.

Jamaica or in the South American countries of Surinam and Guyana, ore is shipped at relatively low cost by freighter to Gulf Coast ports. There, abundant local fuel is used to refine the ore. Only at two other locations, Arvida, Quebec, and Kitimat, British Columbia, are accessibility and cheap electrical power equivalent to the Southern Coastlands'. Thus, these are North America's three sites where imported bauxite ore is refined in significant quantities.

The Region's Two Halves

The distinctive character of the Southern Coastlands contains strong contrasts. The entire region is subtropical, but its eastern half has developed by emphasizing the environment's attractions and advantages, while the western half gets its greatest return from mineral exploitation and consequent industrialization. Recreation and retirement development occur along the Gulf Coast west of the Mississippi River. Although the possibilities there have not been exploited as strongly as have sites east of New Orleans, Texas is beginning to extol the recreational virtues of its offshore islands south of Corpus Christi.

The Southern Coastlands' internal contrasts stand out in its three major urban centers. Both Houston and the Miami area have grown rapidly in recent decades, but Houston's economy is based on manufacturing and trade, and Miami's on recreation and travel. In recent years, Miami's Cuban Americans have stimulated strong financial and trade connections with Latin America, broadening the city's economic and cultural base. Located between the two, New Orleans can claim both industrial and recreational importance. An uneasy mix, New Orleans is not as strong in either economic area. Consequently, New Orleans has not grown as rapidly in the past 50 years as Houston or Miami.

Each half of the Southern Coastlands is under economic and political pressure to become more like the other half. Fluctuations in petroleum prices encouraged areas in the western half to diversify economically, thus developing more recreation sites. After 1975, the western Gulf's industrial and mineral exploitive character began to encroach on the eastern Gulf's recreational sites.

Whenever petroleum costs are high, pressure on the recreational character of coastal Florida increases. Whenever the economy is strong, Florida's recreation economy flourishes. If alternative fuels for cars and trucks are adopted widely within the next several decades, the economic and political demand to exploit offshore petroleum sites in the eastern Gulf may not generate conflict between the Southern Coastlands' two distinctive approaches to land use.

CULTURAL-ECONOMIC CASE STUDY

Latin America in Florida[2]

Florida is a finger pointed at the Caribbean. A peninsula reaching southward approximately 350 miles (560 kilometers) from the main body of North America, Florida extends along a chain of islands known as the Florida Keys to within 95 miles (152 kilometers) of Cuba's north coast. Miami's sprawling metropolitan cluster along the peninsula's southeast coast between the Everglades and the Atlantic Ocean is closer to Havana, Cuba, than to Florida's state capital at Tallahassee. The main themes characterizing the Southern Coastlands are uniquely represented in the greater Miami region's unparalleled cultural transformation after 1959.

Miami became a valued recreational destination after Henry Flagler's railroad was constructed south from Jacksonville in 1896. The region's subtropical location was especially attractive to people living in the northeastern seaboard's major urban centers; cities along Florida's southeast coast grew rapidly for the first half of the twentieth century.

Through the 1950s, Miami's environment attracted many visitors. The city's weather was warm throughout the winter; its coastal location provided easy access to long beaches and warm tropical waters; and the harbor required only modest improvement to bring cruise passengers from North America seeking travel to attractive and culturally distinctive Caribbean islands. These characteristics of Miami's subtropical location were exploited, but it was not until the 1960s that the region's

[2] Many of the data in this section have been taken from Thomas D. Boswell and James R. Curtis, *The Cuban-American Experience* (Totowa, NJ: Rowman & Allenheld, 1984), and Antonio Jorge and Raul Moncarz, *The Political Economy of Cubans in South Florida* (Coral Gables, FL: North-South Center, University of Miami, 1987).

location at the southern extremity of the continent led to Miami's full integration into a hemispheric system of finance and commerce.

Key to the change in Miami is the 1959 Cuban immigration to southern Florida following the successful Cuban revolution led by Fidel Castro. Cubans had lived in Miami for generations; it is a natural point of immigration for people from the island, if only because of proximity. Many who came from Cuba during earlier decades, however, moved to the dynamic labor markets in and around New York City. At the 1950 census, more than 45 percent of all Cuban-born residents of the United States lived in New York, with only about half that number living in Florida.

Beginning in 1959, Cuban refugees in unprecedented numbers moved to Miami and had a tremendous impact on southern Florida's economic and cultural landscapes. Fluctuating policies in Cuba generated three major waves of immigration. Between the revolution and the 1962 missile crisis, more than 200,000 Cubans migrated to Miami, and many settled there. Following a 3-year hiatus during which another 56,000 arrived from third countries, more than 300,000 additional Cuban refugees arrived during the nearly 7.5-year airlift from Havana to Miami. Then following another 7-year period of restricted emigration, the Mariel boatlift brought another 125,000 Cubans to Miami during a single 5 month period in 1980. Overall, during the 22-year period between 1959 and 1981, the greater Miami area's Latino population (about 85 percent of whom are Cuban) grew from 25,000 to almost 700,000, or 40 percent of the area's total 1981 population.

The numbers represent a staggering increase for so short a time, but a broader economic impact on the region is grounded in the immigrants' characteristics. A large share of the Cuban refugees arriving during the initial wave were in the legal

Although the Cuban presence in southern Florida was significant before the 1950s, it catapulted after Fidel Castro assumed power in Cuba in 1959 and sparked an exodus from that island country. The influx of Cuban immigrants to southern Florida transformed parts of its cities. This street scene in Miami vividly portrays the city's Latinization.

and white-collar professions.[3] It would be easy to overstate this occupational tilt, for the immigrants possessed a diverse set of skills.

Subsequent waves were also dominated by professional occupations, but this pattern became less and less pronounced as time went on. Contributions to the local economy made by Cuban immigrants to southeastern Florida, therefore, were very broad, though weighted toward professional, managerial, and skilled occupations.

The immigrants have had a considerable effect on Miami's economic geography. Cubans were absorbed quickly into financial, commercial, and retail activities. The rapidity with which large numbers settled in Miami created an instant market. Businesses located elsewhere in the United States but wishing to expand business in Latin American markets moved at least part of their domestic operations to Miami and employed Spanish-speaking Cubans. Trans-Caribbean business

contacts brought by Cuban refugees were also attractive to American business interests.

Multiplied many times in many ways and over several decades, this pattern opened Miami's natural geographic orientation toward the south. Southern Florida's major domestic competitor for Latin American business, Los Angeles, is oriented mostly toward Mexico and is almost 2000 miles (3200 kilometers) farther from South America than is Miami. Even Houston is almost 750 miles (1200 kilometers) more distant from South American contacts than is southern Florida.

Today, southern Florida is more culturally mixed, with a stronger Latino presence, than ever before. By adding hundreds of thousands of Cuban Americans, the area's economic geography has been transformed. Still subtropical in climate and still attractive for its recreational potential, southern Florida has also become an active commercial and financial hinge on the continent's margin. In a smaller and culturally distinctive fashion, greater Miami's urban complex is beginning to serve U.S. cultural and trade interests with Latin America similar to that which Megalopolis serves through interactions with Europe.

---

[3] Boswell and Curtis, *The Cuban-American Experience*, p. 45.

Review Questions

1. What are the potential climatic hazards found throughout the Southern Coastlands?

2. Examine the positive aspects of New Orleans' site and situation.

3. Describe the general agricultural patterns that prevail along the southern coasts.

4. What factors contribute to the distinctiveness of the Southern Coastlands?

5. Describe the Latinization of South Florida. How did this process evolve?

6. What are the long term implications of Florida's aging population?

THE AGRICULTURAL CORE

North America's interior plains, a gigantic, undulating swath of land, rolls gently from the Appalachians to the Rocky Mountains and from the Gulf of Mexico to the Arctic. Across this vastness, sporadic and far-flung hills and low mountains add scant texture. Geographers subdivide this expanse into numerous regions, some of which overlap thematically. The characteristics they use include topography, average temperature, growing season, and related environmental factors.

Geographers employ two key characteristics to identify regions within the interior plains: *longitudinal* differences in annual precipitation and the way people respond to them in the landscapes the differences create. The interior plains' western portion, for instance, is dry and dominated by grasslands. Its grand openness, seemingly boundless and stark, is what people call the Great Plains and Prairies (see Chapter 12). This western portion is used in ways that keep population densities low.

The interior's eastern plain, by contrast, is quite a different place. More rain falls, and the air is humid throughout the growing season, spawning different vegetation. The blending of these characteristics gives the eastern plains a higher *carrying capacity*, and, coupled to the benefits accruing from numerous natural transportation routes, has lent it the power to become the continent's agricultural core (Figure 11.1). The agricultural core overlaps with other thematic regions (see Figure 1.2). Its eastern half occupies the same area as the western half of North America's manufacturing core. This is no geographic accident. Much of the manufacturing core's urban development and manufacturing capacity was initially stimulated by the demands of agricultural producers and the tremendous volume of food they produced. Cities like Buffalo, Detroit, and Cincinnati in the east and Minneapolis-St. Paul, Omaha, and Kansas City in the west were early agricultural marketing and shipment centers. Businesses arose in these cities to process farmers' products. In tandem, other

Land in the agricultural core is intensively used. Here, corn is harvested in Iowa.

Geographic Name Calling

One of the most familiar names for this region is the *Middle West* or the *Midwest*. Widely used and probably a permanent fixture among America's geographic place-names,[1] this term reflects the fact that Europeans settled North America from east to west. Midwest implies that the West begins at the Appalachians, with the closer Midwest gradually merging into the Far West somewhere in the general area of the Rocky Mountains.

[1] James R. Shortridge, *The Middle West: Its Meaning in American Culture* (Lawrence, University Press of Kansas, 1989).

The Great Plains are clearly as much in the continent's east-west "middle," and many students, when asked to identify the Midwest, point to Kansas, Nebraska, and the Dakotas rather than Illinois, Wisconsin, and Michigan. Only habit and the sequence of European settlement prevent Oregon residents from referring to the region as the Middle East. Nonetheless, "Midwest" will probably continue to be the most widely used and understood name for the U.S. portion of the continent's agricultural core region.

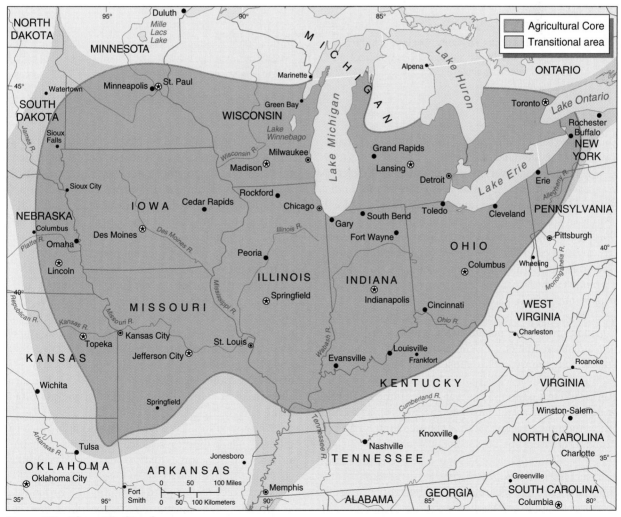

Figure 11.1   The agricultural core.

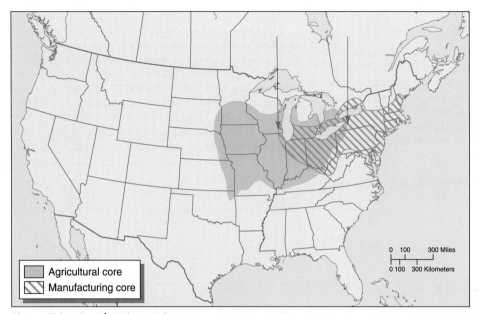

Figure 11.2   **Overlapping regions.** Occupying some of the same territory on the earth's surface, the agricultural core is a small town and farm landscape from the Great Lakes to the Great Plains, while the manufacturing core is an urban and industrial landscape between the Great Lakes and Megalopolis.

businesses joined in to supply the technological hardware farmers demanded.

This regional overlap can be visualized by picturing each thematic region as a separate plane, or flat surface, lying over the appropriate part of North America (Figure 11.2). The manufacturing plane wraps around the lower Great Lakes and extends east toward Megalopolis. The agricultural plane floats at a different level over much the same portion, but it extends farther west and does not reach as far east as the manufacturing plane.

There is another facet of this visualization. The agricultural plane appears to have holes in it where urban and manufacturing functions crowd out agricultural production. But closer inspection shows this to be an illusion and largely a matter of scale and definition. Cook County, Illinois, for example, is almost entirely filled by Chicago and its suburbs and is the second most populous county in the United States. Yet in 2002, there were still 23,836 acres of farmland in Cook County, although this represents a loss of 19,000 acres between 1997 and 2002.

The agricultural core is defined by its aura of small-town, rural America. Though often idealized, especially by outsiders, the cultural tones underlying it are inherently strong and indisputably agrarian. The people are politically and socially conservative, independent, and often friendly to strangers. They cling to what has proven successful and yet will incorporate new technology in their work if the benefits are clear.

Most of the agricultural core's residents are white, descendants of waves of foreign-born migrants who settled in the region until the late nineteenth century. Most of those migrants came from northwestern Europe—Germany, the Netherlands, Scandinavia, and the British Isles. Upon arrival, they laid claim to the prime agricultural lands, and their heirs still dominate the Midwest's rural and small-town areas. Immigrants arriving after the late 1800s tended to come from eastern Europe and the Mediterranean countries. Finding the best agricultural land already occupied, they settled in the manufacturing core's cities.

The agricultural core is rural, and so its character is reflected in its stability, resistance of residents to change, and isolation from forces that create change. It therefore has strong similarities to the agrarian South. Also like the South, recent forces are disrupting the agricultural core's stability. Modifications to the way of life there are now as insistent as elsewhere in North America.

THE ENVIRONMENTAL BASE

The agricultural core's mix of environmental characteristics—rainfall, length of growing season, relief, and soils—allows certain kinds of farming to excel. Initially, European settlers grew wheat. But by the 1830s, they had switched to a variety of corn-based farming schemes that have endured for more than a century.

The schemes may not have altered much, but almost everything else has, especially how farmers plant, run their business, and prepare crops for market, and even the nature of the seed they use. Technology has modified people's relationship to the natural environment just as it has reshaped their urban and industrial activities. Farmers are no exception to this trend. Development and use of industrial fertilizers, herbicides, irrigation, and new plant varieties have expanded areas where crops can be grown and have augmented yields in old ones. Technological innovations have also translated to new tools and farm machinery that have boosted farmer productivity. Some approaches such as controlled environments in greenhouses or research in *hydroponics* show that—for a price—agriculture can be almost as independent of nature as manufacturing.

Climate

Geographer J. F. Hart made a striking remark about the climate in the agricultural core. "For anyone whose aesthetic requirements transcend those of a cornstalk," he said, "the climate is pretty darned miserable. . . . It's no place to have to live, unless you consider making a living the principal purpose of living."[2]

Hart's caustic assessment would probably not have fazed early American and European settlers moving across the Appalachians and onto the eastern interior plains. They were less concerned with a pleasant summer and winter than with survival and pursuit of a livelihood. The environment they found was supportive of their goals.

At the time they struck out, mixed hardwood forests covered most of Ohio, Indiana, and lower Michigan. The trees were familiar, and, to experienced easterners, they indicated moisture and the best soils to grow crops. Aside from being keys to fertile locations, the forests were also sources of fuel and building material. As the

Farming in the agricultural core, as here in northeast Iowa, requires careful application of new methods with traditional crops. The alternative strips of corn and hay reduce erosion while maintaining productivity.

European settlers pushed into Indiana's western margin and on into Illinois and southern Wisconsin, they found openings and glades in the forests that became larger and more frequent in an area people subsequently called the Prairie Wedge. They found, too, that except along rivers and in hillier country, Illinois, Iowa, parts of southern Minnesota, and northern Missouri had equal portions of open grassland and forest.

When later settlers reached north-central and western Iowa, they left dense woodlands miles behind. There, clusters of trees grew only along river courses and in valleys. As people watched the setting sun sink on the horizon, they saw the disc disappear over an open prairie that stretched westward in endless cadence. Unwittingly, they had reached the western margin of the agricultural core and stood at the edge of the Great Plains whose grasses clothed the land to the Rocky Mountains.

[2] John Fraser Hart, "The Middle West," *Annals of the Association of American Geographers* 62 (June 1972): 267, n 14.

In geographers' analytical divisions, the gradual east-west transition from heavily forested land to grassland reflects the annual precipitation settlers could expect at each location. But geographers know, and the settlers probably also knew, that this gauge based on a relationship between vegetation and precipitation is not perfect. Soil and groundwater conditions also govern tree growth. This is why small areas of prairie existed in tree-filled Ohio and Kentucky, for example. The Prairie Wedge, a zone of grassland in Illinois where trees should be growing on purely climatic criteria, was probably caused by human agents— American Indians who set fires from time to time to increase the amount of open land for game.

Contemporary climatology shows that except for the agricultural core's northwest corner and a few sections of Michigan, Ontario, and eastern Wisconsin, the entire region receives an average of more than 75 centimeters (30 inches) of precipitation each year. The core's southern portion can expect in excess of 100 centimeters (40 inches) each year. Important to agricultural core farmers, most of this precipitation falls between the end of April and the beginning of November, consistent with the beginning and the end of the growing season.

It is also important to core farmers that annual rainfall variability be low and its delivery relatively uniform. This makes them less likely to suffer devastating droughts than those raising crops in North America's drier areas. It also means that, even though summer rains often come in intense thundershowers that may bring damaging hail and high winds, agricultural core farmers are not likely to be crippled economically. In contrast, their counterparts in the Great Plains are plagued by wild storms accompanied by high winds, heavy rains, and hail that ravage a percentage of each year's crops.

Rainfall amounts, frequency, and timing, and degree of weather severity are, therefore, all implicated in the health of the agricultural core's economic structure. Also implicated is length of growing season. The spring's last killing frost typically comes in mid-April across the region's southern margins and in mid-May along the northern ones. Throughout, the first autumn frost does not usually arrive until late September, and frost may hold off for several more weeks in southern Illinois, Indiana, Ohio, and northern Kentucky. Regularly extending across 4 to 5 months, the agricultural core's growing season is of sufficient length to permit flexibility in planting and harvesting if late spring rains postpone cultivation and planting by a week or two.

Along with an ample frost-free period, crop growth is stimulated by adequate growing season temperatures. In geographical parlance, the agricultural core has a *continental climate*. The region's wide average annual temperature range, which is the difference between the mean cold month and the mean warm month temperatures, indicates the continental character of the climate (Figure 11.3). Because the agricultural core lies in North America's interior between 38°

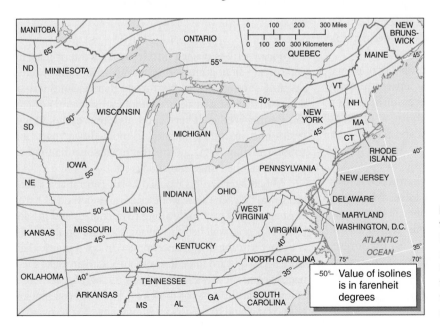

Figure 11.3 Average annual temperature range. The effects of interior continental location and the modest moderating influence of the Great Lakes are evident on this map of the difference between the average temperatures during the coldest month and the warmest month.

and 45° North latitude, its coldest winter temperatures are often as low as those occurring much farther north at sites close to a large water body, such as an ocean. Summer temperatures can climb as high as those found at more southerly latitudes. At Peoria, Illinois, for example, located near the region's center, the average temperature in January is −4°C (25°F), while the average in July is 24°C (76°F).

Although high summer temperatures encourage rapid crop growth, the agricultural core's average non-farm resident is less likely to appreciate summer's miserable combination of hot days, warm nights, and high humidity or winter's long, often gray and uncomfortably bone-chilling weather.

Relief

The agricultural core's landscape rolls gently with few areas of flat or hilly terrain. This low, moderate relief means that farm machinery can be used throughout the region and large fields can be cultivated without risking erosion. Scattered hillocks and stream courses that break up the unending land swells are obvious locations for farmers to maintain woodlots or pasture. The persistently rolling landscape also permits good soil drainage and, in most cases, restricts swamps to small areas. Where land is especially flat, as in the area around the Great Black Swamp of northwestern Ohio, or especially hilly, as in the plateau footlands of the

Appalachians and Ozarks, farm settlement was postponed until the abundant, easier land was taken.

This highly practical, though aesthetically unimpressive landscape is a consequence of the same glaciation that created the harbors of Megalopolis (Chapter 4). During the Pleistocene, as the thick ice mass spread out from its Canadian Shield center, soft sedimentary hilltops were ground down by the glacier's weight and movement and bulldozed before it. The debris, called glacial till, gradually filled valleys between decapitated hills. Left at the glacier's pushing front edge, long and low mounds of glacial till—called terminal moraines—remained as the ice sheet melted and retreated. Today, these moraines offer people who live there relief from the monotony of the interior plain's flatness. Only the southern margins of the agricultural core lie just beyond the glaciers' reach and are the hilliest section of the region, aside from south-central Wisconsin's small Driftless Area.

The tremendous quantity of meltwater released by the glacial retreat was another force creating the region's present landscape. The outflow turned streams into immense scouring pads that eroded valleys for today's major rivers, such as the Illinois River west and south from Lake Michigan and the Mohawk-Hudson rivers east and south from Lake Ontario. The Great Lakes were considerably larger at the end of the Ice Age, submerging extensive areas of what is now dry land south of Chicago, south of Saginaw Bay in Michigan, and the Black Swamp lake plain extending from

The lifestyle of some horse farm owners in the Bluegrass Basin of Kentucky offers an image of Upper South aristocracy in modern trappings.

Toledo, Ohio, to Fort Wayne, Indiana. Much of the land in these places, especially the last, is a *lacustrine plain* (former lake bed). The Black Swamp area is extremely flat and not far above the local water table. While European farmers coveted its deep and rich soils, they forestalled settling there until the techniques needed to drain the area were developed.

Immediately south of the Ohio River in north-central Kentucky is a large basin lying beyond the limit of glaciation. Kentucky's Bluegrass Basin, or Bluegrass Plains, extends the agricultural core's low relief and highly productive farms into the margins of the Appalachian Plateau. Residual *karst* terrain dominates this area, having developed over thick limestone bedrock. Its formation occurred as moving water gradually dissolved the limestone, permitting many major surface features to wear away. Small solution holes (sinkholes) are often scattered on the surface. Underground, the eroding limestone formed stalactite- and stalagmite-columned caves reaching for miles. Mammoth Cave, just southwest of the Bluegrass Basin, is probably the best known complex. This erosional process still imperceptibly sculpts the Bluegrass Plains, but its visible effect on the countryside is to create the same prevalent rolling appearance found throughout the agricultural core's glaciated portions.

Soils

Agricultural core soils fall into two basic groups. With the major exception of central Illinois and south-central Wisconsin, soils east of central Iowa are *alfisols*—almost all of which are in the udalf subcategory (see Figure 2.6). Western soils within the region and those through much of Illinois are *mollisols*.

Alfisols form under conditions of moderate moisture and usually are associated with coniferous or mixed forests. Although the thin, surface soil horizon (the A-horizon) is deficient in humus, the B-horizon has not been heavily leached of agriculturally important minerals (see Chapter 2). In general, for the eastern agricultural core's land to remain productive, the udalfs found there require only that farmers plow in a way to avoid erosion, rotate crops, and apply agricultural lime to neutralize the soil's mild acidity.

Mollisols, superbly suited to grain production, are the most fertile major soil group in the agricultural core. Such soils form under grasses rather than forest cover. They range in color from dark brown to almost black, indicating very high organic content. Mollisols

also tend to be deep, with the surface horizon between 50 and 150 centimeters (about 1.5 to 5 feet) thick. The soil's high organic content comes from the grasslands' dense, tangled root network, with dead roots and grass leaves regularly replenishing it.

Alluvial soils, found in main river valleys and former lake beds, and swamp soils are major exceptions to the two broad soil types in the agricultural core. Alluvial and swamp soils are highly fertile but often require more work by farmers prior to cultivation, such as building deep drainage ditches around fields.

The glaciation that altered prior natural drainage patterns in the agricultural core also left a jumbled landscape of low relief and poor drainage. Small swamps, lakes, ponds, and bogs are scattered across the region, especially in Wisconsin and Minnesota. Most of these wet areas have not been drained because of the expense and the more than adequate quality of the rest of the region's soils.

Waterways

The natural network of waterways permits easy and inexpensive shipment of farm goods to markets and to the countries' main international trade ports.

Movement of European and American settlers into the region occurred earliest along the large waterways (see Figures 3.1 and 3.2). The southern Great Lakes and the Ohio, Illinois, Wabash, and Wisconsin rivers located east of the Mississippi and the Missouri River west of it were the first major routes for settlers as they moved in. Subsequently, the waterways became major routes for shipping produce out.

There are two reasons behind such fortunate accessibility. First, except for the eastern Great Lakes, which offered more direct shipment of goods through the Mohawk-Hudson routeway to New York City, the entire interior river network funneled into the Mississippi system. Second, because topographic relief is low and rainfall regular, the river system is navigable by small boats and barges with only a few interruptions.

The combination of excellent water transportation and productive agriculture contributed to the early growth of several cities in this part of North America. Detroit grew as a military control point and focus for farm products. This city, whose name literally means "the narrows" in French, sits at the best crossing point between Ontario and Michigan and is near where the northern lakes enter Lake Erie. Cincinnati, located at the Great Bend of the Ohio River, became the main collecting and shipping center for products from the agricultural core's southeastern portions. It secured this dominance as early as 1820, and it has never relinquished primacy over that section of the river's trade. Kansas City, at the junction of the Kansas and Missouri rivers, also experienced early urban growth by handling large quantities of agricultural products in river transit. Chicago's location on the southern end of Lake Michigan and close to the upper Illinois River gave the city transshipment opportunities that were supplemented by land connections built west and south across the rich agricultural core.

Initially, the industrial growth opportunities in these and other urban places in the region were direct consequences of high farm productivity on high-quality agricultural land and the natural accessibility provided by the region's waterways. These characteristics are still important in maintaining the cities' vitality.

THE AGRICULTURAL RESPONSE

As settlers gradually moved west across the agricultural core during the early nineteenth century, they created in their wake a wave of wheat production for eastern markets. Wheat was a high-value crop with reliable market demand and familiar to both producer and consumer. Because continuous wheat farming was hard on soils, however, its primary zones of production moved west with the expanding line of settlement.

Shipping raw wheat was not a great problem as long as water transport was continuous. But where the wheat required handling, flour milling quickly became established. Flour mills sprang up at points of embarkation, such as Cincinnati on the Ohio River, or where grain shifted from one mode of transport to another, such as Buffalo, the lake terminus of the Erie Canal. Even after wheat farming shifted onto the plains, these flour milling sites often remained active because they were on the waterways used for shipping.

Farmers who remained behind as wheat production moved on shifted their efforts to their next best agricultural product—meat from domestic livestock. They raised cattle and hogs, with the importance of one over the other varying from year to year. Hogs grow to market size quickly if supplied with sufficient feed; grain-fed cattle also reach market stage more quickly than animals raised on pasture alone. The intensive hog-raising activities of southern Ohio

farmers earned Cincinnati the nickname "Porkopolis" by the 1830s.

Livestock raising and feed grain farming, called mixed farming, became so reliable in the agricultural core that it supplanted wheat production as the dominant farming system. Unlike wheat farming, the mixed farming system was well suited to the natural environment's limitations over the long term. It, too, could use the region's natural waterway network. While alternative farming systems were tried occasionally, none was as persistently successful as feed grain and livestock mixed farming.

Corn, the grain best suited to the agricultural core's environmental conditions, became the region's feed grain of choice. Corn thrives during the region's long hot days and warm nights. But even corn cannot be grown year after year on the same soil, so a 3-year crop rotation scheme became common as early as 1820 in southwestern Ohio; a small grain (usually wheat or oats) and a hay crop (such as clover or alfalfa) are alternated with corn in each field over the rotation period. If a fourth year is needed in the rotation, the land can rest in *fallow* under a recuperative grass, like timothy, and perhaps be pastured lightly. With crop rotation, the land is always contributing to livestock production or cash crop production.

Corn has other advantages beyond its ecological suitability. Yields are high since individual plants can be grown close together, and each plant produces two or more ears of grain. In addition, the large quantity of vegetative matter produced by each plant can be used as feed by adding supplements and cutting it appropriately. Stalks and leaves, chopped finely and stored in large *silos* to permit fermentation, provide winter feed for farm animals. No other plant yields such quantities of usable feed, and corn's dominance has spread to the limits of environmental support (Figure 11.4). This means that along the northern margins of the Corn Belt, as the region came to be called, where summer temperatures do not permit full ripening of the grain before the end of the growing season, the plant is cut green and chopped for *silage*.

Throughout the continent's agricultural core, the mixed farming system of crop–livestock production provides farmers with economic security beyond that found in any other major agricultural region. Certainly, localized specialty crop production elsewhere in the United States and Canada can yield higher profits. But the fact that so many farmers achieve consistently high returns is true only for those in the agricultural core.

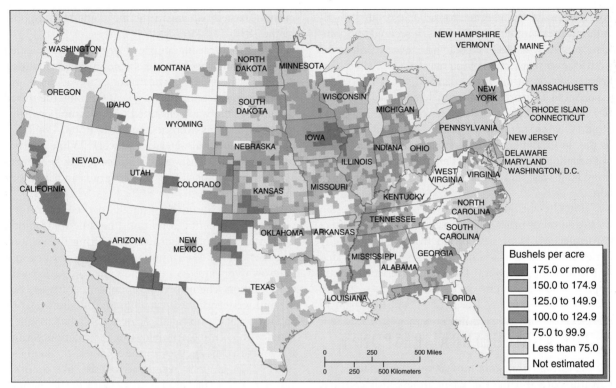

Figure 11.4   Corn for grain, 2002. *Source:* U.S. Dept. of Agriculture, National Agricultural Statistics Service.

The Agricultural Core's Landscape

The rectangular field pattern is distinctive to the central agricultural core's landscape. Although not as rigid as the pattern found across Great Plains states, the area's rural geometry is stubbornly consistent.

The original 13 states developed internal administrative and ownership boundaries in an unsystematic manner, an indication that settlement preceded governmental organization. The *metes and bounds* system of lot designation found in these original states relies on visible landscape features, compass directions, and linear measurement to define land parcels. Irregularly shaped, the land plots were often subject to confused interpretation and litigation.

Through the Land Ordinance of 1785, land north of the Ohio River and west of Pennsylvania, known as the Northwest Territory, was delimited according to a regular, rectangular *township and range* survey before it was opened to settlement by the newly independent United States. A series of east-west baselines and north-south meridian lines were drawn, and township squares 6 miles (9.6 kilometers) on each side were laid out with respect to these lines.

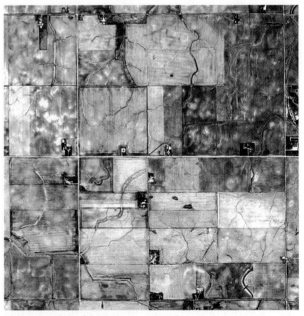

The square township and range land survey system is portrayed dramatically in this vertical aerial photo. Roads were constructed along the edges of many of the 1-square-mile sections, reinforcing the regular survey pattern. This pattern also remains little affected by transfers of land ownership.

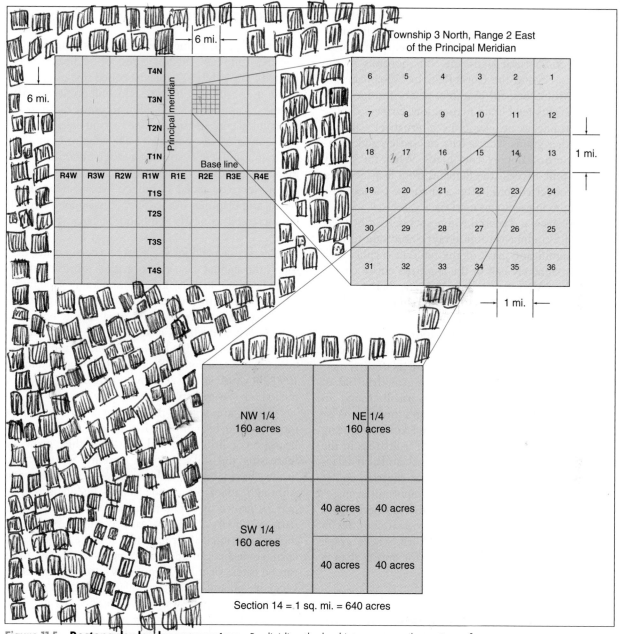

Figure 11.5   Rectangular land survey system. By dividing the land into square-mile sections of 640 acres each, the federal government hoped to standardize the identification of parcels so that land ownership and land transfer would also be more rational and result less often in legal disagreements.

Each township was further divided into 36 sections of 1 square mile (2.59 square kilometers) each (Figure 11.5). The 640 acres (259 hectares) in each square mile was subdivided into quarters of 160 acres (64.75 hectares). These quarter-sections were originally designated the minimum area that could be purchased for settlement (at $1.24 per hectare—50 cents per acre), though the minimum was later reduced to 80 acres (32 hectares) and, ultimately, to 40 acres (16 hectares). The irresistible logic of the township and range system remains visible in the predominantly rectangular road network in the United States between the Appalachians and the Rockies.

The land survey system and the ecological and economic realities of corn-based feed and livestock production lend an inevitable homogeneity to the agricultural core's rural landscape. But portions of the region lie beyond the Corn Belt. In Wisconsin and central Minnesota, for example, where the local climate prevents feed grain maturation, farmers choose dairy farming as an economic alternative. Corn in silage, other grains such as oats and barley, and abundant hay crops support dairy herds. When the supply of fresh milk exceeds the demand from nearby cities, milk is converted into butter and cheese for leisurely shipment to more distant markets.

Wisconsin and central Minnesota, settled by German and Scandinavian immigrants, continue to produce a significant share of the nation's surplus milk and approximately half of the country's entire cheese output (see Map Appendix). A similar belt of dairy farming is located on the Ontario Peninsula north of the province's corn and mixed farming region.

Another distinctive extension of the agricultural core's boundaries beyond the Corn Belt occurs around the western Great Lakes. Just as in the Niagara Peninsula (Chapter 6), farmers grow fruit far to the north of the latitude where it is usually found. In a narrow band along Lake Michigan's shores in Wisconsin and especially Michigan, the lake's moderating influence retards the blossoming of fruit trees in the spring, usually until after the last frost, and also delays the first killing frost in the fall. Sour cherries, apples and, to a lesser extent, grapes are important crops. A similar lake effect is found along the southern shore of Lake Erie, especially the few coastal counties in Pennsylvania and western New York, where grape production has been significant for more than a century.

CHANGES IN THE PATTERNS

By 1890, the mixed farming system based on corn and livestock found to work well in southern Ohio had been carried west to the edge of the Great Plains. Early technological improvements, such as the reaper (1831), the steel plow (1837), and other devices suited to the agricultural core's chief economic activity, ensured the system's success. More recent changes, however, have challenged traditional ways of farming and are creating new geographic patterns in the region.

Soybean Substitution

As late as 1925, farmers harvested less than 200,000 hectares (one-half million acres) of soybeans in the United States. By 1949, they increased soybean acreage to about 4.5 million hectares (11 million acres), with almost 3.7 million (or 9 million acres) of those in the agricultural core and most of the rest immediately south in the lower Mississippi River valley (Figure 11.6). During the next 20 years, soybean acreage exploded to 16.1 million hectares (41 million acres); plantings in the agricultural core exceeded 10 million hectares (25 million acres), and plantings in the lower Mississippi River valley increased to nearly 4 million hectares (10 million acres).

Market demand for soybeans remained greater than supply throughout the period of acreage growth, and soybeans began to be described enthusiastically as the future miracle food. However, during the 1980s, competition from growers in other countries, most notably Brazil, led U.S. farmers to cut back on acreage to keep the total area of soybeans harvested to between 20 and 24 million hectares (50 to 60 million acres). Production increased in the 1990s, however, and by 2000 over 28 million hectares (70 million acres) were harvested nationally.

There are several reasons for the tremendous increase in soybean production. First, as a *legume*, soybeans recondition the soil in which they grow by increasing the soil's nitrogen content. Second, soybeans have broad environmental requirements. Although vulnerable to wide climatic fluctuations, they may be grown throughout the eastern United States, parts of Ontario, and in areas receiving less than 20 inches of rainfall if irrigation is feasible. Third, like corn, soybeans can be used in many ways. The bean can be eaten, or it can be milled to produce an edible vegetable oil and a meal low in fat and high in protein. In the past, the meal was used primarily as a livestock feed supplement, but an increasing amount has now entered people's diets as milk substitutes, meat helpers, cooking oils, and vegetable protein. Fourth, the world food and feed situation maintained export demand for soybeans at high levels. This kept prices relatively stable, encouraging farmers to incorporate soybeans into their crop rotations. Only increased production in other parts of the world tempered soybean farming's expansion in the United States.

Soybean's advantages concentrated a significant share of national production in the agricultural core. Farmers' traditional 3- and 4-year rotations gradually

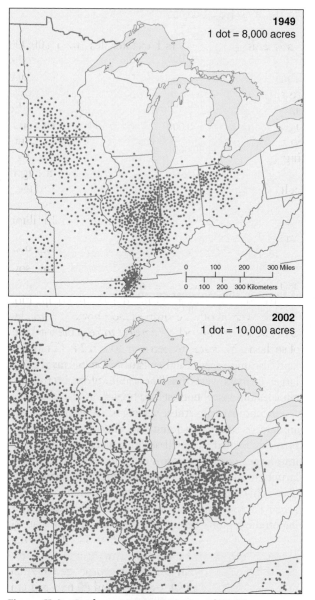

Figure 11.6   Soybean acreage, 1949 and 2002. Expansion of soybean production is clearly evident in what had traditionally been a farming region dominated by corn. *Source:* U.S. Census of Agriculture.

is not necessary. In addition, the crop's high returns per work-hour contributed to its early acceptance by the region's farmers even before the market fully developed.

By 1971, harvested acreage of soybeans exceeded that of corn in southern Illinois and northern Missouri and was greater than two-thirds of the corn acreage in western Ohio and the Grand Prairie region of east-central Illinois. In 1979, for the first time ever in the United States, more acres within the traditional corn-producing sections of the agricultural core were planted in soybeans than in corn. Although hay remains part of crop rotations in the agricultural core, soybeans have proven their worth and will remain significant in the region. The old name of Corn Belt, popularly used to describe the region, should be changed to Corn-Soybean Belt.

Mechanization and Farm Size

A more complex set of changes in the agricultural core's geography grew out of new levels of mechanization and shifts in average farm size. A 1972 study by geographer J. F. Hart found that the amount of land in various farm-size categories remained stable during the first third of the twentieth century.[3] In 1900, about one-third of the farms in agricultural core states were of 73 to 202 hectares (180 to 499 acres), another third were of 40 to 72 hectares (100 to 179 acres), and most of the remainder were smaller than 40 hectares (100 acres). But after 1935, large farms increased across the region, and small farms disappeared rapidly (Figure 11.7).

By 1964, more than 50 percent of the farmland in these states was in farms larger than 105 hectares (260 acres), and by 1992, this share exceeded 75 percent. The largest farms were located in a wide band across central Illinois and northern Missouri and generally along the region's western margins. Smaller farms were more eastern in location—primarily in Ohio, southern Michigan, and eastern Indiana and Wisconsin.

Changes in the core's farm size were caused by the economics of farming and related to ever-greater mechanization. Agricultural core farmers traditionally have taken advantage of mechanical innovations to increase output per work-hour. The region's large fields and gentle terrain permitted farm machinery

gave way to a 2-year corn-soybean rotation. In some southern portions of the core, early maturing varieties of soybeans can be planted in the late spring after a harvest of winter wheat, giving the farmer 3 crops (corn, wheat, and soybeans) every 2 years without significant loss of productivity in any year.

The agricultural core's large, almost level fields are well suited to soybean cultivation. Specialized equipment

[3] Ibid., p. 273.

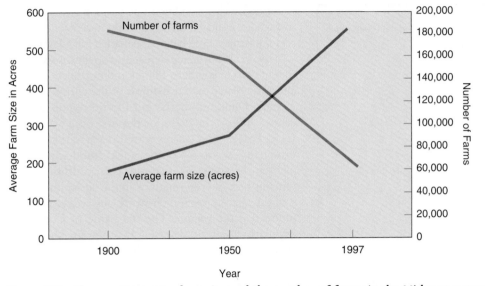

Figure 11.7   Changes in average farm size and the number of farms in the Midwest census region from 1900 to 1997. (*Source:* U.S. Census Bureau, 1997 Census of Agriculture). Small farms were predominant in the agricultural core until the 1930s, but large farms are now most prevalent.

that would have been too large on small, erosion-prone hill farms.

After 1935, various government-backed rural development schemes helped stimulate the transformation. Rural credit was loosened, price support and cropping controls were legislated, and public services were extended into the countryside. Because these programs made it possible for farmers to invest in their operations, tractor-powered machinery became more common on all but the smallest, marginal farms.

The infusion of capital and application of technology to agricultural production had the same results as when capital and technology were applied to manufacturing. Economies of scale favored medium and large farms. A war-related labor shortage during the early 1940s accelerated farm mechanization, and innovations increasingly favored large-scale operations. Two- and four-row equipment gave way to six- and eight-row equipment. Storage and shipment operations became mechanized and more attuned to the requirements of large-volume producers. The main benefits from these changes were garnered by the largest farm operations with fewer benefits going to medium-size farmers. Small-scale farmers were also squeezed by credit limits, since only large-scale farmers could afford the equipment that would give their operation cost efficiency and allow repayment of loans. Consequently, as farmland was gradually accumulated into large acreages, the demand for

land also increased. Higher land prices meant that the larger farmers were those best able to obtain credit or afford additions whenever land became available.

Accompanying these changes in farm size, the amount of land farmed gradually declined. Not all land formerly owned by small-scale farmers was aggregated into larger farm operations. The greatest reduction in farming occurred around the margins of the agricultural core. This suggests that the overall decline in cultivated acreage accompanied a consolidation of successful farming on the region's better land.

Large, nearly flat fields in the agricultural core can be managed by small numbers of farmers as long as they have enough equipment.

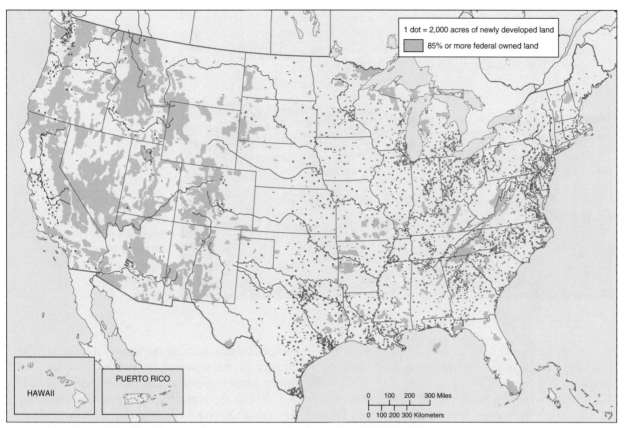

Figure 11.8   Acres of prime farmland converted to developed land, 1982–1997.
Source: U.S. Dept. of Agriculture, Natural Resources Conservation Service.

During the early 1980s, many farm families were forced to leave what had been a way of life for generations. This trend has continued since, although at a slower pace (Figure 11.8). In the late 1970s, some farmers borrowed heavily—at the very high interest rates of the time—to finance new equipment or to lease or buy additional land. Then, as the national economy slowed and the interest rates and prices for farm products fell in the 1980s, farmers could not meet their loan payment schedules and lost their equipment and farms. The early 1980s experience showed that even farmers in the heart of the agricultural core could not ride out wide economic fluctuations while attempting to adapt to the necessary long-term structural changes underway in North American farming.

The Family Farm and Ownership Changes

Deep in American folklore, and to an extent in Canadian, is the almost sacred notion of the family farm. Individually owned and family operated, the idealized farm is passed down from generation to generation with only a succession of hired hands included in what remains a family operation. This image has not been disturbed by the gradual increase in rural nonfarm residents. Brushed over with shades of independence and nature's splendor, the image is embraced by the large numbers of urbanites who either move to the country in pursuit of its idealized version of life or who long to.

Despite exaggerations, the image fit farmers in the agricultural core prior to World War II as well as it fit farmers in any major region. But these days, the fully owned, individually operated, midsize family farm is in swift decline. Relatively self-sufficient dairy farms in Wisconsin—perhaps with an orchard, a vegetable patch, a few hogs and chickens to complement the cows, and a rollicking mixture of children and dogs—approach the nostalgic portrayal better than do farms in most areas. If the current ownership trend

Dairy farms in the agricultural core, such as this one in Wisconsin, paint a bucolic landscape that contains all the idealized elements of the region's culture even as it disappears elsewhere in the region.

continues, however, only a small proportion of the productive farms in the core will soon fit the cultural image.

The decline of the modest family farm is associated with demands for greater farm efficiency. When farm efficiency was defined by hard work, alertness to detail, and rigid adherence to traditionally reliable methods, success was controlled directly by individual farm families. But as large-scale farming became more important, national and international economic swings intruded. A farmer's ability to obtain credit on favorable terms, to survive fluctuations in bank interest and crop prices, and to use large equipment on larger acreages shifted the definition of farm efficiency and agricultural success. Personal effort and individual integrity still contribute to a farm's success, of course, but scale is increasingly critical. Larger operations do better on the average, and midsize family farms are challenged not to follow smaller farms into extinction.

Agricultural core farmers, like those in other sections of the United States and Canada, are adapting to these pressures in several ways. Some farmers rent or lease additional land rather than purchase it outright. Rented land may be part of a nearby farm or several sections of farmland in scattered locations (Figure 11.9). Rented land appears in the census as a separate small farm operated by what is called a part owner, thereby masking the pace of small operation decline. About one-third of those renting farmland lease it from a relative, thereby keeping the land in the family.[4] Other farmers are full tenants, choosing to work for a landowner through one of several arrangements.

Within the agricultural core, tenure characteristics reflect the new economic reality. Of the eight agricultural core states, the three at the region's center have the lowest proportion of their farmland operated by full owners (Table 11.1). Approximately three-fourths of all farmland in Illinois and Indiana, and almost that proportion in Iowa is used under part-time or tenant-farming arrangements, a share greater than that found in any other state. Average farm-size differences also illustrate tenure patterns; farm-size averages are much lower for full-owner operatives in every state.

The same pattern holds for farm product market values per farm. The accumulation of invested capital through part-ownership agreements raises the average market value of part-owner farm combinations to three-to-four times that of single-owner farms, even in Wisconsin. Part owners can use immense equipment over vast farm acreages, justifying the additional thousands of dollars required by large-scale farming.

[4] Bruce B. Johnson, *Farmland Tenure Patterns in the United States* (Washington, DC: U.S. Department of Agriculture, Economic Research Service, 1974.)

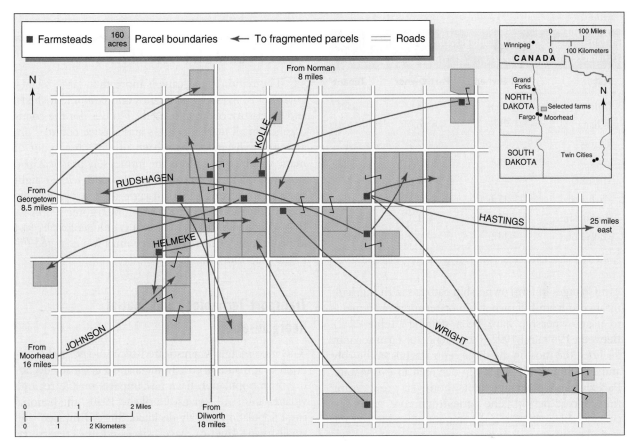

Figure 11.9 Associated parcels on fragmented farms. (From E. G. Smith, Jr., *Annals of the Association of American Geographers*, Vol. 65, 1975, with permission.) This example clearly demonstrates some of the scattered, noncontiguous land parcels owned and worked by individual farmers. The pattern of fragmentation has increased since this study was done.

TABLE 11.1

	Land in Farms			Average Farm Size			Market Value of Goods Sold		
State	**Full Owner**	**Part Owner (percent)**	**Tenant**	**Full Owner**	**Part Owner (acres)**	**Tenant**	**Full Owner**	**Part Owner ($1000)**	**Tenant**
Ohio	32.7	59.9	7.4	91	419	226	30.1	121.5	74.0
Indiana	23.4	68.2	8.4	90	584	350	36.3	179.5	119.5
Illinois	21.6	65.7	12.7	149	706	426	43.8	205.7	137.5
Iowa	27.2	60.7	12.1	173	636	355	77.8	247.5	137.2
Missouri	47.0	46.9	6.1	186	545	330	31.6	96.2	68.5
Michigan	32.8	62.3	4.9	90	445	246	29.7	177.2	147.5
Wisconsin	40.6	54.8	4.6	122	407	202	32.9	179.9	93.3
Minnesota	36.1	56.3	7.7	193	643	386	64.3	202.1	139.9

Title row: *Farm Tenure of Principal Operator: Tenure Characteristics of Farms in Agricultural Core States, 2002*

Source: U.S. Department of Commerce, *2002 Census of Agriculture.*

TABLE 11.2

Changes in Percentage of Land in Farms by Tenure Characteristics, 1969–1992			
State	Full Owner	Part Owner	Tenant
Ohio	−7.8	+13.3	−5.5
Indiana	−9.7	+16.0	−6.3
Illinois	−7.2	+17.1	10.3
Iowa	−10.5	+18.8	−8.3
Missouri	−7.0	+9.6	−2.7
Michigan	−17.6	+17.1	−1.3
Wisconsin	−21.1	+21.5	−0.3
Minnesota	−13.0	+14.9	−1.9

Source: Calculated from U.S. Census of Agriculture, 1969 and 1992.

Changes in farm ownership patterns in the agricultural core from full ownership or tenant arrangements to part ownership have been striking (Table 11.2). Between 1969 and 1992, the practice of farm tenancy declined the most in the three central states in the core and in Ohio. Full ownership declined in every state. Part-ownership arrangements, however, grew during this 23-year period. The agricultural core states saw considerable increases in the amount of land leased and rented to others to farm.

SETTLEMENT PATTERNS

The region's landscape is more than large rectangular fields, livestock pens, silos and equipment sheds, and a network of straight roads. Houses dot the countryside. Small market centers and service counties are at regular distances from each other. The layout of many small towns reflects the immensely practical grid survey system used to parcel out the land, with straight streets intersecting others at right angles. Historically etched, this settlement pattern is undergoing a process that will change the agricultural core's geography just as profoundly as increased mechanization changed the fabric of farming.

Transport Technology and Spatial Reorganization

The increasingly sophisticated technology at the disposal of agricultural core farmers and a gradually declining farm population have had impacts on the region's villages and small towns. Until the 1950s, the national rural population steadily declined. Since then, rural and small-town America have witnessed the reverse as

Barns, such as these in Central Iowa, are an integral part of the cultural landscape of the agricultural core.

The adaptability of the Agricultural Core is seen in the rapid increase in ethanol production plants, such as this one near Goldfield, Iowa.

increasing numbers of city-dwellers move to the countryside and commute (or telecommute) to work. This pattern of rural growth holds true everywhere except in the agricultural core and the Great Plains.

Both the national growth in rural residents and the core's stasis are related to changes in transport technology. During the 1930s and 1940s, as rural roads were paved, farmers began to afford reliable trucks to carry livestock or farm produce to market. In economic terms, transport costs declined as transport efficiency increased. The improved road network allowed more rapid movement, so farmers carried produce to markets farther than they could before. Farmers began trucking produce past the local village to larger market centers or livestock yards. Deprived of a portion of their consumers, the smallest communities felt consequences of their own; they did not grow, and many lost the small number of residents they had.

A study of population change between 1910 and 1960 in 15 rural counties in northern Ohio and Indiana[5] found a direct association between village size and population increase. The study concluded that the larger the town in 1910, the greater the town's population increase over the next 50 years. Small communities with fewer than 1000 residents in 1910, however, lost population across this period.

The decline of rural areas began to change nationally in about 1970 when urban residents found it easy to move to the country and yet still work in town. Rural areas and nonmetropolitan places in the eastern United States and Canada began growing as city dwellers moved their homes beyond the urban fringe. Improvements in Internet access will accelerate this re-peopling of rural landscapes.

The agricultural core lagged behind other parts of the country in this trend. Early in the period, demographers Fuguitt and Beale found that "most resurgence of growth in very small places appears to have taken place in the South." And "it is essentially in the North Central states, where there are hundreds of very small incorporated places, that the notion of 'dying' small towns comes closest to reality."[6]

The efficiency that made agricultural core farming so successful, together with the successful adaptations its farmers made to part-time operation, may have slowed the flow of urbanites to village and country settings in the region.

---

[5] S. Birdsall, unpublished manuscript, 1962.

[6] Glenn V. Fuguitt and Calvin L. Beale, *Population Change in Nonmetropolitan Cities and Towns, Agricultural Economic Report* no. 323 (Washington, DC: U.S. Department of Agriculture, Economic Research Service, 1976), pp. 4, 6.

A VILLAGE CASE STUDY

Clarksville, Iowa

The Clark family traveling west by wagon with several other families arrived in northern Iowa in July 1852. The Clarks and the others had left their homes in north-central Indiana to seek new and better land across the Mississippi River. The small group decided on a site in Butler County, and it is from this modest beginning that Clarksville, Iowa, marks its origins. There are hundreds of small towns and villages like Clarksville in the agricultural core (Figure 11.10). The fortunes of such settlements represent the region's broad characteristics.

Clarksville's first railroad arrived in 1871, connecting town markets to Cedar Rapids, Chicago, and the rest of the growing country. A second rail line was completed in 1879, adding further insurance to the town's future.

As Iowa became more fully settled in the decades following the Clark family's arrival, Clarksville grew slowly but steadily. Its population reached 895 in 1910, 1328 in 1960, 1424 in 1980, but increased only slightly to 1441 in 2000. While the village is incorporated and has its own banks, schools, and government buildings, it remains basically a service and market center for nearby farmers.

Clarksville is not served by any of the region's main highways. This kept the village's growth slow, even for one in the agricultural core, and this contributes to the citizens' open curiosity about any strangers. Individuals arriving in town by road are recognized immediately and are expected to be there for a reason; few would drive through town when traveling between major destinations. A recent visit to Clarksville elicited several friendly inquiries of, "Who in town are you here

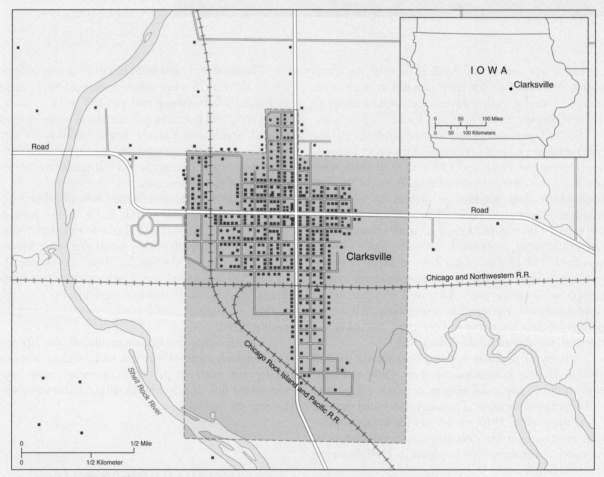

Figure 11.10   Clarksville, Iowa. This representative agricultural core town has changed little despite the economic forces altering the farming community that supports it economically.

Clarksville, Iowa, like other small towns of the agricultural core, derives much of its quiet, steady character from its history of slow—or nonexistent—growth. Clarksville's downtown is only a few blocks long, and its small, square residential blocks have only six or eight homes. From anywhere in town, the surrounding countryside is less than a 10-minute walk.

to visit?" and clear surprise at the response, "No one; just passing by."

Although Clarksville has continued to grow slowly, the potential for decline is suggested by the ratio of females to males in town and the age structure of that population. There were 778 females in Clarksville in 2000 compared to only 663 males. The median age of Clarksville's population (about 38.8 years in 2000) was also almost 5 years older than the median age for the entire state (34 years). And 21.6 percent of the town's residents were 65 years old or older in 2000. (The state proportion was only 14.9 percent.) It appears that a significant number of young people leave after high school—probably in search of greater employment opportunities. A somewhat elderly population remains behind.

Clarksville's urban form is typical of agricultural core farm communities. The square and rectangular residential blocks are laid out around the two through streets, Main Street and Superior Street. The pattern's regularity ignores the area's modest topographic variation. Clarksville's small business district clusters about the intersection of Main and Superior,

but neither road is so busy that the intersection requires a traffic light. Homes on large lots retreat outward from this center at low densities. With a small six-block exception to the north, the village boundaries form a neat $2 \times 2\frac{1}{2}$-mile rectangle, although Clarksville's incorporated area is far from filled by the village's home lots. The square land survey system also extends well beyond the village's limits. Only the railroads and short sections of the two main through roads deviate from straight, north-south/east-west section boundaries. A large sign pointing to Clarksville's amenities, including the newly constructed golf course, was recently erected on State Highway 3 just south of the village. Both the golf course and the sign are modest attempts to widen Clarkesville's reach.

Like thousands of other small communities in the agricultural core, Clarksville is being affected in small ways by changes in technology and the region's economy. A peaceful, attractive, friendly place in which to live, Clarksville offers few of the locational advantages of communities with greater access to local centers of economic activity.

TECHNOLOGY CASE STUDY

The Dorr Farm, Marcus, Iowa

Computers have infiltrated every conceivable element of agriculture, influencing what technology-savvy farmers like [Tom] Dorr grow, how they grow it, and how they market the fruits of their labor.

The terminal beside Mr. Dorr's desk, for instance, links him to DTN, a nation-wide agricultural and weather data network. There is also his personal computer and printer, which is part of a local area network connecting five computers and a server in this small clapboard building. Formerly the home of a tenant worker, the office is now the information hub of 3,800 acres of northwestern Iowa prairie where Mr. Dorr and his 11 full- and part-time employees raise corn, soybeans, and hogs, sell seed and run a grain elevator that serves his and neighboring farms.

With gross revenue of about $2 million in most years, the Dorr operations rank among the 4 percent of the largest commercial farms that account for 50 percent of the nation's agricultural output. Such commercial-scale farmers are usually among those most active in experimenting with new equipment and management techniques.

. . . Mr. Dorr is doing what thousands of other American farmers are doing: using machinery laden with electronic controls and sensors to achieve pinpoint seed spacing, analyze soils for moisture and nutrients, track weather, and manage the rates at which fertilizer and pesticides are applied. He has experimented with global positioning via satellites to track exactly where each machine is as it carries out these functions.

And come harvest season, still other devices will calculate crop yields in real time.

Mr. Dorr . . . has a love of the soil . . . But [he] does not let agrarian sentimentality befuddle his business acumen. The family farm he grew up with was part of an agricultural enterprise that besides livestock and crops, included a feed store and turkey hatchery.

[Mr. Dorr's vision is] of a 225,000-acre operation made up of three "pods," each with its own manager but sharing an information system back at farm headquarters . . . a farm sprawling over thousands of individual fields—many of which might be only partly owned by Mr. Dorr and his relatives, while others could be rented, either for money or for a share of the crop.

His information system would know what was grown in each field in the past and how much it yielded under different growing conditions. It would also know about crucial characteristics of the field like irrigation, drainage, and soil.

The system would also have constantly updated information on available labor, machinery, and supplies. Operations like storage, marketing, and distribution would be tied in, so that the past and the projected profitability of each field would be constantly visible. . . .

Assembling this digitally enhanced megafarm would require . . . at least a $2 million technology investment. Put it all together, though, and one can envision a farm that rearranges planting or harvesting on the fly as weather changes or new sales opportunities arise.

Source: Barnaby J. Feder, "For Amber Waves of Data," *New York Times*, May 4, 1998, pp. D1, D4. Copyright © 1998 by The New York Times. Reprinted by permission.

Review Questions

1. What are the major physical geographical factors that enhanced the establishment of an agricultural core in North America?

2. Why is the population of the agricultural core predominantly of European heritage?

3. Describe the rural landscape that dominates the region.

4. What has prompted the tremendous increase in soybean production during recent decades?

5. How have farm ownership patterns changed over the past several decades?

6. What makes Clarksville, Iowa, typical of farm communities in the agricultural core?

7. Describe the impact of recent technological innovations on farm operations, as in the case of the Dorr Farm in Marcus, Iowa.

CHAPTER 12

THE GREAT PLAINS AND PRAIRIES

Preview

1. Over the past 450 years, views of the Great Plains have shifted between a dry wasteland to a land of rich agricultural opportunity.

2. The region has some of the least variation in vegetation and topography of any region in North America.

3. Severe weather, including tornadoes, thunderstorms, and hail, is widespread across the Great Plains.

4. Agriculture dominates the Plains economy and is characterized by large-scale operations and a high reliance on heavy machinery.

5. The control and management of water is the most important natural resource issue in the region.

This is a land of large landscapes, where gentle topography and limited tree cover open distant vistas. Farmers in the Great Plains often plant wind breaks of trees and bushes around their cluster of buildings to protect against wind and blowing snow.

Historian Walter Prescott Webb suggested in his book *The Great Plains* that the great, continuous grasslands located astride North America's center were so unlike the home regions of northwest European settlers that before they could live in the Great Plains, they had to develop new crops, new land-use patterns, and new technologies. The agricultural crops and settlement patterns they carried with them from Europe were inappropriate (Figure 12.1).

European settlers encountered several environmental surprises on the Great Plains. Rainfall in the grasslands was far less than in the forested East, and the Plains' western margins were much drier. Violent storms accompanied by high winds, hail, and tornadoes were common. Blizzards' wintery blasts intensified the season's worst cold and mounded snow into immense drifts. In summer, the soil was parched by hot, dry winds, sometimes carrying dirt away in huge, billowing clouds of dust. Small, intermittent streams were the only source of water able to support the region's sparse trees. Farmers newly arrived from eastern states had to adjust. No longer were plentiful water for crops and animals and ample wood for building, fencing, and heating available. They faced conditions on the Great Plains for which they had little reliable information.

Europeans and Americans had formed opinions about the Great Plains and prairies. Though strongly held, their notions were contradictory, and like many opinions, they were only modestly useful to European settlers who had to deal with practical problems in an unfamiliar environment. What were those contradictory images of the Great Plains? How have the images changed? What were the real problems facing European settlers as they moved onto the plains, and how have their solutions fared? What are the Great Plains and prairies like now?

IMAGES AND IDEAS OF THE PLAINS

In 1540, when the Spanish explorer Francisco Vázquez de Coronado wandered into south-central Kansas in search of Cíbola, the mythical cities of gold, he had no more success in finding mineral wealth there than elsewhere. Even so, he was struck by the richness of the Kansas land he encountered: "The country itself is the best that I have ever seen for producing all of the crops of Spain," he wrote, adding that the land was "very flat and black" and "very well watered by the rivulets and springs and river."[1]

Kansas was a great distance from Spain's colonial core in central Mexico, however. The lack of gold and

[1] Walter Prescott Webb, *The Great Plains* (New York: Grosset and Dunlap, 1931), p. 107.

Figure 12.1   The Great Plains and Prairies.

silver meant this land with its attractive environment, like the rest of the Great Plains, was never occupied by Spain.

Centuries later, Americans exploring land west of the Mississippi River found an endless, almost flat land covered with grass. They formed an image of the Great Plains that contrasted sharply with Coronado's view. Americans tended to call this land a desert, so different was it from the East's wooded countryside. "Almost wholly unfit for cultivation, and of course, uninhabitable for a people depending upon agriculture for their subsistence" was the reaction of Major Stephen Long during his 1821 expedition across the plains.[2]

The Oregon Trail crossed the Great Plains west of Kansas City. For many settlers who took the long trek to the Willamette valley in Oregon, the trail was a way of getting past the Plains' unfriendly lands for the more familiar demands of the Pacific Northwest's wooded territory. American geographies published during the middle decades of the nineteenth century labeled the Great Plains "the Great American Desert"; for many this definition was acceptable.

Both Coronado and Long were describing the same part of the continent, the area now known as the southern Great Plains. Their differing experiences and environmental backgrounds influenced their perceptions of the area they were visiting.

Coronado was raised on the dry, treeless plateau of central Spain. He then spent additional time in the equally treeless and drier lands of central and northern Mexico. To him, Great Plains vegetation seemed lush. Compared to his previous experiences, the streams in Kansas were abundant and flowing. The absence of trees was irrelevant to his assessment of the Great Plains' agricultural potential.

Major Long was raised in the humid, forested eastern United States and had a different perspective. To Long, the streams seemed sparse and inadequate. To a person accustomed to well-wooded countryside, the Great Plains' lack of trees was shocking. Without trees, how could anyone appraise which land was good for agriculture and which was not?

Such variation in perceptions of environmental quality is not strange. Numerous studies of human perception emphasize the importance of an individual's experience on his or her perception of an environment. New information about a place is filtered and evaluated; previous experience is the filter's framework.

[2] Ibid., pp. 156–157.

In a related way, a person's reaction to natural hazards is directly related to how frequently the hazard occurs and how accurately the individual perceives the frequency. In a study of floodplain development, for example, geographers found that where flooding was rare, residents paid little heed to its likelihood. Where flooding was frequent, on the other hand, residents paid special attention to prevention measures when planning new development. Similarly, in a study of drought perception in counties extending from central Kansas to eastern Colorado, geographers found that farmers in the eastern part of the study area, where drought is less frequent, consistently underestimated how often it occurred. Toward the west, the discrepancy between image and reality declined; in eastern Colorado, farmers' expectations of drought frequency was matched by its occurrence.

Given that previous experience contributes to perception in a major way, it is not surprising that Spaniards saw the Great Plains in a favorable light. These explorers were familiar with the physical environment found on the Great Plains; it mirrored that of their homeland. The differences—a little more water, a little more grass—were small and positive. For typical eastern Americans, such as Major Long, the difference between the Great Plains and the landscapes they had experienced was more radical, and their reactions were strongly negative.

THE GREAT PLAINS AS A REGION

The Great Plains are a recent addition to popular geographies. Even today, the idea of the Great Plains is vague for most of its residents. Asked to name the physical region in which they live, residents usually give a more local area than "the Great Plains." Thus, a Texan may say he lives on the Staked Plains. A Kansan will reply that she lives in the Gypsum Hills. A Nebraskan may reply "the Platte Valley" or "the Sand Hills."

The Great Plains as a broader region is substantially an academic invention of the twentieth century. It is an idea used to frame responses to the widespread economic and environmental problems that developed in this part of the United States during the Depression years of the 1930s. Built on the powerful image of "the Great American Desert" and reinforced by literature and film, the Great Plains entered Americans' conceptual framework of regions.

One major result of the development of the Great Plains idea as a broad umbrella term has been extension

The plains and prairies landscape is more varied than commonly imagined. Badlands, such as these in South Dakota, result from the erosion of sedimentary rock by the action of water and wind.

of the belief that the Great Plains environment is uniform and monotonously flat. This verdict is reinforced for many travelers crossing the region by car. Two of the most traveled east-west routes across the United States, I-40 and I-80, pass through the Staked Plains and the Platte River valley, respectively. These areas are flat and arguably the least attractive parts of the Great Plains.

In truth, the Great Plains and prairies possess substantial variation in landscape. The Sand Hills of Nebraska, the Badlands of South Dakota, the Alberta uplands, the Manitoba lowland—these are a few examples of Great Plains landscapes that are different both geologically and in their appearance.

Another misconception of the Great Plains and prairies is that all its residents share the same ethnic background. This, too, like images of a homogeneous physical environment, is not the case. Many parts of the region are strikingly diverse, especially in Canada where recent immigration has played an important part in the prairies' population growth. Scandinavians settled in the northern U.S. plains. Eastern Europeans were important in many areas of the plains, as illustrated in Willa Cather's fine novels of pioneer settlement in Nebraska. Mexican Americans are a major part of the population of Colorado's Piedmont. Today, Winnipeg is one of Canada's more ethnically diverse cities. Large numbers of American Indians also live on the Great Plains, especially in South Dakota and

Oklahoma. In spite of its size and shared landscape features, the Great Plains and prairies is a region of great diversity, both physically and culturally.

GRASSLANDS LITERATURE

Yes, we buried him there on the lone prairie,
Where the owl all night hoots mournfully,
And the blizzard beats and the winds blow free
O'er his lonely grave on the lone prairie.

From the folk song,
"The Ballad of the Dying Cowboy"

The strongest images of a place are often portrayed in the songs, folktales, and literature set there. The environment may be used as an ever-changing stage for the novelist's characters; the landscape is scenery or a backdrop for the story's human actors. However, many tales do not depict a place's landscape as passive. The land influences and molds the characters' thoughts and actions and may be altered by those actions. Thus, the landscape becomes a vibrant, even a dominant, element in regional fiction.

Authors from the Great Plains, like those from the South, have produced an uncommonly large volume of works in which the grasslands environment is strikingly painted. Hamlin Garland, raised in South Dakota in the late nineteenth century, introduced one representative theme in his novel *Main-Traveled Roads* (1891):

> How poor and dull and sleepy and squalid it seemed! The one main street ended at the hillside at his left and stretched away to the north, between two rows of the usual village stores, unrelieved by a tree or a touch of beauty.

The treeless, desolate landscape and its impact on individuals is a theme repeated in many Great Plains novels. Beret, the wife in O. E. Rölvaag's *Giants in the Earth*, a magnificent tale of a Norwegian frontier settlement on the Minnesota–Dakota border, would look at the treeless horizon of her new home and realize "the full extent of her loneliness, the dreadful nature of the fate that had overtaken her." She wondered: "How will human beings be able to endure this place? Why, there isn't even a thing one can hide behind." In the story, her husband eventually dies in a blizzard. The isolation and barrenness of her life drive Beret insane.

In *The Grapes of Wrath* (1939), John Steinbeck viewed an Oklahoma wracked by a decade of economic depression and years of severe drought.

> The right of way was fenced, two strands of barbed wire on willow poles. The poles were crooked and badly trimmed. Beyond the fence, the corn lay beaten down by wind and heat and drought, and the cups where leaf joined stalk were filled with dust.

To emptiness, then, is added the theme of a mercurial, often devastating climate.

Marie Sandoz developed the idea of climatic hazard in *Old Jules* (1935). In this biography of her father in the Sand Hills of Nebraska, Sandoz describes her father's pride in an orchard he had planted. Then a hailstorm struck.

> Suddenly the hail was upon them, a deafening pounding against the shingles and the side of the house. . . . One window after another crashed inward. . . . Jules came in from the garden and sat hunched over a box before the house. All the trees were stripped, bark on the west and south, gone. Where her garden had been Mary planted radishes and peas and turnips as though it were spring. The corn and wheat were pounded into the ground, the orchard gone, but still they must eat.

Despite such random malevolence, some writings set in the Great Plains and prairies offer portaits of opportunity and stark beauty in the land's emptiness. Per Hansa, Beret's husband, in Rölvaag's *Giants of the Earth*, rushed to his new field to begin breaking sod with a "pleasant buoyancy" and felt "the thrill of joy that surged over him as he sank the plow into his own land for the first time." In Willa Cather's *My Ántonia*, a parallel is drawn between the plains and the sea.

> As I looked about me I felt that the grass was the country, as the water is the sea. The red of the grass made all the great prairie the colour of winestains, or of certain seaweeds when they are first washed up. And there was so much motion in it; the whole country seemed, somehow, to be running.

THE PLAINS ENVIRONMENT
Relief and Vegetation

The Great Plains topography and vegetation are among the least varied—some would say most boring—in Canada or the United States. The region lies entirely within the physiographic province called the interior plains (see Figure 12.1). The land's elevation increases gradually, almost imperceptibly, from east to west. Along the Great Plains' eastern margin, the land lies only 500 meters (1600 feet) above sea level. Denver, located at the region's western edge, is known as the "Mile-High City."

The apparent uniformity of the plains and prairies is partly a function of the great extent of the region's physiographic subdivisions. The High Plains is the Great Plains' largest subdivision, occupying the Great Plains' western margin between the Edwards Plateau in south Texas and southern Nebraska. Covered by a thick mantle of sandy, porous sediments, the High Plains is rather flat. Some parts of it, such as the Llano Estacado (Staked Plains) in west Texas and eastern New Mexico, are extremely flat. Only along streams, such as at Scottsbluff on the North Platte River in western Nebraska or at Palo Duro canyon on the Red River in the panhandle of Texas, has erosion resulted in substantial local relief. Formerly occupied by the largest of the Pleistocene lakes (see Chapter 2), the exceptionally flat Lake Agassiz Basin includes much of southern Manitoba, eastern Saskatchewan, and the valley of the Red River of the North in North Dakota and Minnesota.

Not all portions of the plains and prairies are flat. The Black Hills, associated both geologically and topographically with the Rocky Mountains to the west, are an obvious exception in South Dakota and

Wyoming. Other, smaller outliers, such as the Cypress Hills in southern Alberta and Saskatchewan, dot the northern plains' western margins. The Sand Hills in central and northwestern Nebraska offer a dense, intricate pattern of grass-covered sand dunes, many of which are well over 30 meters (100 feet) high. Badlands topography—extremely irregular features resulting from wind and water erosion of sedimentary rock—is widespread on the unglaciated Missouri Plateau from northern Nebraska northward to the Missouri River. The best examples of this desolate, irregular topography are found in the Badlands National Park in western South Dakota and the Theodore Roosevelt National Park in North Dakota.

Not surprisingly, the characteristic vegetation of the grasslands is grass. But this is a deceptive oversimplification ignoring the variety and distribution of natural grasses. A continuous tall-grass prairie, with grasses 30 centimeters (1 foot) to 1 meter (3 feet) high, covered the moister eastern portions before agriculture destroyed much of the original grasslands vegetation. Big bluestem was perhaps the most characteristic grass. Grasses were shorter in the drier climates toward the west, until continuous prairie grasses gave way to bunch grasses, such as grama grass, along the dry steppe margins of the western Great Plains.

The prairie grasses developed deep, intricate root systems commonly extending deeper into the soil than the grass blades reach above. Although the tangled root system benefited soil quality by adding decayed organic matter, it made the prairies exceptionally difficult to plow. The first settlers broke the sod by employing "bonanza teams"—heavy plows pulled by as many as 20 animals. Prairie sod could also be cut into large bricks; during the early period of Plains settlement, Europeans used them to build sod houses.

The Cyclical Pattern of Plains Precipitation

Most precipitation on the Great Plains is caused when cool, dry air masses from the continent's northern interior conflict with warm, moist air moving onto the Great Plains from the Gulf of Mexico. The tropical maritime air flowing in from the Gulf of Mexico commonly curves up the Mississippi valley before moving northeast, missing much of the western and northern Great Plains entirely. Thus, the mechanism producing precipitation is less strong in the region's western half and toward the north, and less rain falls there. In Kansas, for example, average annual precipitation varies from a moist 105 centimeters (42 inches) in the southeast to a semiarid 40 centimeters (16 inches) in the

Few regional boundaries are as obvious as that between the Plains and Prairies and the Empty Interior along the eastern front of the Rocky Mountains.

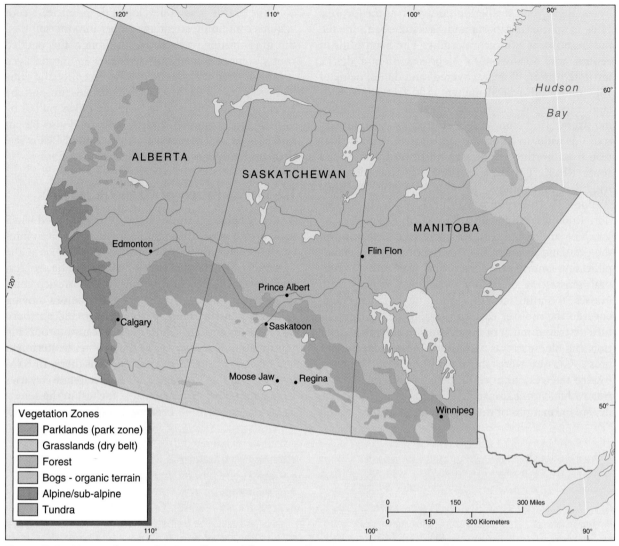

Figure 12.2   Vegetation on the Canadian Prairies. The southwestern corner of the Canadian prairies is a dry grassland, and ranching is the primary focus of the rural economy there. The better-watered parklands to the north and east of the grasslands are a major focus of the prairie agricultural economy.

southwest. Air masses crossing the American Southwest also interact with continental air masses, but air arriving from the arid Southwest usually carries less moisture and creates much less precipitation on the plains.

The Canadian grasslands' precipitation pattern is markedly different (Figure 12.2). The Great Plains' east-to-west zonation of southern latitudes is replaced by one oriented northeast-southwest. The dry grassland to the southwest is limited to use as pasture. But a mixed tree-grass area to the north and east, called the Park Zone, can be cultivated.

Periods of higher than normal precipitation on the Great Plains occur when tropical air masses move northwestward from the Gulf of Mexico. This provident current is far from dependable, however. Annual precipitation amounts vary widely, although usually falling between 80 and 120 percent of the average (Figure 12.3). During severe drought years, such as during the mid-1930s, precipitation dips to less than 76 percent of the long-term average. Fortunately for Great Plains farmers, about three-fourths of a year's precipitation generally falls between April and August when crops need it.

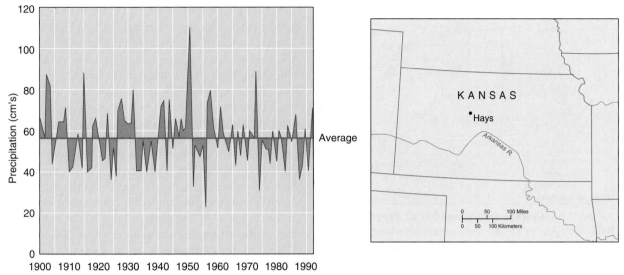

Figure 12.3 Precipitation variability in Hays, Kansas. (Courtesy of Charles Konrad.) Wetter-than-average periods are regularly followed by drier-than-average periods in Hays. Note especially the 1930s drought, which was preceeded and followed by stretches of above-normal precipitation.

Major droughts have occurred in approximate 20-year cycles, with the last six centering on the 1890s, 1910s, 1930s, 1950s, 1970s, and 1990s. For many Great Plains farmers, widespread optimism and expansion during above-average rainfall periods was followed by economic calamity during drought years.

During World War I, when high prices for wheat coincided with abundant precipitation, farms were expanded by putting marginal land in grain. Farmers moved into the dry, southwestern edge of the main wheat-growing area in Kansas and Oklahoma. When wheat prices collapsed in the early 1920s, economic recession hit the Wheat Belt. Farmers responded by planting even greater acreage to wheat on more marginal land in an attempt to maintain overall income. Then, when the Depression of the 1930s coincided with a succession of severe droughts in the Great Plains, economic ruin hit farm families. Crop yields, not high even in good years, shrank to nearly zero.

One environmental consequence was the infamous Dust Bowl. The soil-binding quality of the land was destroyed by years of repeated cropping. Dunes 3 meters (10 feet) or more high were piled up by the dry winds. The sky was often heavily laden with dust. As a result, large numbers of destitute farm families left their land on the southern Plains and headed to the new promised land, California.

Extreme Weather

Not all instances of catastrophic weather are as long term as the Plains' 20-year drought cycle. The warm, moist air masses from the south and the cool, dry air from the north are uninterrupted as they cross the treeless, level land. When this contrasting air meets full force, extreme weather events can occur. And they do, with great regularity, in the Great Plains.

Spring and summer precipitation often arrives in violent thunderstorms. If the storm is strong enough, hail accompanies the wind, rain, and lightning. Hail measuring more than 5 centimeters (2 inches) in diameter and looking like golf balls or even baseballs can devastate a crop of mature, top-heavy wheat. The frequency of hailstorms is high in the southern and west-central Great Plains, with parts of western Nebraska and southeastern Wyoming leading the continent in average hail frequency Figure 12.4).

Wind has been a mixed blessing on the Great Plains. In the past, steady late spring and summer winds on the central and southern Plains—the highest in interior North America—served to maximize windmill efficiency in the region. However, the high winds also mean that moisture evaporation and transpiration are high across the Great Plains. Maximum potential evapotranspiration rates vary from about 165 centimeters

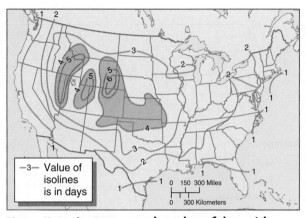

Figure 12.4   Average annual number of days with hail. Each year, hailstorms destroy crops valued at many hundreds of millions of dollars. Damages are especially heavy on the High Plains.

(60 inches) annually in Texas to around 75 centimeters (30 inches) at the northern edges of the Great Plains in Canada. Because these rates are so high, precipitation effectiveness is limited.

Tornadoes are another violent result of Great Plains storm systems. Although tornadoes occur throughout North America, the clash of contrasting air masses characteristic of Great Plains weather makes them a storied but real regional hazard (Figure 12.5). The area affected by any one funnel is small, but funnel wind speeds are in excess of 350 kilometers (200 miles) per hour, so damage is severe in a tornado's path. The frequency with which they occur in the central Plains has given the area the dubious honor of being named "Tornado Alley."

The *chinook*, a winter wind, takes place when dry, relatively warm air from the Pacific Coast pushes over the Rocky Mountains and descends onto the Great Plains. As it descends, the air warms further. A chinook is much warmer than the cold, continental air commonly found over the Great Plains in the winter. As the Pacific air pushes the cold air from the western Plains, a rapid, dramatic temperature rise results. Partly because of this phenomenon, winter temperatures along the higher altitude, western Great Plains are slightly warmer than along the eastern edge of the Plains. Cities such as Denver and Calgary occasionally bask in abnormally balmy, midwinter days with temperatures over 10°C (50°F).

Winters are usually very cold and summers very hot on the Great Plains. The annual temperature

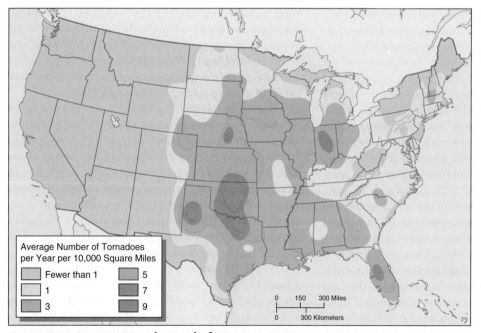

Figure 12.5   Average annual tornado frequency. Tornadoes are far more common on the Great Plains than in any other area of the world. Tornado warning sirens are a normal part of spring and early summer in communities in Tornado Alley between northwest Texas and northeast Kansas.
Source: Oklahoma Climatological Survey.

This tornado swept across the prairies near WaKeeney, Kansas on June 9, 2005.

ranges are large and increase northward. Both North Dakota and Texas have recorded high summer temperatures of 49°C (121°F), the only locations in the United States outside the desert southwest ever to experience temperatures in excess of 120°F. North Dakota has also recorded a winter low of −52°C (−60°F), the only place in the United States east of the Rocky Mountains ever to record so low a temperature.

Snow, wind, and bitter cold characterize one of the most devastating weather events on the Plains: the blizzard. A blizzard begins in winter when a very cold, polar air mass pushes south along the Rocky Mountains to meet moisture from a Pacific storm that has moved inland. High winds, intense cold, and considerable amounts of snow are typical of Plains blizzards. Ordinarily, regional snowfall amounts are not high, averaging about 100 centimeters (40 inches) annually on the northern Plains and less than 25 centimeters (10 inches) in most of Texas. However, a blizzard can last for days, bringing half or more of an entire winter's usual snowfall in one storm.

A blizzard's effects can be harsh and widespread. Plains ranchers usually leave their stock outdoors during the winter. A severe blizzard can block animals' access to food, resulting in high stock mortality. During a series of unusually difficult blizzards in 1887 and 1888, as many as 80 percent of northern Plains cattle died. A decade of overgrazing exaggerated the impact of these storms. The quality of steppe grasslands had diminished, so the animals were not nutritionally buffered. Today, a blizzard may block highway travel for days and require air drops of hay to cattle on the range.

EARLY SETTLEMENT OF THE PLAINS

Plains Indians

The pre-European occupation of the Plains by American Indians was selective. Hunting, particularly for buffalo, was the primary economic activity, and most tribes lived along the streams in semipermanent settlements. With no means of rapid long-distance overland movement (the dog was the only domesticated animal in pre-European North America), the Indians could not leave the dependable water supplies of the streams for long periods. This was a substantial problem, for the migration of the great buffalo herds often took this food source far away from settlements for many weeks.

When Spanish explorers departed from the southern Plains in the sixteenth century, they left some of their horses behind. This action dramatically altered the Plains Indians' lifestyle. When Euro-Americans reached the Great Plains in the early nineteenth century, they found what many have called the finest light cavalry in world history. Horses had spread throughout the grasslands. Plains Indians were no longer restricted to the region's sparse waterways and freely followed the buffalo migration. Horses enabled the Indians to thrive as never before. A few tribes, like the Dakota (Sioux) in the north, the Apache and the Comanche, dominated large sections of the Great Plains.

Native Americans were gradually pushed from most of the Plains during the nineteenth century (see Chapter 14). To some extent, this occurred after white trophy hunters caused the near-extinction of the Indians' main source of livelihood—the Plains buffalo. But effectively, Indians were swept aside by a flood tide of Euro-American settlement.

Euro-American Settlement

The early Euro-American perception of the Great Plains as an unpromising and difficult place to settle was not totally wrong. The widespread lack of trees meant that farmers had none of the traditional material needed to build houses and barns, install fencing, or provide fuel. Water sources were also scarce, and most local rivers and streams dried up each summer.

The earliest Euro-American settlers chose home sites along the best waterways, excluding later arrivals from ready access to flowing water. The existing system of water law gave all riparian rights to the landowner. This use of English common law in North America had been satisfactory in the more humid eastern United States, but it was totally out of place on the Great Plains.

The American settlement frontier hesitated along the eastern boundary of the Great Plains as a result of these and related problems. Settlers bypassed the Plains for Pacific Coast lands until technological and land ownership changes made widespread Plains settlement feasible and more inviting.

Grasslands Ranching

As the expansion of American farming settlement hesitated, an alternative economy swept across the region. Cattle ranching had been introduced into south Texas by Spaniards and into east Texas by frontier settlers from the U.S. South. But it was not until 1867 that this use of the land spread north from Texas.

A series of events and changing conditions permitted this rapid expansion of ranching. During the American Civil War, as south Texas ranchers went off to fight, a large number of unbranded cattle spread across the grasslands. The war ended in 1865, the South's economy was in shambles, and south Texas ranchers were broke. But millions of unclaimed cattle grazed the Texas plains.

Also coinciding with the war's end, railroad companies were given the right to construct lines to the West Coast. When they found little profitable freight in the territory between the Mississippi River and the Pacific, the United States government subsidized the railroad companies with land grants totaling tens of millions of acres along the railroad rights-of-way. Railroads, on their own behalf, encouraged agricultural settlement along their routes, hoping to generate demand for transportation.

Opportunity and need coincided as great herds of cattle were driven north from south Texas to railheads in Kansas. Some cattle were shipped east; other herds were driven farther north to stock the huge, almost empty Plains. Within a decade of the Civil War, open-range cattle ranching spread north across the Great Plains and into Canada. By 1880, an estimated 5 million head of cattle had been moved from south Texas to the Kansas railheads or to central and northern Plains ranges.

The open-ranching economy collapsed even more rapidly in the late 1880s than it had spread 20 years earlier. Overgrazing, new cattle-raising operations in the Midwest, a slipping national economy, the disastrous winter of 1887–1888, and an influx of farmers into the Plains ended this short period in American history. Unimproved, open-range ranches were pushed to the drier western side of the Plains or forced into more restrained, fenced-in operations.

As short as this period was, it fired Americans' imaginations and continues to fascinate millions. Countless novels, films, and songs have portrayed the experiences of open-range ranching. At least one television series, *Gunsmoke*, had a longer tenure than the period it attempted to depict.

The Agricultural Frontier

A series of innovations developed after the Civil War also gradually eliminated, or at least ameliorated, the problems encountered by new Great Plains settlers. Barbed wire, for example, was developed commercially in the 1870s and became an effective fencing material, compensating for the local lack of wood. Dwellings constructed of sod provided damp and insect-infested but adequate housing for a time. Prairie sod was difficult to plow, but when lifted off the soil in thick bricks and used as a building material, it resisted erosion for years. When lumber was brought in by the railroads, most settlers replaced their sod houses as soon as possible with frame homes.

Other technology was imported. The technology for deep-well drilling and windmills had long been known, but windmills had never been used much in northwestern Europe or in the eastern parts of North America. Together these devices provided a means of obtaining water far from surface streams and rivers. Grain farming also became increasingly mechanized during this period. By enabling an individual farmer to cultivate more land than previously, farms could be larger and, thus, compensate for lower yields per acre.

Finally, new crops suited to the Great Plains' growing conditions were introduced as farmers improved their understanding of how to use the Plains environment. Hard winter wheat is probably the best example of such a new crop. First brought to the United States by Mennonite immigrants from the Ukraine, it was better adapted to the dry growing conditions of the Great Plains than were the wheat strains grown there earlier.

Settling the Canadian Prairies

In 1670 the Hudson's Bay Company was given a royal charter to trade through Hudson Bay. It received a title to Rupert's Land, a vaguely defined territory regarded as Hudson Bay's drainage basin.

For the next 200 years, the Hudson's Bay Company was the sole agent of British authority across a broad area, including all of what is today the Canadian prairies. Company policy discouraged settlement. Its goal was to collect and sell animal furs, and agricultural settlement was seen as incompatible with this goal. The company established trading posts at Fort Gray (near present-day Winnipeg) and Fort Edmonton (near Edmonton). These posts and the transport routes between them seeded regional urban development when Rupert's Land was acquired by the newly independent Dominion of Canada in 1870.

The Canadian government fostered settlement of the prairies after they were acquired. But the uninviting Canadian Shield lay between southern Ontario and central Manitoba, and few Canadian settlers undertook the long trip. When the Canadian Pacific Railway was completed in 1885, the move was much easier. Still, many of the early prairie farm settlers came from the American Midwest. By World War I, more than a half-million Americans had moved to the Canadian prairies and were joined by immigrants from central and eastern Europe as well as Canadians.

By moving onto the Canadian prairies later than the Great Plains had been occupied by American farmers, Canadians learned from the Americans' mistakes. Farmers were encouraged to increase their landholdings by purchasing railway land grants or other privately owned land. Ranchland was leased by the government, not sold, and could be reacquired if needed for agriculture. The government encouraged (and still encourages) mixed farming in the prairies' Park Belt, hoping diversification will cushion the economic blows caused by climatic fluctuations.

PLAINS AGRICULTURE

Today, Great Plains agriculture is large scale and machinery use is intensive. It is also dominated by a few crops that are especially suited to the Plains climate. Wheat is by far the most important crop grown in the region, combining crop tolerance of dry conditions with large, reliable market demand (Figure 12.6).

Winter wheat is planted in the fall and becomes established before the winter dormant season sets in. Its major growth comes in the spring when precipitation is plentiful and the summer's desiccating winds have not begun. Winter wheat is harvested in late May and June.

Cattle raised on the grasslands of the Great Plains are often taken to feedlots for fattening on a special feed before being shipped to market. Feedlots, such as this one in Lubbock, Texas, dot the region and reflect the relocation of the beef meatpacking industry from major Midwest cities to small Plains facilities.

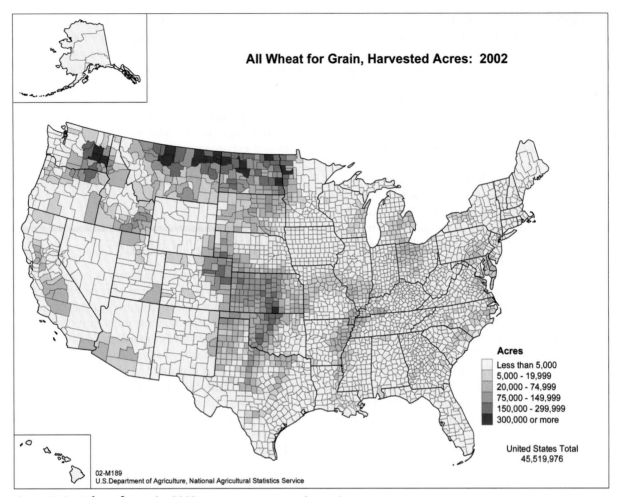

All Wheat for Grain, Harvested Acres: 2002

Acres
Less than 5,000
5,000 - 19,999
20,000 - 74,999
75,000 - 149,999
150,000 - 299,999
300,000 or more

United States Total
45,519,976

02-M189
U.S.Department of Agriculture, National Agricultural Statistics Service

Figure 12.6   **Wheat for grain, 2002.** *Source:* U.S. Census of Agriculture.

Winter wheat is grown across much of the United States, but it is concentrated in the southern Plains between northern Texas and southern Nebraska. As new strains of winter wheat have been developed to better tolerate cold winter weather, its area of production has expanded north as far as Montana.

Spring wheat, grown primarily from central South Dakota to the south-central Prairie Provinces of Canada, is planted in early spring and harvested in late summer or fall. It is suited to areas where winters are so severe that newly germinated winter wheat is killed by the cold.

The Great Plains is the continent's premier wheat-producing region. Canada's three Prairie Provinces grow most of the country's wheat, with Saskatchewan producing more wheat than its two neighbors, Alberta and Manitoba, combined. Kansas usually leads wheat production in the United States, followed closely by North Dakota and then, in distant order, by Oklahoma and Montana. Of the top seven wheat states in the United States, only Washington (ranked fifth) is not a Plains state. Thus, Plains agriculture is the basis for the United States and Canada ranking first and second, respectively, among the world's wheat exporters.

Most grasslands wheat is grown using dry farming techniques—that is, methods that do not use irrigation. Most conspicuous, especially in the northern Plains, is the widespread use of strip fallowing, where strips of land—plowed and tilled but not planted—are alternated with cropped land. The strips are then rotated each year. The result is a striking pattern of long, narrow, alternating rectangles of brown soil and either green or gold wheat.

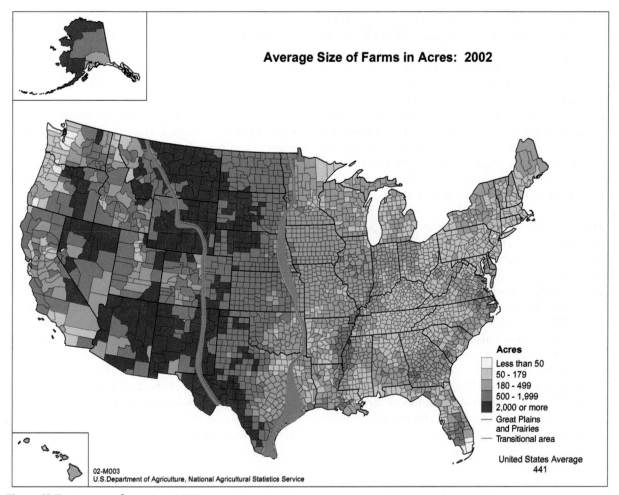

Average Size of Farms in Acres: 2002

Acres
- Less than 50
- 50 - 179
- 180 - 499
- 500 - 1,999
- 2,000 or more

— Great Plains and Prairies
— Transitional area

United States Average
441

02-M003
U.S.Department of Agriculture, National Agricultural Statistics Service

Figure 12.7 **Average farm size, 2002.** Farmers on the dry western margins of the Great Plains need large farms to compensate for low per-acre economic returns. *Source:* U.S. Census of Agriculture.

A form of migrant labor has long been important in harvesting Great Plains wheat. Crews operating huge combine equipment begin each year in the south around June 1 with the winter wheat harvest in Texas. The crews gradually follow the harvest north, finishing some 4 months later with spring wheat in the Prairie Provinces.

Unlike migrant farm laborers elsewhere in the United States, each large crew uses many combines and trucks, and they have traditionally been well paid. Custom combining is less prevalent now than it was before World War II. Because wheat farms are larger, more individual wheat farmers can afford their own combines. Still, these custom combining crews harvest probably a third of all Great Plains wheat and a higher share of winter wheat.

Pressures to increase farm sizes and to purchase land as it becomes available, as in the agricultural core (Chapter 11), have led to more noncontiguous farms in the Great Plains (Figure 12.7). This, in turn, has fostered a growing number of "sidewalk farmers" who live in town and travel to their various acreages. Although demanding, fieldwork on a wheat farm is required during only two or three periods a year. Farm families do not have to live on their land. Another variation are so-called suitcase farmers who live even farther away from their fields and visit them only occasionally.

In addition to the freedom to live in town if they wish, wheat farmers see yet another advantage in fragmented land ownership. A hailstorm usually affects only a small area. By dividing their wheat crop among several widely separated fields, growers are gambling that in any one year their entire crop will not be lost to hail or to a particularly severe thunderstorm.

Although farms are often fragmented, the crop is almost all wheat. This crop uniformity poses a challenge at harvest time. When everyone is growing the same crop, how can it be moved quickly and cheaply to safe storage? Many Canadian prairie towns, for example, are dominated by small, often wooden, grain elevators to which grain is brought by truck before being sent on to larger storage facilities by rail. The small elevators can hold only part of the local crop. If wheat is to be moved from fields to the large storage facilities before rain destroys the harvest, railroads must make many grain hoppers available at the right times and in the right locations. Occasionally, the harvest outstrips the railcar supply, and wheat is piled in large, temporary mounds near the elevators.

Most Canadian wheat goes first to Winnipeg, and then to Thunder Bay on Lake Superior for transshipment on the Great Lakes. An increasing amount of Canada's wheat harvest now moves through Vancouver on the Pacific Coast. Some is also shipped north to Churchill on Hudson Bay, usually for export to Europe. Most U.S. exported wheat moves through the Great Lakes or in barges down the inland waterway system and the Mississippi River.

A soft international market for wheat during the 1960s led to crop diversification in the U.S. Plains states and Canada's Prairie Provinces. Today, Alberta has more acreage in barley than in wheat. Total wheat area in the Prairie Provinces, which reached 12 million hectares (30 million acres) in 1967, declined to 7.6 million hectares (18.7 million acres) in 1971 before increasing again when demand rebounded. However, only in Saskatchewan does wheat still dominate total farm production. Overall, the Province grows 45% of Canada's grain.

Various other crops have been successfully adapted to Great Plains conditions. Sorghum emerged as a major crop on the southern Plains, especially in Texas and Nebraska. This African grain is able to withstand the Great Plains' dry growing conditions. Used mostly as stock feed, sorghum now equals winter wheat in importance on the Plains' hot, dry southwest margins. On the northern Plains, barley and oats are major second crops, with most of the continent's barley crop coming from the Lake Agassiz Basin. In addition, nearly all flaxseed produced in North America is grown in the northern Plains. Sunflowers are of rapidly increasing importance in the Red River valley of Minnesota, North Dakota, and Manitoba. Canada is now the world's dominant producer of canola, a leading vegetable oil and an important protein ingredient in many livestock feeds.

Throughout dry-farming areas of the High Plains, farmers seek to preserve the limited moisture available for growing crops by alternating strips of wheat with plowed but fallow ground.

Large and expensive, center-pivot irrigation systems throughout the Great Plains draw on subsurface water. Each circle shown is 1/2-mile (805 meters) in diameter. How might the remaining nonirrigated areas be used?

WATER CONTROL AND IRRIGATION

Water, aside from the land itself, is arguably the single most important resource in the subhumid Great Plains. Even though precipitation is generally sufficient for wheat farming, many High Plains farmers find it worthwhile to irrigate their crop as a way of assuring a satisfactory harvest even in drought years. But where is the water?

Much of the High Plains is underlain by the Ogallala aquifer, a vast, natural underground reservoir beneath 250,000 square kilometers (100,000 square miles) and containing an estimated 2 billion acre-feet of water. The aquifer's water is "fossil" water, much of it deposited more than 1 million years ago. About a quarter of the land covering the aquifer is irrigated with Ogallala water.

The High Plains is a major agricultural region. It provides, for example, two-fifths of the country's sorghum, one-sixth of its wheat, and one-fourth of its cotton. Its irrigated lands produce 45 percent more wheat per acre, 70 percent more sorghum, and 135 percent more cotton than do neighboring non-irrigated areas.

There is an indirect cost associated with such productivity, and the future is being borrowed against to pay the cost. Groundwater withdrawals have more than tripled since 1950 to more than 20 million acre-feet annually. Natural recharge from precipitation replenishes only a small percentage of what is withdrawn from the aquifer each year.

The area that centers on Lubbock, Texas, provides an example. Early in the twentieth century, irrigated cotton farming gradually replaced the dry-farming approach. Using water from wells drilled into the water-bearing sands underlying the southern High Plains, the region became the most important area of cotton production in the United States. More than 50,000 wells currently supply local farmers with irrigation water.

But the subsurface water reservoir is being depleted. Groundwater levels have dropped more than 50 feet in a 15,000-square-mile area north of Lubbock and more than 100 feet in almost half of that area (Figure 12.8). The average well depth, currently more than 50 meters (155 feet), is increasing. Because deeper wells are much more expensive, a recent estimate suggests that early in the twenty-first century irrigation will be too expensive on half of the Texas High Plains.

New Mexico's laws concerning the conservation of underground water are stricter than those in Texas. Thus, land in New Mexico adjacent to Texas remains in ranches, whereas Texas land is cropped. This legal

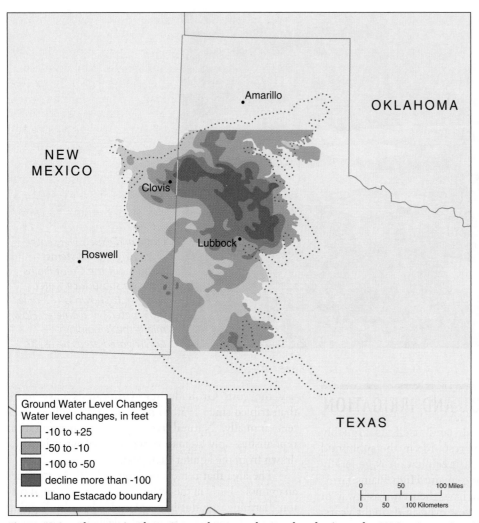

Figure 12.8 **Changes in Llano Estacado groundwater levels since the 1930s.** (Reproduced by permission from the *Annals of the Association of American Geographers*, Volume 85, 1995, Jacques Emel and Rebecca Roberts.) Agriculture in this dry section of Texas and eastern New Mexico use groundwater pumped from the region's large subsurface aquifer. As a result, the groundwater level across a large area between Lubbock and Amarillo, Texas, has declined by more than 50 feet since the 1930s.

difference is clearly visible from the air; Texas's subdivided farms stand in sharp contrast to the extensive, unbroken landscape of New Mexico's ranches. Also possibly reflecting the different state laws is the higher water table on the New Mexico side of the boundary.

Several other, governmentally funded projects also increased water availability on the United States Plains. The National Reclamation Act of 1902 was passed largely at the instigation of Plains and western farmers and provided for the development of federal irrigation districts. This act is one part of a complex pattern of federal dams and water movement systems

scattered widely over the West and constructed at a cost that has reached billions of dollars.

Even so, these waters are no longer adequate to meet demands. In recognition of this demand, the Big Thompson River project was built between 1938 and 1957 to carry water from the west slope of the Rocky Mountains' Front Range to the east slope and the irrigated lands beyond. The most striking technological feature of this massive project is a 33-kilometer (13-mile) tunnel, lying 1200 meters (3950 feet) below the continental divide in Rocky Mountain National Park.

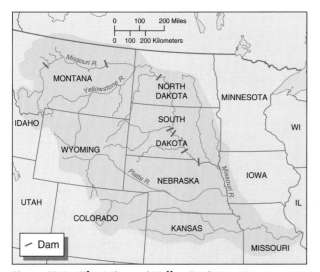

Figure 12.9   The Missouri Valley Project. This project meets the need for irrigation water in the western portion of the Missouri River basin and reduces flooding downstream; it is an example of large-scale federal intervention intended to help address regional problems.

The largest water impoundment project on the Plains, however, is the Missouri valley project, originally called the Pick-Sloan Plan (Figure 12.9). People living at the lower end of the Missouri River valley, including those in Kansas City and St. Louis, wanted an effective system of flood control for the river. People living in the upper Missouri valley, especially the Dakotas and Montana, hoped for ample water for irrigation.

The resultant multibillion-dollar system, composed of a series of large earth-fill dams on the upper Missouri as well as numerous dams in many tributaries of the river, is meant to satisfy both desires. Irrigation expanded considerably in many upstream areas, and major flooding is less likely in the lower valley. An added benefit has been a boom in water recreation throughout an area previously containing almost no standing water.

OTHER PLAINS RESOURCES

If agriculture is the Great Plains' past, energy production may be its future. Many Plains residents view this prospect with a mix of anticipation and concern.

Great Plains sediments contain major reserves of petroleum, natural gas, and coal (see Figure 2.7 and 2.8). The southern Plains' panhandle field, encompassing western portions of Texas, Oklahoma, and Kansas, is the world's leading supplier of natural gas. The same three states, plus Wyoming, are also major petroleum producers. Alberta, long Canada's predominant producer of both petroleum and natural gas, currently produces over 45 percent of the country's coal. Coal has been mined for many years at scattered locations in the region from southern Alberta and Saskatchewan to Texas.

Increases in energy demand and energy prices have created a boom economy in the Plains' rich coal and petroleum fields. The resources were not mined earlier because of their distance from major markets, technological problems, and abundant, alternative U.S. sources. When prices increased after the mid-1970s, however, economic barriers to exploitation disappeared. Plains coal is often found in thick, easily mined seams, but it is expensive to ship to distant eastern markets. The passage of the Clean Air Act in the early 1970s provided an important boost for extraction of the West's low-sulfur coal, which is used for electrical power generation.

As a result of these changes, Wyoming, for example, is now the leading coal-producing state in the United States. Its mines provide almost 20 percent of total U.S. coal output. Coal accounts for nearly 70 percent of the state's tax base and provides over 90 percent of all Wyoming railroad freight. Wyoming coal production in 2006 was over 446 million tons, roughly 348 million tons more per year than in 1980. Gillette, the largest town in the center of Wyoming mining activity, saw its population increase by a factor of five in a decade. Emerging from years of economic stagnation, scores of other communities are experiencing similar growth spurts. Even with an 8.9 percent increase between 1990 and 2000, however, Wyoming's population is still growing more slowly than the United States' population, and it remains the country's least populated state with just 515,004 people in 2006.

Technological problems limit petroleum production from Alberta's Athabasca tar sands. Located about 400 kilometers (250 miles) north of Edmonton, these sands may contain at least 300 billion barrels of petroleum, an amount equal to the entire world's known reserves of conventional oil. Because tar sand petroleum saturates the sands, it must be mined somewhat like coal, and refining technology is still in the development stage. However, as energy prices rise and the applicable technology improves, this reserve is

When completed, the Crazy Horse Memorial in the Black Hills of South Dakota will be the largest sculpture in the world, and will honor a dominant figure in Native American History.

becoming of tremendous economic importance to Canada.

The scale of potential energy developments on the Great Plains is almost incomprehensible. At least 100 billion tons of low-sulfur sub-bituminous coal—whose burning meets strict antipollution laws—lies near the surface in the northern Plains of the United States. This is an amount equivalent to U.S. needs for the next 125 years at current levels of consumption. Within 2000 meters (6000 feet) of the surface, the total is perhaps 1.5 trillion tons.

The structure of the Great Plains regional economy is already shifting, with agriculture and ranching less important than they once were, and mining much more important. Despite strict regulations, environmentalists fear that future exploitation will erode the soil, pollute the Plains' water and air, and stimulate excessively rapid population growth.

Wind energy can be added to the array of mineral fuel resources so abundant across the Great Plains. Winds can be strong and persistent in this region and were harnessed on late-nineteenth and early twentieth century farms to draw needed water from deep wells.

The use of wind as a major source of electrical energy has taken decades to realize, but is now well underway in many parts of North America. The large-scale adoption of wind energy technology has required changes in the cost and anticipated availability of fossil fuels; growing concern about emissions caused by burning coal, petroleum products, and natural gas;

wind turbine technologies; and public attitudes toward the visible presence of large wind turbine arrays. The earliest large-scale wind "farms" were erected in California, but Texas has far exceeded California's wind power capacity. By late 2007, wind power capacities in place or under construction in Texas alone were greater than the combined equivalent capacities in California, Oregon, and Washington. Wind turbine clusters are certain to become a widespread feature of Great Plains and Agricultural Core landscapes.

POPULATION CHANGE AND REGIONAL ORIENTATION

For the past 50 years, changes in Great Plains and Prairies farming guaranteed that local economic opportunities for young people would be limited, and many left when they could. Some went to the region's larger cities. The Edmonton and Calgary metropolitan areas, for example, accounted for three-fourths of Alberta's population increase between 1941 and 1986. Their share of the provinces' population grew from 23 percent to 61 percent.

Much of the Great Plains is served by major urban centers located beyond the peripheries of the Plains or having dual regional orientations to the Plains and elsewhere. Kansas City and Minneapolis-St. Paul, Denver, Dallas-Fort Worth, and San Antonio, the

A Buffalo Commons?

Authors of a 1989 essay suggest that most forms of agricultural activity should be abandoned in a broad swath of territory between the 98th meridian and the Rocky Mountains from Texas to Montana and North Dakota.[3] They think that U.S. settlement of the area is the longest running environmental mistake in the country's history. The best use of most of the area, they argue, is as a Buffalo Commons, where large herds of buffalo may once again be allowed to roam freely. To back up their idea, they identified more than 100 counties that had lost more than 50 percent of their population since 1930, with more than 10 percent of people leaving in the 1980s. These sparsely settled counties have fewer than 11 people per square kilometer (4 per square mile) and a high median age due to outmigration of young adults seeking opportunity elsewhere. Twenty percent or more of all residents in these counties have incomes under the poverty level, and annual new construction per county is minuscule, valued at less than $50. The authors suggest that if current trends continue, two-thirds of the area's farms and towns will vanish by 2020.

The idea of the Buffalo Commons has been met on the Great Plains with varied reactions ranging from anger and disdain to expressions of pride in their region as a continuing home for millions of Americans.

[3] Deborah E. Popper and Frank J. Popper. "The Fate of the Plains," in Ed Marston (ed.), *Reopening the Western Frontier* (Washington, DC: Island Press, 1989), pp. 98–113.

largest American cities on the Plains, are all at the region's margins. Denver is a regional office center as well as the focus of financial activity for energy resource development on the northern Plains as well as the Interior West. The Dallas-Fort Worth consolidated metropolitan area, with a population exceeding 5.2 million in 2000 and easily the region's largest, straddles the boundary between the Plains and the South. Dallas is more a city of the humid East, whereas Fort Worth—50 kilometers (30 miles) to the west—is a ranching and stockyard center and clearly part of the Great Plains. San Antonio, third in size on the Plains, is more than 50 percent Hispanic, placing it in the Southwest Border Area as well (see Chapter 14).

All of Canada's major Plains cities are located great distances from other heavily populated areas by broad expanses of sparsely settled territory. Winnipeg, for example, is nearly 1600 kilometers (1000 miles) from southern Ontario. Calgary is separated from Vancouver by almost 1000 kilometers (600 miles) of rugged mountains. As a result, all of Canada's prairie cities focus on one exit center, Winnipeg.

Winnipeg, with a 2006 metropolitan population of 694,668, had long been the largest prairie city until the recent energy resource boom stimulated the growth of Calgary (population 1,079,310) and Edmonton (population 1,034,945). Winnipeg's location at the southeastern corner of the region gave it considerable locational advantages as the focus of transport

As farm sizes grow and population numbers decline slowly across the Great Plains, an open landscape dotted with abandoned dwellings is created, such as with this relic farmstead in eastern Colorado.

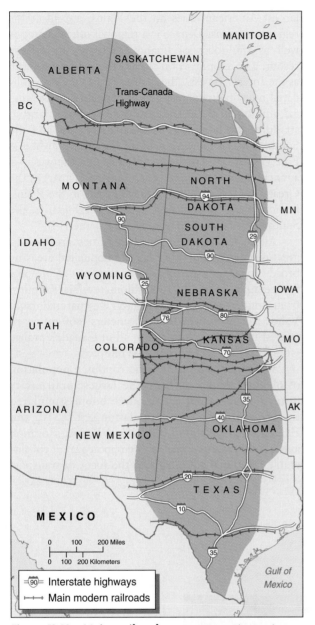

Figure 12.10   Major rail and auto routes. The north-south and, especially, east-west linearity of major overland transportation routes on the Great Plains can be frustrating for anyone wanting to travel by land diagonally across the region.

highways passing through substantial parts of the region in the United States, six are east-west routes. In Canada, the only substantial north-south highway follows the western edge of the prairies in Alberta.

An important consequence of this pattern for the Plains' residents has been a continuing economic and perceptual orientation toward places off the Plains instead of toward areas within it. North and South Dakotans, for example, focus strongly on Minnesota's Minneapolis-St. Paul area. The financial and trade institutions of the Twin Cities play a major role in the two Dakotas; the principal Twin Cities newspaper is read widely across the states; the Twin Cities professional sports teams are followed fervently by Dakotans, as well. Typical residents of the Dakotas are likely to be more aware of Minneapolis-St. Paul than they are of the bordering Plains areas to their north and south. The same orientation pattern is repeated elsewhere in the United States across the Plains.

It is too early to be certain how much of the past Great Plains experience will change as the economy becomes more diverse and energy resources play a greater role. Exporting coal or petroleum from the Plains in the future could have as little effect on job prospects of young people living there as exporting wheat or barley had in the past. The Great Plains and Prairies is an immense region located far to the interior of the continent and, except for its southern margin, far from urban centers on the Atlantic and Pacific coasts.

Review Questions

1. Why was the region initially perceived as the "Great American Desert"?

2. Discuss the challenges of the physical environment that early settlers encountered. What were the technological innovations that helped overcome these obstacles and facilitated long term settlement?

3. Differentiate between the winter and spring wheat belts. What are the physical geographic factors that contribute to this pattern?

4. What are the impacts that the major transportation routeways have on the Plains?

5. Identify the principal factors that contributed to the "Dust Bowl."

6. What is the Ogallala aquifer and what is the vital role that it plays in support of agriculture throughout the Plains? Describe the current issues stemming from water supply and use.

between the prairies and eastern Canada. Most of the bulky products leaving the Canadian grasslands funnel through the narrow outlet occupied by Winnipeg.

The pattern of transportation routes on the Plains also does little to integrate the region. Most routes were originally built to cross the area rather than to serve it (Figure 12.10). Thus, major highways and railroads in both countries pass east-west across the Plains, with few lines running north-south. Of the seven interstate

THE EMPTY INTERIOR

Preview

1. The physical geography of the Empty Interior is dominated by arid plateaus and dramatic mountain ranges.

2. The federal government is a major landholder in the region and conflicts often arise over use of public lands.

3. The Empty Interior is the core area of the Mormon religion in the United States.

4. Although agriculture is present in the region, irrigation is often a necessity, causing struggles for use of limited water resources.

5. Tourism, mining, ranching, and lumbering play an important role in the region's economy.

Columbia (Figure 13.1). Population density is so low across this space, geographers sometimes call the region "the Empty Interior."

Although relatively singular in the sparseness of its population, the territory is marked by variations in other elements. Some portions, notably the Rocky Mountains in the region's east, the Sierra Nevada- Cascades-Coast Range-Coast Mountains systems in the west, and the Alaska and Brooks Ranges in the north, have some of the most rugged terrain in the United States and Canada. On the other hand, many plateaus containing extensive flat areas lie between these mountain chains. Annual precipitation varies from more than 125 centimeters (50 inches) in northern Idaho's Bitterroot Range to less than 25 centimeters (10 inches) in the plateau country.

Most of the region's people are of northern European origin, although Hispanics and American Indians form significant proportions in the south (see Chapter 14). The region's residents support themselves in several ways. Irrigated agriculture and ranching are important

N orth America's largest area of sparse population south of the Arctic stretches from the Rocky Mountains' eastern slopes west to the Sierra Nevada of California, the Cascade Range of the Pacific Northwest, and the Coast Mountains of British

The Empty Interior is a land of stark contrasts where aridity and lushness intermingle across wide desert valleys and high, rugged mountains. This photo shows a portion of Nevada's so-called Extraterrestrial Highway where the human imprint on the landscape is especially sparse.

Figure 13.1 **The Empty Interior.**

in many areas. Where these pursuits are not feasible, lumbering, tourism, and mining dominate.

The Empty Interior's massive expanse of land extends nearly 1600 kilometers (1000 miles) from east to west in its wider southern sections and some 600 kilometers (375 miles) in its more narrow, northerly parts in British Columbia. Some of the most strikingly scenic portions of the continent exist within its boundaries. This vast and varied scenic resource has molded the perception most Americans have of "the West." With few exceptions, human impacts in the region, though locally important, are overshadowed by the varied splendors of the natural environment.

A SPECTACULARLY SCENIC LAND

The Mountains

The mountain ranges of the West present dramatic elevation changes. From base to peak, mountains commonly sweep up 1000 meters (3000 feet) or more. Typical are abrupt, almost vertical slopes, with peaks frequently appearing as jagged edges pointing skyward. This appearance is due partly to age. Most western mountain ranges, though by no means all of them, are substantially younger than their eastern counterparts. Thus, they have not been subjected to the long-term erosion that smoothed the land and eliminated many of the jagged edges of eastern mountains. This undulating softness is eons away in the West.

Yet erosion has started its shape-shifting chore. A drive across the mountains on Colorado's Trail Ridge Road, for example, affords striking scenes of its effects. The road bisects the alpine country of Rocky Mountain National Park's higher elevations. Much of this area was arched upward during several different geologic epochs. Interspersed periods of relative quiet allowed erosion to smooth the surface and gouge the flanks of the uplifted surface. Today, the effects of a trip along Trail Ridge Road are dramatic, with steep increases in elevation and breathtaking vistas of deep canyons and jagged ridges. Along the highest parts of the road, elevations hover above 4000 meters (12,500 feet), and the terrain gently rolls for 18 kilometers (12 miles). Here, far above the mountains' base, is the last area to be affected by the latest cycle of erosion.

Climatic conditions, both historical and current, have contributed to the rugged character of Western mountains. During the most recent geologic epoch, the Pleistocene, substantial parts of the Empty Interior were affected by the actions of ice. The massive ice sheets that were part of continental glaciation enveloped portions of British Columbia and Alaska. Smaller areas of ice, called *alpine glaciation*, existed as far south as the mountains of northern New Mexico.

Cultural Carryovers

Within North America, much nomenclature for various physical features is borrowed from sections of Europe where the feature is widespread. Thus, most features of alpine glaciation have French names, reflecting similarities to the French Alps. Many of the names for arid landscape features, such as *canyon* and *arroyo*, are from the Spanish language and reflect the earlier Spanish and Mexican presence in the arid lands of the U.S. Southwest.

The carving done by alpine glaciers did much to form the topography of the Empty Interior. Remnants of the glaciers still exist in parts of the region, although climate warming in recent decades is shrinking them quickly. Most widespread in the Pacific mountains of southern Alaska, small glaciers linger as far south as the central Rockies in Colorado and the Sierra Nevada of California.

Alpine glaciers form in higher elevations and gradually flow downhill as the volume of ice increases. The moving ice is a powerful agent for erosion. Where this erosion continues for a sufficiently long time, a deep U-shaped valley is created with almost vertical sides and a relatively flat bottom. If two glaciers flow in parallel valleys, a narrow, jagged ridge line forms with small peaks called *arêtes*. Where several glaciers form in the same general locale and flow outward in different directions, a sharp-faced pyramidal peak called a *horn* results. The Matterhorn (in Europe) is the best known such horn, but Mount Assiniboine in the Canadian Rockies is also a fine example of this feature.

The general result of alpine glaciation is a rugged landscape. Yosemite valley in the Sierra Nevada in California, an almost classic, glacially carved valley nearly a mile deep, is probably the region's most photographed example of alpine glaciation. The consequences of such glaciation are visible in higher ranges throughout the entire Empty Interior.

The Plateau Country

Most of the Empty Interior is filled with plateaus rather than mountains. With some exceptions, plateaus are the basic feature of all the area between the Rocky Mountains and the Sierra Nevada-Cascades-Coast Ranges-Coast Mountains complex from Mexico

The Canadian Rockies contain remnants of the glaciers that covered more than half of North America during the last Ice Age. The Athabaska Glacier is a favored site of visitors to Banff and Jasper National Parks.

through Alaska. The term *plateau* is somewhat misleading, however, because parts of a plateau can have great changes in relief.

The Colorado Plateau along the middle Colorado River in Utah and Arizona is one of the most spectacular examples. Although there are large structural changes in relief, most of the Colorado Plateau is underlain by gently dipping sedimentary rocks. Its major landscape features result from erosion by *exotic streams*, most notably the Colorado River and its tributaries. In this arid environment, streams are easily the predominant erosive influence. Thus, in combination with geologically recent, substantial uplift over much of the plateau, streams and rivers cut deeply into the land.

The Empty Interior's erosion-carved canyonlands are some of the best known scenic resources in North America. One recent major travel survey conducted in the United States identified the Grand Canyon of the Colorado River in Arizona as the most widely recognized natural attraction in the country. River erosion deeper than the Grand Canyon's exists within parts of the Snake River Canyon. But the Snake River Canyon's typically narrow gorges have not caught the public's eye and imagination. The greater attractiveness of the Colorado River canyon, particularly in its visually sweeping form in Grand Canyon National Park, results from a couple of features. Many deep tributary canyons branch off the Colorado gorge, creating a canyon system that is, at places, more than 16 kilometers (10 miles) wide. In addition, the layered sequence of strongly and weakly resistant rocks in the canyon's sedimentary formations has created an angular pattern of scarps and benches, each with its own hue that together yield a kaleidoscope of colors.

The Colorado Plateau is not the only plateau in the Empty Interior, or even the largest. The Basin and Range region fills the plateau country from the Colorado Plateau in the south across southern New Mexico and Arizona, west into Death Valley and the Mojave Desert in California, and as far north as Oregon and Idaho. This vast area is composed of a series of about 80 broad, flat basins containing more than 200 north-south trending, small linear mountain ranges. Each mountain range is generally no more than 120 kilometers (75 miles) long and typically rises 1000 to 1600 meters (3100 to 5000 feet) from its base. North and west of the Colorado River Basin, most of the Basin and Range has interior drainage. Streams flow into the area and end there. The Basin and Range has no drainage outlet to the sea. As a result, much of the area has received vast quantities of alluvia eroded from the surrounding mountains.

Figure 13.2 **Pleistocene lakes.** Large sections of the Empty Interior were filled by lakes during the Ice Age. The Great Salt Lake is a remnant of the largest of these. The flat surface of Bonneville Salt Flats in northwestern Utah is part of the former lakebed.

During the late Pleistocene, the Basin and Range was cool and moist; alpine glaciers draped the mountains. Because of the colder, wetter climate and slow melt of mountain ice, substantial parts of the area were covered by lakes (Figure 13.2). The largest, Lake Bonneville, is spread over 25,000 square kilometers (9500 square miles) in northern Utah.

Most of the Pleistocene lakes are gone now or are greatly diminished because the stream flow feeding them draws on lower annual precipitation and because evaporation is higher in today's warmer temperatures. Many of the lakes that remain, such as Nevada's Pyramid Lake or Utah's Great Salt Lake, are heavily saline. Flowing water always picks up small quantities of dissolvable salts. Normally, these salts make a minor contribution to the salinity of the world's oceans. But because the Basin and

Range's streams lack an outlet to the ocean, lakes there have high salt concentration. The Great Salt Lake, covering about 5000 square kilometers (1900 square miles), is what's left of Lake Bonneville and has a salt content much higher than that of the oceans.

North of the Basin and Range, the Columbia Plateau formed from a gradual buildup of lava flows. Contained by surrounding mountains, repeated flows, each averaging 3 to 6 meters (10 to 20 feet) thick, accumulated to depths up to 650 meters (2000 feet). A few small volcanoes or cinder cones dot the area, stark evidence of ancient volcanic activity. On the Columbia Plateau, too, streams have eroded deep, steep-sided canyons. Beautiful in their rugged way, they lack the Grand Canyon's stepped appearance because the lava beds cut by the streams are a more homogeneous erosive material.

With some gaps, the pattern of eroded plateaus continues north into the Yukon Territory, occupying the space between the Rocky Mountains and the Pacific Ranges. In central Alaska, the Yukon River's drainage basin fills the territory from the Alaska Range to the Brooks Range. Surface materials there are mostly sedimentary rocks. Large portions of the middle and lower Yukon valley are quite flat and marshy in summer.

Climate and Vegetation

Much of the Empty Interior has a precipitation pattern that leaves it arid or semiarid. The dryness becomes increasingly apparent toward the region's southern zones. Nearly all of the area in the United States that can be classified as desert is found in the Empty Interior or in the Southwest's borderlands. In the Empty Interior's north, especially in northern British Columbia, the Yukon Territory, and Alaska, a long cold season is the dominant climatic theme.

Aridity is of fundamental importance to the Empty Interior's landscapes and geographic definition. Indeed, water deficiency has played a greater role in the Empty Interior's limited population growth than has the region's rugged terrain.

A strong, direct association exists between precipitation and elevation throughout the Empty Interior. In fact, a precipitation map of the region serves as a reasonably accurate topographic map. Low-lying areas throughout the region are generally dry, with heaviest precipitation usually found on the midslopes of mountains. Surface water across the entire Empty Interior is almost totally dependent on exotic streams that flow down and away from the mountains.

The association between topography, temperature, and precipitation results in a striking altitudinal zonation of vegetation throughout the Empty Interior. The lowest elevations are generally covered with desert shrub. Sagebrush is the region's dominant low-elevation species. One subspecies, big sagebrush, can grow 5 meters (15 feet) high and live 100 years, and it may be America's most abundant shrub. In the region's far south, precipitation increases modestly in late summer. Locally called the monsoon season, the light rains allow a sagebrush/grasslands combination. This combination is found at elevations above the desert shrub in other Empty Interior districts.

Upslope from the sagebrush is a treeline where precipitation is sufficient to support tree growth. At the lower treeline, forests are a transitional mix of grass, piñon pine, and juniper along with other small trees. At higher elevations, this blend transforms into extensive forests of majestic ponderosa pine, lodgepole pine, or Douglas fir. If the mountains are high enough, a zone of smaller trees such as subalpine fir exists. And then a second treeline (technically, the "upper treeline") is crossed, beyond which the combination of high winds and a short, cool growing season render tree growth impossible. Above the upper treeline, conditions are suitable only to tundra.

The actual elevations at which these vegetation changes occur vary by latitude. Both treelines are found at substantially lower elevations in the Empty Interior's north. The lower treeline may also have a lower elevation on west-facing slopes, which receive more precipitation from the eastward-moving weather systems. It can be lower, too, on north-facing slopes where shade lowers temperatures and retards water loss due to evapotranspiration. The boundary between forest and tundra is above 3700 meters (11,500 feet) in New Mexico, whereas in southern British Columbia it occurs at elevations less than 2000 meters (6000 feet).

Wildlife

The Empty Interior, with much of its territory publicly owned, supports a diversified, growing, and sometimes controversial wildlife population. The region, like most of the West, has shared in a wildlife population explosion. Since 1935, the bison (buffalo) population has grown from little more than 10,000 to 65,000 today. The North American elk population grew from 225,000 to 500,000; pronghorn antelope from 40,000 to 750,000; white-tailed deer from 5 to 15 million; and wild turkeys from 30,000 to 2.5 million.

Within the Empty Interior today, Colorado and Idaho each have over 100,000 elk; Montana and Wyoming nearly match that number. California, Idaho, Montana, and Oregon each have more than 10,000 wild bear. A substantial deer population lives everywhere. Mountain lion, although elusive, are believed to have spread throughout the region. These wild animals attract large numbers of tourists, as anyone caught in an elk-viewing, bear-watching traffic jam in Wyoming's Yellowstone National Park will attest. Wild animals have also attracted big game hunters, and big game hunting is now big business. A 2001 U.S. Fish & Wildlife Service survey indicated that hunters spent more than $2 billion on equipment, trips, and other expenses, such as licenses. Much of this money is used in state conservation efforts. Wildlife watching is also on the rise, and in some places the needs of hunters are being challenged by the needs of the wildlife watchers.

The overwhelming reason for the swelling wildlife ranks is a dramatic change in public attitudes and a resultant improvement in management and conservation practices. But these practices have critics, and the future of some programs—and species—is unresolved. Reintroduction of gray wolves to Yellowstone was adamantly opposed by some livestock ranchers who feared the wolves will roam beyond the park's borders to prey on domestic herds. Some of these ranchers are among a contingent who want the area's small buffalo herds removed as well, for they are concerned the buffalo may spread disease and infect cattle.

Economics drives these concerns. The management of national parks and conservationists are trying to alleviate ranchers' worries by offering compensation to those who lose animals. So far, actual incidence of wolf predation is low, as is risk of disease spread by bison. Management officials use their tools in tandem with compensation packages. Educational programs aimed at supplanting myth with fact are key.

THE HUMAN IMPRINT
Public Land Ownership

Most land in the Empty Interior in both the United States and Canada is in governmental hands (Figure 13.3). In Nevada, for example, various branches of the U.S. government control almost

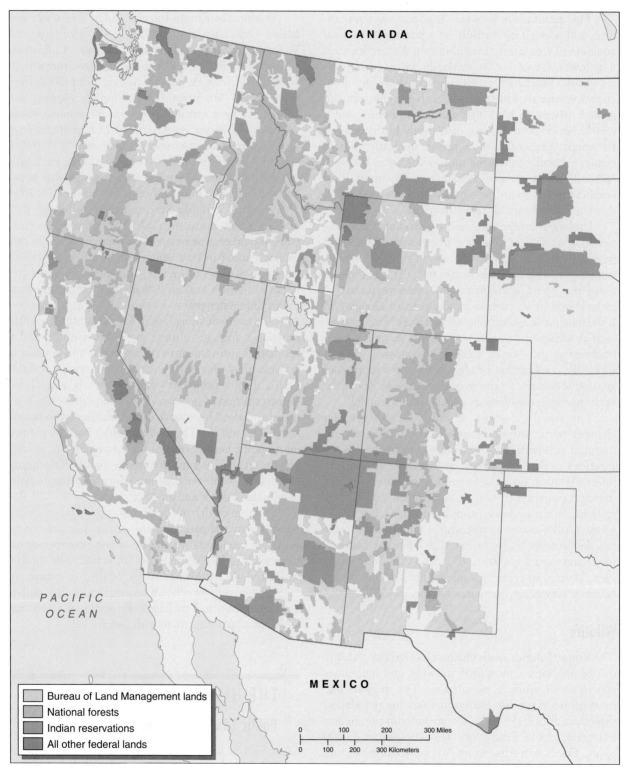

Figure 13.3 **Federal lands.** Some Westerners question the wisdom of widespread federal land control in the region.

90 percent of all land. Although percentages are lower elsewhere in the region, the basic pattern of predominant government ownership is universal. And this affects how the land is used.

That so much of the land in the Empty Interior is in governmental hands is a residue of history and policy. In both countries, this region and the far Northlands were the final lands that a substantial number of people occupied. In both cases, federal programs of land distribution designed to encourage agricultural use were not relevant because the regions held little or no agricultural promise. The United States Bureau of the Census proclaimed the end of the country's settlement frontier in 1890, a time when much of the Empty Interior was still not settled. Although settlers expanded onto Canada's prairies for several decades after 1890, they, too, ignored most of the Empty Interior.

By the time other interests, such as lumbering and mining, pushed for greater private land ownership, the federal governments of both countries were seriously reevaluating earlier programs in which they distributed land almost for free. In the United States, Theodore Roosevelt created the U.S. Forest Service in the early 1900s largely to counter a threatened transfer of Empty Interior land into private ownership. This act marked the beginning of an effective conservation movement in the United States. In Canada, settlement of the Empty Interior lagged behind that of the United States. Federal and provincial governments kept in hand an even larger portion of the region's nonurban, nonagricultural land.

Public lands have been put to a wide variety of uses in both countries. In each, but especially in the United States, a substantial part of the national park system is located in the Empty Interior (Figure 13.4). The basic

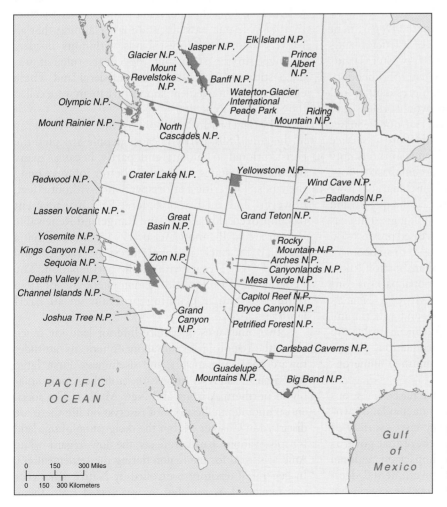

Figure 13.4 Major national parks. The interior West, with its dramatic and rugged topography, is the location of many national parks in Canada and the United States.

task of both the United States and Canadian national park systems is to preserve the unique or highly unusual sections of their country's natural environment. Much of the physical landscape satisfying this requirement exists in the Empty Interior. Some of the continent's most famous national parks—Yellowstone, Grand Teton, Glacier, Banff, Grand Canyon, Bryce, Zion—are found there. Yellowstone, acknowledged to be the world's first national park, was set aside in 1872 as the first nature preserve in the United States.

The national parks represent the best known use of this vast governmental land resource, but they are only a small portion of the Empty Interior's total public land area. The largest share in the United States is held by the Bureau of Land Management (BLM), an agency of the Department of Interior. Grazing is easily the most important land use administered by BLM. But the BLM has also been the main agent in constructing irrigation and hydroelectric dams in the region.

The U.S. Forest Service, based in the Department of Agriculture, is the second largest U.S. federal landholder. Much of the Forest Service's western lands were set aside during a few frantic days of Theodore Roosevelt's administration when the president feared impending congressional action would release to private interests federal lands that had not been set aside for other purposes. Today, the Forest Service controls a majority of the forested lands in the Empty Interior. The service traditionally has emphasized logging and grazing under its multiple-use charge. However, in recent decades, it has increased the quality and quantity of recreational uses in its holdings. This has been, at least partly, in response to increased pressures on the limited facilities in national parks as tourism boomed.

In the Empty Interior's Canadian portion, most governmental land not in federal national parks is controlled by provincial governments. Most of Canada's provincially owned land is forested, and Alberta and British Columbia have used their provincial ownership to control and expand lumber production. British Columbia, in particular, has also established numerous provincial parks.

Some residents of the West have raised questions about the extensive nature of governmental land ownership in the Empty Interior. They cite a variety of reasons for their concern. Some believe that governmental ownership has a negative impact on regional economic growth. Others think governmental decisions on land use do not address local interests. No property tax is levied on governmental land, and some argue that federal payments fail to cover that tax loss. Still others are concerned about what they see as poor management of governmental lands. Research strengthens concern on certain points. A recent study, for instance, indicates that misuse has damaged much of the federally owned grasslands in the West, damage that is difficult to reverse in arid and semiarid environments.

Most basically, Westerners critical of governmental control of Empty Interior land claim this ownership pattern harms their ability to plan and control their own, and their region's, future. Long-range economic and environmental plans depend largely on decisions made at the federal level; local and national authorities often disagree on planning goals (see the discussion of Alaska in Chapter 16). In 1970, state governments in Nevada and Arizona proposed that the federal government donate 2.4 million hectares (6 million acres) of federal land to each state. That proposal was rejected. In 1978, Nevada passed a state law claiming that all BLM and national forest lands within its borders belonged to the state, but the courts ruled it was unconstitutional for states to claim federal land. These efforts and the movement supporting them were called the Sagebrush Rebellion.

The Sagebrush Rebellion lost much of its energy in the late 1980s, partly because people accepted the constitutional argument and partly because many doubted that states would do a better job managing the lands. The critics' efforts were not without effect, however. The quality of federal land management has improved. The Forest and Rangeland Renewable Resource and Planning Act of 1974 mandated that a use plan for each national forest be created and updated every 5 years with public participation. The Federal Land Policy and Management Act of 1976 requires a similar planning program on all BLM lands.

Two other uses of Empty Interior land say much about the region's past and about Americans' attitudes toward the land's quality and usefulness. First, large American Indian reservations are set in the region, notably in northern Arizona and New Mexico. The social and economic implications of reservation life there are discussed in Chapter 14, but the designation of this land for reservations partly reflects the low regard white Americans had for the region during the settlement era. In general, if territory controlled by Native Americans was perceived as valuable, it was seized from them.

Monument Valley on the Navajo reservation in northern Arizona was made famous by countless films. But the valley was for much longer—and still is—an area where Navajo live and let sheep graze among the dry buttes and mesas.

Second, some of the United States' largest bombing and gunnery ranges, as well as its only atomic bomb testing facility, are in the Empty Interior. Critics of this use condemn it because of its ecological effects. But government officials cite need and national security precedence over alternative demands on the land.

The Mormon Presence

The combination of rugged terrain and widespread aridity limits the scope of agriculture. As the agricultural frontier moved west in the late decades of the nineteenth century, it swept past the Empty Interior. Were it not for minerals, transportation, and the Latter-Day Saints, few Americans or Canadians would have chosen to live in the region until well into the twentieth century.

Members of the Church of Jesus Christ of Latter-Day Saints, more commonly called Mormons, strongly affected settlement patterns within the Empty Interior. Established in upstate New York in 1830, the Mormon church and its followers were attacked repeatedly, both verbally and physically, for what were considered their "unusual" beliefs. To escape such assaults, the Mormons moved several times in search of a place to practice their religion in peace. In 1844 a mob killed the

church founder, Joseph Smith, in the Mormon town of Nauvoo, Illinois (which at the time was vying with Chicago for the honor of being the largest city in the state). Over the next few years, many Mormons pushed west to settle far from their persecutors and where they hoped to create an independent Mormon state.

The locale the Mormons selected for their initial western settlements was the Wasatch valley, tucked between the Wasatch Mountains and the Great Salt Lake in northern Utah. This spot, which would later become Salt Lake City, must have seemed an unlikely one to begin an agricultural settlement. The climate was dry, the lake saline and useless, and the landscape barren. Nevertheless, the Mormons quickly began their agricultural operations.

Their settlements grew as new arrivals came, for the Mormons were, and are, fervent missionaries. High birth rates also pushed their population numbers up. Because they dreamed of founding a country they would call Deseret, stretching north into what is now Oregon and Idaho and southwest to Los Angeles, Mormons established communities at greater and greater distances from Salt Lake City.

Mormons failed in their hope to create Deseret. American expansion moved through and beyond the Mormon-held area, and Mormons again found themselves under the will of the United States. Ironically, the initial site the Mormons found so satisfying in its isolation lay directly astride what would become the California Trail. The discovery of gold in California in 1849 initiated a large movement of non-Mormons through their settlements within a few years of their arrival. What would have been Deseret was eventually divided among a half-dozen states.

This failure did little to lessen the Mormons' impact on the region. They were the first white people to face the problems of life in the Empty Interior, and their solutions were shared with later non-Mormons. No solution was more important than their innovative irrigation techniques. The techniques and careful central control necessary to collect and move water to agricultural users were almost unknown to other Americans. The Mormons, with their theocratic government providing a strong central organization, built a large number of storage dams on the western slopes of the Wasatch Range along with miles of canals to move water to users in the valley below.

Contemporary benefits of these efforts are agricultural crops, trees, and green lawns covering much of the Wasatch valley. In early summer, the multihued greens of irrigated fields contrast dramatically against the browns of the valley's non-irrigated portions. But the primary legacy of Mormons' early efforts at large-scale irrigation was the start of an irrigation boom in the Empty Interior. Today, most agricultural crops grown in the region are raised on irrigated lands.

Mormons still have a substantial impact on the Empty Interior. Of the region's roughly 14.5 million people, about one in seven are Mormons. They constitute more than 70 percent of Utah's population and a majority of south-central Idaho's. Substantial numbers of Mormons also live in Nevada, northern Arizona, and western Wyoming.

Temple Square in Salt Lake City, Utah, provides visible religious focus amid evident Mormon success in settling this arid land. With the Wasatch Mountains immediately to the east, Temple Square contains both the Temple (the building with the spires) and the Tabernacle (the low, oval structure just beyond the Temple).

A DISPERSED ECONOMY

Irrigation and Agriculture

Irrigated crops dominate agricultural production in the Empty Interior (Figure 13.5). Without irrigation, a stable agricultural economy would be impossible in this dry region. Today much of the flow of several important rivers is diverted for various uses, with irrigation claiming the lion's share. This massive diversion would be prohibited under the water laws in effect in most parts of the United States, which require that usage not noticeably diminish stream flow. The unique water problems of the Empty Interior led to the development of new standards called the Doctrine of Prior Appropriation. This doctrine grows out of a "first come, first served" approach as long as the initial user continues beneficial use of the water. Under this doctrine, water is treated like a commodity and can be bought and sold independent of the land through which it passes. The Doctrine of Prior Appropriation is in contrast to eastern water law that divided the water among all owners of the land through which it

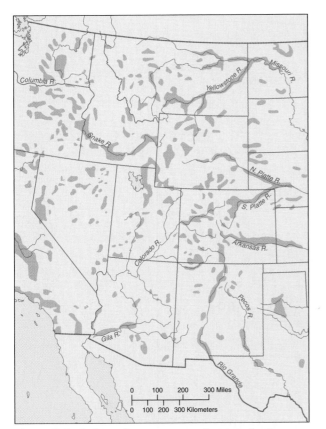

Figure 13.5   Irrigated areas. Irrigated areas in the Empty Interior support most of the agriculture and also provide a focus of settlement in this sparsely populated region.

passes unhindered. Clearly, in the Empty Interior water law permits total diversion of a stream unless there have been alternative contractual arrangements.

The Reclamation Act of 1902, passed with strong Western backing, provided federal support for the construction of dams, canals, and eventually hydroelectric systems for 17 states west of the 100th meridian. Today, irrigation projects encompassing 4 million hectares (10 million acres) claim more than 80 percent of the water from these projects. Although most of this irrigated land is in California, large irrigation projects are scattered throughout the Empty Interior. Agricultural water users have been required to pay only a small portion of the real cost of supplying irrigation water. Urban users, often to their consternation, pay far higher rates. Contested equity issues aside, much of the cost of development is borne by the federal government.

The use (or misuse) of the Empty Interior's longest river, the Colorado, offers perspectives on conflicts over water ownership (Figure 13.6). By 1915, nearly the entire flow of the Colorado River had been allocated, mostly downstream in Arizona and California. This meant that the upper basin states of Wyoming, Colorado, Utah, and New Mexico had little right to the water that originated largely in their states' mountains. In 1929, the Colorado River Compact reallocated these water rights, giving half of the river's total allocated flow of 15 million acre-feet to the upper basin and half to the lower basin states of Nevada, Arizona, and California.

Most stream flow in the region originates as spring and summer melting of the previous winter's mountain snows. Thus, winter snow accumulation is quite important, but it is also quite variable. The winters of 1982–1983 and 1983–1984 were two of the snowiest in history. Water spilled over the top of Boulder Dam, one of Salt Lake City's main streets turned into a temporary river, and the Great Salt Lake, historically shrinking due to irrigation withdrawals, reached its largest areal extent in a century while its salinity halved. Six times saltier than the oceans in the summer of 1982, the Great Salt Lake had only three times the oceans' salt content two years later. In the late 1980s, however, winters were dry in the Sierra Nevada and central Rocky Mountains, so resultant water deficiencies left consumers parched the following summers.

Competition among the states for the Colorado River's water is intense. Until recently, the states of the upper Colorado used relatively little of their allotted share. The lower states, California and Arizona, increased their withdrawals by including most of the water "surrendered" by their upriver neighbors, although the northern states, especially Colorado, have no desire to relinquish permanently any of their allotted share. But, in recent years growth in Colorado and Nevada has increased demand for water once sent to California and Arizona. In October of 2003, the six states that share the Colorado River's water signed an agreement that will allow California to continue to draw more than its share of the water while reducing its dependence gradually over the next 14 years.

In a separate issue, as Arizona was confronted with water supplies inadequate to support its population and economic booms, it argued that California's share was too large and Arizona's too small. Arizona pressed the issue in the courts several times and gained increases in its share of Colorado River water. In 1973, it began construction of the Central Arizona Project, designed to carry water from Parker Dam on the

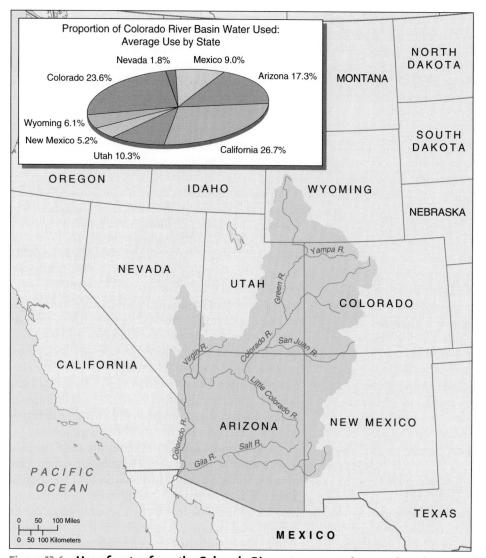

Figure 13.6   Use of water from the Colorado River. Competition for water from the river is fierce. It is a scarce, vital resource in this generally dry region. *Source:* Southern Nevada Water Authority.

Colorado River to Phoenix and Tucson. Water first reached Phoenix through the Project in 1985.

The Colorado River projects are the most notable of a number of examples of intensive irrigation development in the intermountain region. Irrigation projects along the Gila and Salt rivers in south-central Arizona were using virtually the entire flow of the Gila by the early 1930s. Farmers and the area's growing urban population then turned to wells for additional supplies.

Today Arizona depends on subsurface waters to supply several times as much water as the Gila-Salt

River system. However, this solution is short term. Hundreds of millions of cubic meters of water are taken from the ground and not replenished by natural discharge. The U.S. Geological Survey reported that between 1952 and 1978, thousands of square miles of southern Arizona sank as much as 7 feet because of groundwater withdrawal. In short, the pressures on a deficient water supply are growing more acute throughout the drier southern half of the Empty Interior.

Future water resources are a matter of pressing concern. Politics and economics have discouraged the federal government from proposing major new impoundment and diversion programs. Urban and industrial water users demand a larger share of available water. Environmental problems can also affect politics. A recent drought in Colorado, for example, has reduced the amount of water flowing into the Colorado River. But, Colorado is obligated to supply a certain amount of water to the river each year. This places stress on Colorado's relationship with its neighbors and on the officials within Colorado trying to provide water to a growing population.

By some estimates, Arizona could support 20 million people with its current water supplies instead of its present 3.5 million people if none of the state's water went to agriculture. In an innovative response to its water problems, Arizona in 1980 passed a comprehensive water planning law. All large wells are now metered, and withdrawal fees are charged. Water rights can be sold, with the expectation that they will shift to urban and industrial users. If that does not happen, the state will eventually buy farmland with water rights and retire it from agriculture. The goal is to reduce the state's 415,000 hectares (1.025 million acres) of irrigated farmland to 324,000 hectares (800,000 acres) by 2020.

A succession of years with below-average snowfall beginning in 2000–2001 sharply increased concerns among the region's local water managers. Because snowmelt is the primary source of water in the Empty Interior, inevitable tensions between agricultural and non-agricultural users have grown. Lower stream and river flow means that reservoirs do not reach capacity each spring, and the total amount in storage has declined steadily. Water storage on the Colorado River, for example, the region's primary water source in the south, shrank each year. By 2007, Lake Meade in Arizona was filled to less than 50 percent of storage capacity. Lake Powell was also at its lowest level since the Glen Canyon Dam was completed. Meanwhile, the region's urban population continued to grow rapidly. Abnormally high precipitation years in the region, as occurred in 1982–1984, could go a long way toward refilling the reservoirs, but such extreme events are, by definition, uncommon. As supplies diminish, it is agriculture that is most vulnerable to greater restrictions on water use.

Although irrigated land is scattered thinly across much of the Empty Interior, a few areas are sufficiently large to merit special mention. The 1 million hectares (2.5 million acres) irrigated in the Snake River Plain make Idaho the region's leader in terms of the number of acres irrigated and enables the state to be among the country's leaders in potato and sugar beet production. The Columbia Valley Reclamation Project, supplied by the bountiful waters of the Columbia River impounded behind Grand Coulee Dam in central Washington, contains well over 200,000 hectares (549,000 acres) of irrigated land. It produces water for such crops as alfalfa, sugar beets, and potatoes. Irrigation along the Wasatch valley has expanded little since the first decades of Mormon settlement, with about 400,000 irrigated hectares there devoted primarily to sugar beets and alfalfa. The Grand Valley, along the Colorado River in west-central Colorado, produces alfalfa and potatoes as principal crops, although tree fruits, especially peaches, are important.

In Washington, tributaries of the Columbia River, notably the Yakima and Wenatchee in Washington and the Okanagan in British Columbia, flow eastward out of the Cascades and supply water for both countries' famous apple-producing regions. The Okanagan valley (spelled Okanogan in Washington) has been irrigated since the 1860s. It was one of the first major irrigated areas in the Empty Interior. The combination of an attractive product and extremely effective marketing enables this dry area with an average annual precipitation of less than 25 centimeters (10 inches) to grow a third of each nation's apple crop.

Though bountiful, each of these areas produces a limited set of crops. The short growing season precludes the farming of most long-season crops. Local demand is limited, minimizing the need for dairying or many fresh vegetables. Livestock, sugar beets, apples, and potatoes, however, can be grown locally and shipped profitably to distant markets.

The Empty Interior's southerly irrigated districts, though under pressure from urban and industrial water consumers, nevertheless have one major advantage over their northern counterparts—a far longer growing season. The Imperial Valley, with a frost-free period in excess of 300 days, is one of the United States' premier truck-crop farming areas. Much of the winter head-lettuce supply comes from here, as do grapes, cotton, and alfalfa for fattening beef. What's more, the valley's cattle population is more than 320,000 head. An electric power-generating facility there uses local, abundant cattle manure for fuel. The growing season is long enough to support *double-cropping*, and this increases overall productivity. The Coachella valley north of the Salton Sea produces dates, grapes, and grapefruit. The Yuma valley along the lower Colorado River supplies

Generating electricity and moderating the river's flow downstream, the Glen Canyon Dam holds back the Colorado a few miles above where it enters the Grand Canyon.

principal crops of cotton, sugar beets, and oranges. In the Salt River valley, near Phoenix, winter lettuce, oranges, and cotton are the major crops. These crops, unlike those grown farther north, face little direct competition from agricultural centers in the major market areas of the eastern United States.

Transportation Service

As North America's West Coast attracted settlers and developed economically, the Empty Interior became a broad barrier between the more densely populated areas of both countries. Little traffic is generated within the Empty Interior, so transport developers sought to align rail and highway routes to move people and freight across the region as speedily and

inexpensively as possible. Consequently, most major routes pass through the region east-west, between the urban centers of the Midwest and those of the West Coast, traveling as directly as topography will allow.

Despite this solution, the Empty Interior's great width demands development of many service facilities for the traffic passing through. The Empty Interior's smaller towns, as well as a fair number of its larger ones, began as centers established to service and administer the railroads. They were located wherever railroad personnel were needed, and the region's population density was irrelevant to the towns' existence. Later, automobile traffic provided a demand for gas stations, car-repair facilities, motels, and restaurants for travelers. Although fewer local railroad workers were required as technological innovations were implemented after World War II, this loss of jobs was more than compensated for by the growing need for people to serve truck and automobile traffic. Few regions today have as large a share of the total workforce involved in transportation services as the Empty Interior.

Although transport services were the principal influence on the Empty Interior's early urban growth, cities that have become the largest have usually been aided by some additional attribute. Spokane, Washington, for example, with 446,706 people in its Metropolitan Statistical Area in 2006, has become the principal center for the state's "Inland Empire." This area is geographically defined and half encircled by the sweep of the Columbia River across central Washington State and has long been a place rich in agricultural production. Albuquerque, New Mexico, with a Metropolitan Statistical Area population of 816,811 in 2006, gained a role similar to Spokane's due to its centrality and accessibility in that state.

Phoenix, Arizona, was initially an agricultural center but then boomed as Americans flocked to its warm, dry environment. With its 4,039,182 million people in 2006, Phoenix ranks among the country's urban regions. It has become a retirement center as well as a focus of manufacturing activities. Industries producing small, high-value products such as electronics have been important in the city's growth. The higher transportation costs engendered by locating a great distance away from most of their markets are not significant to these light industries. Amenities, such as outdoor recreation opportunities, attract both labor and management and far outweigh the slight economic locational disadvantage (see also Chapters 5 and 10).

Ogden, Utah, is one community that still exists as a major rail center. Early in its history, it was among the most important of such places in the region, but it has not become a major urban place. Ogden, only about 55 kilometers (35 miles) north of Salt Lake City, is in the shadow of the larger city, one whose continued dominance derives from its early development and its key functions as the capital of Mormon religion.

Tourism

The great variety and appeal of the Empty Interior's scenic wonders attracts millions of visitors each year. The federal government controls most of the major attractions, like the Grand Canyon, but the growth of tourist-oriented industries has been substantial. Visitors to most major parks must first pass a garish strip of motels, snack bars, gift shops, and other sources of ersatz local color. This service industry has undoubtedly grown partly because distances between attractions are great, and services are needed in countless locations.

When Nevada's gambling operations are included as part of the Empty Interior's tourist industry, the overall regional impact of tourism becomes even greater. In 1931, Nevada passed laws legalizing many forms of gambling and simplifying divorce proceedings. The state soon became the gambling and divorce capital of the nation. Liberalized divorce laws in other states have since eliminated part of the state's claim to fame, but Nevada's gaming industry remains the cornerstone of its economy. State gambling taxes account for 45 percent of Nevada's revenues. The industry first centered in the Reno/Lake Tahoe area, but Las Vegas

now dominates. Las Vegas has the advantages of relatively cheap electricity from nearby Boulder Dam to power the amazing display of lights advertising the city's gambling and entertainment activities, along with proximity to southern California. A direct interstate highway and frequent, inexpensive air travel connect Las Vegas to urban California's millions, its most important source of customers. Today, Nevada casinos typically take in over $800 million a month from gaming alone. Hotel, food, and entertainment earnings multiply the gains for state businesses.

The total volume of tourism has grown so much in this thinly populated region that many attractions are overburdened. The U.S. National Park Service, faced with budget reductions and substantial environmental damage to its overused facilities, restricts facility availability in some parks. In Yellowstone, where the summer morning lineup to obtain an empty campsite is as much a part of the landscape as the roadside bears once were, the Park Service has been forced to reduce the number of campsites. Part of the road along the south rim in Grand Canyon National Park is closed to private vehicles, as are sections of the road system in California's Yosemite National Park. Reservations made in advance of arrival are required to secure a campsite in many of the more populated national park campgrounds during the busy summer season.

The tremendous growth of tourism has exerted pressure on the U.S. Forest Service and the Bureau of Land Management (BLM) to provide more recreational

Las Vegas, once a sleepy desert town, is now the fastest growing city in the United States. Between 1990 and 2000, Las Vegas grew 83.3 percent!

Yosemite Valley in Yosemite National Park, California, is one of the United States' premier natural features. The valley is so heavily used in summer that camping is by reservation only and a portion of the valley is closed to private vehicles.

facilities on their lands as a way to decrease pressure on national parks. Both agencies have complied, but they are not agencies primarily intended to serve the touring public. Both encourage multiple-use activities on their land. Although this includes recreation, the growth in tourist demand is stretching Forest Service and BLM resources as well.

Lumbering and Ranching

Ranching, lumbering, and, to a lesser extent, mining depend on governmental lands for many of their basic materials. U.S. Forest Service and BLM lands are open to grazing, and most lumbering in the Empty Interior is carried out on Forest Service lands in the United States and Canada. Productivity per acre is relatively low on these lands for both ranching and lumbering, particularly in the United States, especially in comparison with land held in private hands. Many who oppose the government's land use policies contend that poor management has resulted in widespread overgrazing of federal lands and that the land's carrying capacity has declined as a result.

There are several reasons for the discrepancy between the productivity of privately held and government-held land. Government land is, by definition, multiple-use. Serving several purposes limits developing the land just for one. Another likely explanation is the land's low quality, particularly the land controlled by the BLM. In many drier areas, 40 hectares (100 acres) or more of land per head of cattle are needed for satisfactory grazing.

The Empty Interior's great seasonal climatic variations make this one of the few areas in North America where *transhumance* is practiced. Transhumance is the seasonal movement of animal flocks by those who tend them from the lowlands in winter to mountain pastures and meadows in summer. It is especially important in the sheep-ranching economy. Many Basques, expert shepherds from the Pyrenees of Spain came to this area as contract laborers to manage the herds. Today, descendants of the Basques make up a substantial part of the population of several states, especially Nevada.

In British Columbia, the wood products industry remains the cornerstone of the provincial economy. Wood products and paper and allied products are British Columbia's two most important manufactured goods. More than 90 percent of the province's forest is Crown land—held largely by the provincial government. The government has actively encouraged both cutting and reforestation.

Mining

Some Empty Interior urban places were established because of the region's mining activities. Miners became, after the Mormons, the second largest group of Americans to settle the Empty Interior. The Comstock Lode discovery in Nevada gave rise to Virginia City. It grew to a city of 20,000 people by 1870 before nearly disappearing as high-quality ore played out. Nevada was admitted to the United States in 1864,

Parts of the Empty Interior still reflect the open rangeland life, where horses and cowhands work as teams on roundups.

long before its neighbors, largely on the basis of the rapid population growth that followed the Comstock discovery as gold and silver mining boomed in the state.

As Nevada's mineral resources were depleted, the state's population also declined. Today, the mining economy is not important in the state, although some old, abandoned mining centers, such as Virginia City, are tourist attractions.

Gold and silver deposits still exist in the Empty Interior. But the richest, most accessible deposits have been mined out, and the economic boom that gold generated is long past. Renewed mining activity has been encouraged by continuing high gold and silver prices, but the new interest is at a smaller regional scale than it was a century ago.

Copper leads any list of the Empty Interior's economic contributions from mining. Production today is concentrated in Arizona and Utah. The vast, open-pit mine at Bingham, outside Salt Lake City, is one of the world's largest human-made excavations. It has yielded more than 14 million tons of copper since mining began in 1906. Arizona's most important copper-mining center is at Morenci, in the eastern part of the state, but other important mines are at San Manuel, Globe, and Bisbee, all in the south.

Most copper ore mined in the Empty Interior is low grade, with metal content less than 5 percent. At Bingham Canyon, for example, the ore is about 0.6 percent copper. As a result, smelting or concentrating facilities are located near most mines to reduce costs by eliminating the waste portion of the ore before

Copper mined in the Empty Interior is usually dug from vast open pits, such as the Bingham Canyon copper mine in Utah, one of the largest human-made holes in the world. The mineral content of ore is low in many mines, and this has led some mines to close in the last several decades.

Bodie, California was once a thriving community but is now abandoned except for the tourists who come to see its remains. So-called ghost towns are common in the Empty Interior; New Mexico alone has 2000 of these abandoned towns.

shipment. Thus, refining is a major manufacturing industry in the region, employing over 12,000 people.

Lead and zinc are the next most important metals mined in the United States' portion of the Empty Interior, although they are more important than copper in Canada's portion. The vast mine near Kimberley, British Columbia, dominates Canadian production of both metals. Lead and zinc mining is more widespread in the United States and is usually extracted with other minerals. The Butte Hill mine in Montana, for example, long produced lead, zinc, and copper; the Coeur d'Alene district in northern Idaho produces gold, silver, lead, and zinc; and Colorado's Leadville district produces gold, silver, lead, zinc, and molybdenum.

Other minerals are mined in the Empty Interior, some of them in economically significant quantities. Uranium deposits are widespread, with Utah and Colorado currently the principal producers. Large coal deposits in southwestern British Columbia have been developed, with approximately 25 million tons of coal mined annually, most destined for export to Japan.

A huge, potential resource is spread across thousands of square miles where Utah, Colorado, and Wyoming meet. Vast oil shale deposits underlie the area in the Green River geologic formation (Figure 13.7). As much as one trillion barrels of oil are locked in the

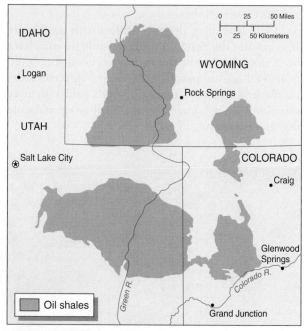

Figure 13.7   Oil shale deposits. Few people presently live in this isolated part of the Empty Interior. How much local population growth would a major development of this resource create?

formation's rocks, an amount far more than the world's entire proven alternative oil reserves, including those in the Athabasca tar sands (Chapter 12).

The issues that need to be resolved before extracting oil from the Green River deposits, however, are daunting. With present extraction technology so expensive, mining is not economically feasible. The process also requires huge amounts of water, a problem in this dry area where access to available water is already disputed. And the process may discard enough waste material to level the area's topography, a concern to environmentalists and others dependent on local recreation opportunities. Because of these and related issues, the oil shale deposits are likely to remain only a potential resource for the indefinite future.

As rich, abundant, and diverse as these mineral deposits were (and are) across the Empty Interior, mining's local economic benefits are precarious. Local deposits are economically ephemeral and can be depleted. Economic recession, such as occurred in the 1980s, also affected some mines due to both increased foreign competition and concern over mining. Smelting's environmental and health consequences can also affect particular mines. Montana's Butte Hill mine, for example, is closed. The large zinc smelter in Kellogg, Idaho, is also closed. And many other mines and smelters have cut production.

Other than temporary boom towns, cities rarely develop because of local mining activity. In the Empty Interior, only copper offers a partial exception, because some mines operate for many years. Even with copper, however, mining towns remain small in population size. Butte has about 33,900 residents, but it is also an agricultural processing center. Globe has a population of 7486, Bisbee has 5123 people, and Morenci has fewer than 2000 residents.

The Empty Interior is a region of vast open spaces, sparse settlement, and many areas where the frontier is close at hand. Most people live in isolated urban nodes. Residents know they live where thousands of others choose to spend their vacations. For its residents, the Empty Interior is a delightful place to live but a difficult place in which to make a living.

NATURAL HAZARDS CASE STUDY

The Yellowstone Fires

The summer of 1988 started like any other in Yellowstone National Park. The park, like much of the northern Rockies, was in a period of long-term drought interrupted by shorter periods of heavy rain. May was a wet month. Natural fires that started in the park through much of June burned themselves out. Human-caused fires were doused quickly, as policy dictated.

Until 1972, National Park Service policy had been to extinguish all fires expediently within a national park. But in that year, the service adopted a "let it burn" policy for natural fires based on studies that showed such burns are necessary to maintain vegetative diversity and, thus, a diverse animal life. The fire management plan for Yellowstone gave park officials latitude to monitor and manage such fires. Fires would be allowed to burn if they did not threaten life or property. Still, by 1988 most of the park's forest was littered with 100 years' accumulation of forest deadfall.

The rainy May that year merged with a summer of no precipitation, very low humidity, high winds, and high temperatures. By mid-summer, much of the deadfall was drier than kiln-dried lumber. By July, one of the worst forest fire seasons in western U.S. history was underway. Before it was over, some 70,000 fires burned 1.5 million hectares (3.75 million acres) in the West. The cost of fighting those fires exceeded $500 million.

In Yellowstone, the first fires that did not quickly burn themselves out started in late June and early July at scattered locations. Lightning caused most of them, although one of the largest, the North Fork fire, was probably started accidentally by a lumberman working a power saw near the park's western border. By mid-August, more than 40 separate fires had been identified.

On July 21, park officials decided to suppress all fires; the situation was clearly more critical than they had expected in early June. By then fires had burned some 6000 hectares (15,000 acres), and still no rain fell. On August 20 alone, 25,000 hectares (62,000 acres) of Yellowstone burned. On September 7, flames lapped much of the area surrounding the Old Faithful geyser, and officials feared the nearby historic inn would be lost. The inn survived, however, and finally, on September 10, it rained. The worst was over, although isolated fires continued to burn into November. When tallies of charred land were complete, officials documented almost 400,000 hectares (1 million acres), or 45 percent of the park's land, had burned that summer.

The 1988 fires in Yellowstone National Park tested the National Park Service's "let it burn" policy and the endurance of the firefighters brought in to attempt to guide the fires' direction. Subsequent regrowth of vegetation was rapid and widespread within the Park.

The fires generated a storm of controversy over the National Park Service's "let it burn" policy. Tourist officials in service industries peripheral to the park feared a loss of business if fewer people chose to visit Yellowstone. Loggers and ranchers who worked near the park argued that fire did not respect park boundaries and that the policy threatened their livelihood unnecessarily. Others were simply saddened over the tarnishing of Yellowstone's beauty. Ecologists countered that the century-long absence of major fires in Yellowstone had resulted in loss of meadowlands and old forest that supported animal and bird life, leading to less diversity and lower populations. The fire led the Park Service to rewrite its policy, which now allows natural fires to burn but calls for more caution when developed areas are threatened.

Apart from such bureaucratic debate, Yellowstone itself began a speedy recovery. Although many large animals had died, no more had succumbed than are lost during a difficult winter. Grasses were greening in some locations by late September. By the following spring and early summer, the most spectacular wildflower display in memory dappled Yellowstone with vivid beauty. Besides the floral resurgence, tree seedlings from the lodgepole pine and other species common to the park were sprouting everywhere. In fact, one type of cone on the lodgepole pine, the park's dominant species, opens only when heated.

The other dominant species, the tourists, returned in record numbers, anxious to see the new wonders of Yellowstone.

SETTLEMENT CASE STUDY

Moab, Utah

The "Invincible Three" were jittery after robbing the San Miguel Bank at Telluride, Colorado, on a June morning in 1889. Seeing a posse in every dustcloud, Butch Cassidy and his sidekicks, Tom McCarty and Matt Warner, spurred their horses toward a favorite northeastern Utah hideout called Brown's Hole. From Telluride, they galloped through Mancos and Monticello in a relentless ride to Moab, a small, isolated town situated at the only sensible crossing of the Colorado River for miles in either direction. There, they grabbed fresh horses and ferried across the river to find refuge in rugged Utah terrain.

Outlaws passing through southeastern Utah late in the nineteenth century had an instinctive appreciation for Moab's geographic attributes. Sitting astride a natural transportation route in a region of irregular terrain and harsh climate, Moab was remote from populated places. Today, the attributes that appealed to Butch Cassidy more than a century ago have kept Moab small but holding a strong hand of economic cards.

Moab exists because it is located at what, until modern times, was the region's only reasonable place to cross the Colorado River. Typically, the Colorado is deeply entrenched in the Colorado Plateau across southeastern Utah as well as much of Arizona. Few side streams and tributary rivers enter the Colorado along this portion, and almost all of those that do have cut steep-walled canyons to join the main river's larger one.

Mormon missionaries made the first, though unsuccessful, attempt to settle Moab in 1855. The site appealed to them for several reasons. The river crossing was a natural transportation control point, and it was also in the Moab valley, a graben lying perpendicular to the course of the Colorado. Modest yet promising grazing lands existed in the valley south of the river, and access to river water meant irrigated agriculture was possible in the floodplain where it cut across the valley.

When the Mormon missionaries set up at Moab, most American settlement was following east-west trails many miles to the north or south. So when the Mormon mission closed in 1855 after an attack by Ute Indians, Moab's site was too isolated to be worth serious attention by other settlers. Decades passed before people returned for another try at permanence, though prospectors, trappers, cattlemen, and outlaws continued using the river crossing.

In 1902, Moab finally had enough people and organization to incorporate as an independent community, even though the residents were few and most of the town's surrounding land had little economic appeal. Minerals existed, but at the time they could be extracted only with difficulty because of the harsh environment. Cattle could be grazed, but only at low densities and at some risk because water availability was limited. Agriculture was a gamble because the area's average annual precipitation was well below 25 centimeters (10 inches).

Moab stayed dormant until it experienced a minor spurt in the 1920s, when an enthusiastic but fruitless search for oil drew hopefuls temporarily to town. Soon disillusioned, they left. The town stayed sleepy for another 30 years until its population tripled due to a successful search for uranium. But this wake-up, too, was relatively short-lived; Moab's population growth stalled and then shrank as the mines played out.

From 1900 until the mid-1970s, Moab underwent several boom-and-bust cycles, an experience typical of many Empty Interior communities. But unlike many others, Moab had

The canyon country around Moab, Utah, is stark and extremely rugged, yet its beauty attracts tens of thousands of visitors each year.

sufficient geographic resources to survive even the difficult times. Though agricultural land was limited and restricted to the floodplain between the town and the river, people took advantage of what existed. Most used the space for personal crops or stock, but some planted fruit trees for commercial markets. Water was reliable, not only from the Colorado River but also from snowmelt streams flowing off the nearby LaSal Mountains. In addition, other local mineral resources surfaced and were exploited. Potash mining gained an economic toehold even as the larger uranium mining operation of the 1950s, 1960s, and 1970s slipped from prominence to closure. Perhaps most important, local roads were improved to make Moab less isolated from major east-west traffic.

In the 1990s, Moab was back on the boom side of the cycle. Ironically, the harsh and broken landscape that provided outlaws haven a century earlier and kept most land in federal ownership is a magnet attracting economic opportunity. Moab's scenic setting is spectacular. Red, soft orange, rust, and mahogany cliffs, buttes, and spires are the town's backdrop. The desert gems of Arches National Park and Canyonlands National Park lie on either side of Moab, while a cool, moist mountainous national forestland is less than an hour's drive up into the LaSal Mountains.

This landscape, now widely known and accessible, lures entrepreneurs and scads of summer tourists who spend several days in the area based in Moab. White-water rafting offers a respite between short hikes and windshield vistas, with trips of one day or several days skimming the lowest undammed portion of the Colorado River. Bikers cruise about in their bike-laden 4-wheel drives, headed for designated biking trails crisscrossing slick rock, desert terrain. Representatives of the motion picture industry regularly take advantage of the picturesque landscape, shooting ads and films.

In economic terms, even as Moab's uranium mining operations went bust, its orientation toward tourism filled the vacuum. Businesses boomed. Touring companies, T-shirt shops, trading posts, jewelry shops, and gas stations now fill every niche in town. Even a small retirement community now exists on the outskirts, basking in the desert climate.

Moab's latest wave of economic growth is a mixed blessing, however. On an ecological scale, the desert landscape surrounding the town is fragile and easily damaged. As the number of tourists increases, so does the number who are careless of the land. On a human scale, Moab's boom is seasonal. In winter, the town shuts down as many businesses close until next year's tourist time. In summer, some long-time residents who remember the 1950s and early 1960s long for solitude and an abatement of the noise accompanying hordes of visitors. But all in all, Moab's economy is more stable now than at any earlier time, and most who choose to live in this town enveloped by beautiful but unforgiving landscape believe the stability is worth the cost.

Review Questions

1. How might one consider tourism to be a mixed blessing in the "Empty Interior"?

2. Why is much of the land of the Empty Interior still in government hands? Discuss the pattern of federal lands in the region.

3. How could one argue that the pattern of governmental land ownership has had a negative impact on regional economic growth?

4. What is the impact of the Mormon church on the cultural landscape of the region?

5. What are the contentious issues associated with irrigation-based agriculture within the region?

6. What are some of the specific issues involved with using the water from the Colorado River? What are some of the proposed solutions to the intense competition?

7. What are the widespread implications of the National Park Service's "let it burn" policy?

CHAPTER 14

THE SOUTHWEST BORDER AREA: TRICULTURAL DEVELOPMENT

The Southwest is a distinctive place to the American mind but a somewhat blurred place on American maps which is to say that there is a Southwest but there is little agreement to just where it is. **Donald W. Meinig**

Preview

1. The Southwest is dominated by three coexisting cultures: Spanish American, American Indian, and European American.

2. Native American populations in the region are large and diverse, but less integrated into the American melting pot than other ethnic groups.

3. The Hispanic presence in the region dates back over 400 years, but populations have increased dramatically in recent decades because of immigration and higher than average rates of natural increase.

4. The economy of the region is highly integrated with the relatively poorer areas on the other side of the border with Mexico.

5. Population in the Southwest Border Area has exploded in recent decades.

The Southwest is a culture region distinguished by coexisting Spanish American, American Indian, and Northwest European American (Anglo) people. The presence of each group imbues the region with different characteristics that, together, make the Southwest one of the most plural and transitionary American regions.

From southern California's Pacific Coast to the Texas Gulf Coast north of the Rio Grande, the Southwest Border Area parallels the U.S.-Mexico border (Figure 14.1). Although most of the international border has been in place for 150 years, Spaniards occupied the American Southwest for centuries before the United States started making claims to this part of North America in the 1830s. Permeable culturally, the border is no hindrance to continuing infusions of non-Anglo influences in the region from Mexico.

Beginning thousands of years before Spaniards arrived searching for gold, a succession of American Indian peoples moved into the Southwest. These first settlers thrived and, as different Indian groups arrived over time, they spread across the region developing various cultures and livelihoods. Native Americans still live in the Southwest in greater numbers than in any other of North America's regions.

Northwest European and American settlers are recent arrivals. Because of its aridity, the Southwest was not generally attractive to them. Anglo disdain for the region changed, however, as definitions of success and satisfaction shifted and as technological developments neutralized many of the Southwest's climatic disadvantages. Cities arose and prospered, water was secured, and the long, warm, dry climate became an attraction for millions of Americans.

Today, the Southwest Border Area's tricultural complexity lends it its special character, and the polychromatic physical environment is like a stage designed to emphasize aspects of each culture while setting the region apart from the rest of the country. American Indian and Spanish populations coexisted there for 250 years after the Spanish arrived at the end of the sixteenth century but before Anglos began trickling into the area in the middle of the nineteenth century. The result was an unceasing period of both forced and inevitable *acculturation*, or cultural borrowing and sharing among groups.

Acculturation in the Southwest Border Area became more complex when Anglos arrived. This third major culture, backed by its government and by a

Although the United States of America and los Estados Unidos Mexicanos are linked by a common border more than 1000 miles (1600 kilometers) long, much shared culture, and a tradition of peaceful relations, contrasts between the Mexican (right) and Californian sides are dramatic at Tijuana because of the fence restricting cross-border movement.

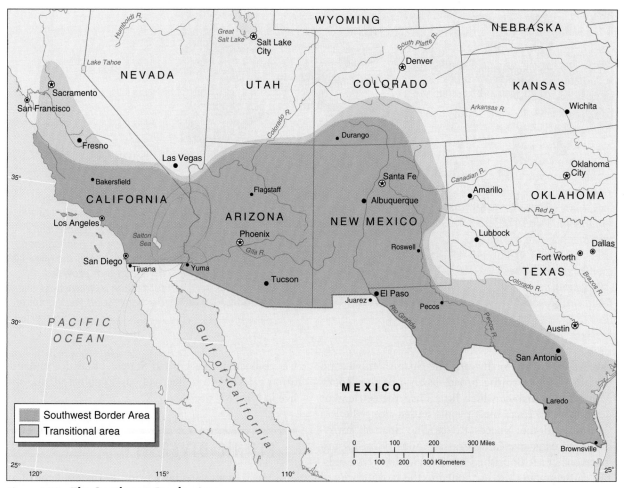

Figure 14.1 The Southwest Border Area.

large, economically thriving and expansionist population, joined the region's mix. The cultural borrowing and sharing that makes the Southwest Border Area distinctive took on a decidedly Anglo tilt.

Today, residents in the tricultural border region are preponderantly non-Spanish and non-Indian. Perhaps 1 person in 4 has a Spanish surname, and little more than 1 in 100 is American Indian. The expectation, then, might be that these minority populations would be engulfed by the larger and relatively homogeneous Anglo one. But this has not been the case. Both minority groups have had a major, sustained impact. No other region in the United States except Hawaii is as clearly tied to non-Anglo ethnic patterns.

The cultural impact of Indian and Spanish groups on the regional landscapes is obvious. Spanish place-names abound, especially along the Rio Grande in Texas and New Mexico and in coastal southern California. American Indian place-names are locally important, especially on the Navajo, Hopi, and Papago reservations in Arizona. Hispanic neighborhoods are sometimes identified by the use of adobe and tiled roofs, outside ornamentation, and yards encircled with high fences. The traditional circular hogans with their dome-shaped roofs on the Navajo reservation and the multisorted pueblos in New Mexico are striking, enduring elements of Indian architecture. Foods rich in chiles, corn products (especially tortillas), and pinto beans are a mainstay of regional fare. Catholicism, a religion held by most Hispanics as well as many American Indians, dominates.

Spaniards expanded into the Southwest from their colonial core in Mexico. When they encountered a difficult or different environment, the expansion halted. The canyonlands of northern Arizona and southern Utah provided an effective barrier to Spanish

Mesa Verde National Park showcases the incredible culture of the region's early Native American inhabitants. Pictured is one of the large cliff dwellings that provided protection from enemies and the harsh Southwest climate.

expansion to the northwest. The Spanish moved up the Rio Grande to the broad expanse of the Rocky Mountains, north of which little Hispanic settlement developed. In Texas, most people settled along the Rio Grande and the Nueces River. The Spanish introduced an extensive cattle-grazing industry into south Texas, where it flourished on the flat, dry grasslands. But cattle grazing was not suited to the moist, forested lands in the eastern section of Texas. As a result, that area, left as a frontier, was largely ignored by the Spaniards.

Thus, the Southwest Border Area is most clearly identified by the extent of Spanish settlement, and its basic outline was manifested well before the arrival of the westward migrating American population in the 1800s.

The aridity of Arizona, New Mexico, and bordering areas in Utah and Colorado discouraged large-scale Anglo agricultural settlement in the nineteenth century. The slow pace of Anglo penetration ensured that substantial numbers of American Indians stayed in those states, although Indian strength was as important as Anglo hesitancy. Pueblo-dwelling cultures like the Anasazi of the upper Rio Grande valley—with multi-storied apartment dwellings, complex water-gathering systems for irrigated agriculture, art, and far-flung trade—had developed one of the technologically most advanced pre-European Indian civilizations in what

was to become the United States. Their descendants, often referred to collectively as Pueblo Indians, still live in New Mexico.

ETHNIC DIVERSITY

Most Americans take pride in the concept of an "American melting pot" in which diverse peoples form a population united by common goals and sharing a broadly common culture. Except for Quebecois separatists, this concept applies to Canada as well. Blended in America's melting pot were 45 million Europeans and millions of Africans who arrived primarily between the early seventeenth century and 1920s, and, more recently, Latin Americans and Asians who moved to the United States in large numbers during the latter half of the twentieth century. Many of their progeny still hold dear the customs of their homelands. In their divergent ways of eating and drinking, religious preferences, and musical tastes, these Americans continue to demonstrate their ethnic heritage. Sometimes they choose to live with others of the same background in closely knit neighborhoods. Yet nearly all of these people now have far more in common with other Americans than with the present residents of their European, African, Latin American, or Asian homelands.

American Indians

Ironically, the residents of the United States (and Canada) who have been least well integrated into the melting pot are those who have been here the longest. Of all the country's major ethnic groups, American Indians—divided into hundreds of tribes with varied languages and customs—are the least integrated into society. For many, this is purposeful. They wish to preserve and restore the integrity of tribal ways or foster pan-Indian ways—that is, customs common to or adopted by various Indian cultures.

Although American Indians live in every part of the United States, today most of them are found in areas that white settlers had deemed undesirable or that were not part of early settlement frontiers (Figure 14.2). Countless Eastern tribes died out, succumbing more often to disease than to the bullet. Others were forcibly moved west, out of the way of European settlement. Still others clung to the land, dispersing or hiding in small numbers until they regrouped or joined together in small and large confederacies to resecure toeholds near ancestral lands. Today, many Indian tribes live in nonreservation communities.

The Sioux and other Plains tribes, as well as the tribes of the Interior West (particularly in the Southwest), fared far better numerically. Because lands there were not easily suited to agriculture, a wanton governmental disregard of Indian rights was replaced with a less overtly hostile approach within a few decades after the first substantial contract. Still, the government's goal was to contain these peoples, and as a result, the great majority of American Indian reservations as well as most Indians live in the Plains and Western states (Figure 14.3).

Within the Southwest Border Area, the American Indian population is culturally diverse. The largest tribes are the Navajo in the Four Corners area, where the states of Colorado, Utah, Arizona, and New Mexico meet; several Apache tribes in Arizona and New Mexico; the Pueblo groups in New Mexico; the Papago in southern Arizona; the Hopi in northwestern Arizona; and the Utes in southwestern Colorado. Though the tribes span the region, Indian people are by no means evenly distributed across it. Most are found on the major reservations, especially those centered on the Four Corners, and in some southern California cities. The Navajo reservation of the Four Corners area has ten times the population of any other reservation. Arizona and New Mexico together are home for nearly 400,000 Indian people. Only Oklahoma outranks Arizona in the size of its Indian population. California ranks third, and New Mexico fourth. Among American cities, the Los

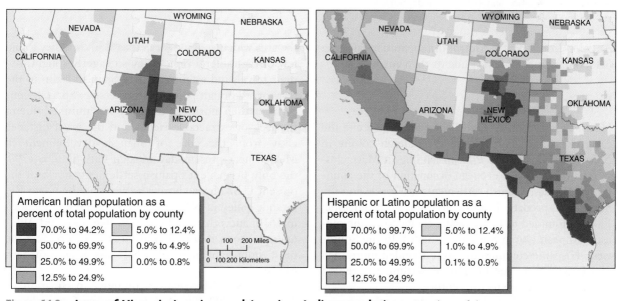

Figure 14.2   Areas of Hispanic American and American Indian populations. Members of these two groups are numerically dominant across much the Southwest Border Area. *Source:* U.S. Census Bureau, 2000.

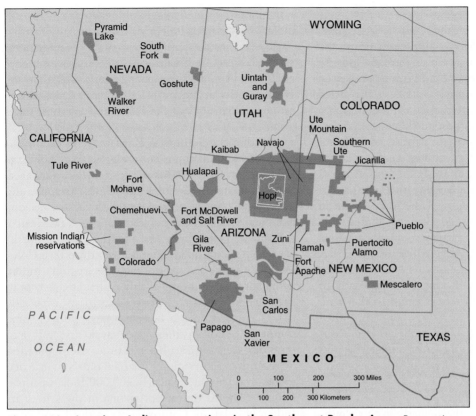

Figure 14.3   American Indian reservations in the Southwest Border Area. Reservation lands were largely ignored by Anglo settlers in the nineteenth century because of the lands' apparent lack of economic potential. Some of these same lands are now valued for their resources, most notably the Navajo Reservation and its large coal deposits.

Angeles-Long Beach Metropolitan Statistical Area (MSA) has more Indian residents than any other.

Hispanic Americans

History texts in the United States commonly cite the early coastal settlements like the Raleigh colony in North Carolina or the Pilgrim landing in Massachusetts as the start of European occupation of the country. But in fact, Spanish settlements in Florida and the Southwest predate English arrivals. St. Augustine, Florida, founded in 1565, validly claims to be the oldest European city in the United States. In the Southwest, Hispanic communities were founded more than 200 years before Anglos started settling the region during the early nineteenth century.

All of what is today called the Southwest was incorporated into the Spanish Empire during the early sixteenth century. By 1550, the Spanish had explored the area widely looking for gold and silver. They found few of the mineral riches they so eagerly sought, but their travels did establish a firm Spanish claim to the territory. Though politically possessive, Spanish concern for their northern territory was minimal due to the lack of extractable riches and the great distance away from the core of Spanish development in Mexico near present-day Mexico City. Before 1700, the only permanent Spanish settlements north of the present U.S.-Mexico border were along the upper Rio Grande valley in New Mexico. Santa Fe was founded in 1610, and other communities, like Taos and Albuquerque, soon followed. This area was to remain the most important part of what was then northern Mexico.

The Spanish began a tentative occupation of Arizona in 1700. The Apache were a constant threat, repeatedly raiding settlements there. Consequently, Arizona remained a small settlement outlier.

Chimayo, a village located north of Santa Fe in New Mexico, is one of the few remaining examples of Spanish influence on town form in the region. Spanish towns tended to be small and unplanned, and their distinctive pattern was usually obliterated by urban growth during the last 100 years.

Colonization of Texas, which began about the same time, had long-range results that were considerably more successful. Nacogdoches was founded in 1716, followed two years later by San Antonio. During the middle 1700s, the Spanish settled the lower Rio Grande valley. Still, by the early nineteenth century, the authorities viewed these and other Hispanic settlements as an inadequate deterrent to an expansionist United States then eyeing the Texas territory. Thus, Mexican authorities allowed foreigners, mostly Americans, to develop settlements there during the 1820s and 1830s, hoping that these welcomed new residents would be loyal to Mexico (which was independent from Spain after 1821) and form an effective bulwark to the American government's designs on Texas. History was to prove this hope false.

California, the most distant of Spain's northern territories, was settled much later by Spain. The Spaniards first established a mission and presidio (military post) at San Diego in 1769. During the next two decades, a string of missions, with a few presidios and civilian communities, spread along the coast as far as Sonoma, north of San Francisco. Spain, and then Mexico, encouraged this thin band of coastal occupation partly in response to growing British and Russian interest in the continent's west coast.

After U.S. acquisition of Texas in 1845 and the Mexican Cession, which ended the Mexican-American War in 1848, the estimated Mexican population of this broad territory was 82,500. Of this number, roughly 60,000 were in New Mexico, 14,000 in Texas, 7500 in California, and 1000 in Arizona.[1]

By 1850, the Mexican population of Texas and California was less than 10 percent of the two states' total residents. East Texas, the new western frontier for Southern settlements, attracted many from that region. The discovery of gold in California in 1848 lured non-Spanish peoples into central and northern portions of the state. Only in New Mexico, southern California, and Texas south of San Antonio did Hispanic people continue to dominate demographically for a few more decades.

The Southwest's originally low Hispanic population greatly increased through immigration from Mexico during the twentieth century. Many live at or near poverty levels, but however difficult their existence may be, most view their present living conditions as a substantial improvement over what is available in northern Mexico. Between 1900 and 1990, roughly 2.9 million people migrated legally from Mexico to the United States; California, Texas, and Arizona easily rank as their leading destinations. Annual Mexican immigration into the United States in the late 1980s averaged between 75,000 and 90,000, or about 15 percent of the United

[1] For a presentation of the settlement geography of the Hispanic population to 1960, see Richard L. Nostrand, "The Hispanic-American Borderland: Delimitation of an American Culture Region," *Annals of the Association of American Geographers*, 60 (December 1970): 638–661.

States' total immigration. Dramatic increases occurred during the 1990s, with 2,249,421 Mexicans immigrating legally to the United States during the decade. Since 2000, the annual number has exceeded 171,000. An increasing number of Mexican people are now moving into areas quite distant from the Southwest, but most remain in the region's border states.

Large numbers of illegal immigrants make it difficult to estimate how many Mexicans enter the United States each year. An accurate figure for the illegal migration total is impossible, as is reflected in the wide-ranging estimates of 2 to 12 million for the number entering. Although legal migration is considerable, general quota restrictions and denials of Mexican requests to immigrate mean far more people have wanted to enter the United States than have been admitted. The illegal movement has existed along the Mexico-U.S. border for decades. While many of these illegal migrants are caught and returned to Mexico, most simply try again until they are successful.

Legal or not, compelling reasons lie behind the migration. A high rate of population growth exists in Mexico, along with widespread unemployment and poverty in the north. The U.S.-Mexican international border, though more closely regulated than that between the United States and Canada, is generally marked by only a fence and can be crossed with relative ease. Despite the border officials' increased efforts to control access and despite attempts by Mexico and the United States to negotiate an enforceable migration agreement, there is scant likelihood that the flow of people will be staunched in the near future. Even the hundreds of miles of more substantial fencing under construction during 2007 and 2008 are not likely to end illegal immigration.

In the past, the United States legally admitted large numbers of temporary workers from Mexico. Many entered the United States to harvest agricultural products in the Southwest for a few months and then returned to Mexico with their profits.

In the early 1960s, the federal government decided this influx had a detrimental impact on the living and working conditions of U.S. migrant farm laborers. Willingly or not, temporary workers tolerated sunup to sundown drudgery and overcrowded, unsanitary housing along with low wages. Organized labor, too, opposed a large, temporary labor migration. Attempting to unionize the U.S. migrant agricultural labor force, labor leaders also argued that temporary Mexican workers held down wages by being a plentiful and inexpensive labor force. In response to these pressures, the United States greatly restricted the number of Mexican temporary workers admitted.

Researchers debate whether this decision had any positive effect on working conditions for U.S. migrant workers. Certainly, it has been a factor in the rapid mechanization of crop harvesting. Tomatoes, iceberg lettuce, peaches, asparagus—these and other types of produce once picked by hand—can now be picked by machine. The decision has also indisputably pushed many workers like those who had formerly entered the country legally to enter illegally.

This East Los Angeles Mexican American neighborhood has decorative and symbolic murals on the housing reminiscent of the great Mexican muralists.

A significant share of the Southwest Border Area's people are Hispanic. In 2000, persons of Spanish surname represented 25.2 percent of the population of Arizona; in California, the proportion was 32.4 percent; in Colorado, 17.1 percent; in New Mexico, 42.1 percent; and in Texas, 32.0 percent. In 2000, the Bureau of the Census estimated the Hispanic population of the United States at 35.3 million, an increase of 142 percent over 1980. One American in eight currently is Hispanic, and more than 60 percent of the Hispanic population is Mexican American.

Hispanic percentages have increased since 1980 in each state in the Southwest Border Area. Hispanics currently represent almost one-third of the populations of the country's two most populous states, California and Texas. Almost a third of California's schoolchildren are Hispanic. In short, the proportion of the Spanish-surname population in the states lying wholly or partly in this tricultural border region increased faster than the rest of the states' populations, even though all, except New Mexico, had a substantial immigration of non-Spanish-surname people as well.

This Spanish-surname population increase is due both to immigration and to a higher fertility rate as compared to that for the total population. Completed family size for Spanish-surname families averages about one more child per family than that for non-Hispanic families. Although part of the explanation for this higher fertility rate is economic, culture also plays an important role. In much of the Southwest, the family—especially a large, extended one of grandparents, aunts, uncles, cousins, and other assorted relations—forms the heart of Hispanic life. Even in the region's larger cities, family events from weddings to graduations bring dispersed family members together to celebrate and share kinship.

Today, Hispanic people concentrate in those areas where most were found in 1850 (see Figure 14.2). Nearly every county in the Rio Grande valley is populated predominantly by people with Spanish surnames. In some south Texas counties, 90 percent of the population is Hispanic.

The Hispanic population is also overwhelmingly urban. More than 3 million Spanish-heritage people presently reside in the Los Angeles-Long Beach MSA, a population that represents roughly 1 out of every 10 Hispanics in the country. Greater Los Angeles is the largest Spanish culture community north of Mexico City.

Socioeconomic Disparities

The cultural and economic gaps between the Anglo population and the Hispanic and American Indian groups are immense in many instances. In most of the Southwest's cities, many Hispanic neighborhoods are comprised of low-income residents with large families living in poor-quality housing. Everywhere across the region, a close association exists between a high percentage of the population living below the poverty level and a high percentage of the population being Hispanic or American Indian (Figure 14.4).

The cultural differences among Anglos, Hispanics, and American Indians surface over much of the Southwest, especially in rural areas. Here the American melting pot has worked to only a limited extent, and a cultural gap remains between these various groups. When the Southwest was annexed by the United States, its Hispanic residents suddenly became part of a country with a different language, a different legal system, and a different economic structure. Anglos quickly and assertively took control of local politics and economics. They also grew rapidly in numbers and, by force of presence, predominated.

The U.S. policy that placed most American Indians on reservations effectively maintained distinctions

---

What Is In a Name?

What is in a name? Sometimes a great deal. For the Spanish-surname population of the Southwest, the issue of what to call themselves has been important but difficult to agree on.[2]

Some younger people, angered over injustices suffered at the hands of Anglo society, call themselves *la raza* (the race) or *Chicanos*. Recent migrants from Mexico also often prefer the latter term. Spanish-surnamed people from the Southwest prefer *Hispano*. Many in Texas call themselves *Tejanos*. *Mexican* is usually reserved for people born in Mexico, although some would also use the term when referring to the older, rural, conservative population. People who are from Latin American countries or identify with one of the Latin American cultures are called *Latino*. Arbitrarily, we use *Hispanic* or, when appropriate, *Latino*.

[2] For an interesting discussion of this problem of nomenclature, see Arthur L. Campa, *Hispanic Culture in the Southwest* (Norman: University of Oklahoma Press, 1979), pp. 6–10.

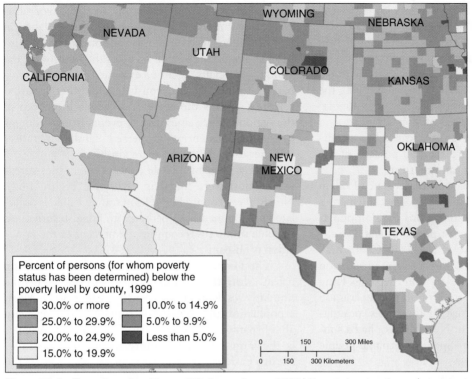

Figure 14.4 **Poverty rate.** (*Source:* U.S. Census Bureau, 2000.) Poverty is extensive and persistent throughout the Southwest Border Area.

between Indian and other cultures. Almost by definition, reservations were in out-of-the-way places that offered little obvious economic opportunity and little contact with Anglos. Until recently, educational opportunities were minimal on reservations and usually emphasized Anglo cultural goals at the expense of Indian perspectives. Local opportunities to use an education were almost nonexistent. In spite of this, most Indians chose to stay. Although deemed undesirable by Anglos, reservation lands often held swaths of Native American ancestral homeland. So ties to the land and to family, coupled with a legitimate concern about coping within the Anglo urban culture, lessened what on economic grounds might have been expected to be a wholesale movement off the reservations.

In small but meaningful ways, politics and conditions for the Southwest Border Area's Hispanic and Indian minorities have changed for the better in recent years. Navajo experiences are indicative. An elected tribal council now makes most economic decisions for the reservation, although final authority remains— much to the dislike of many Navajos—with the Bureau of Indian Affairs. All-weather roads now cross the

reservation, greatly reducing isolation. Health and educational facilities have been improved.

Huge reserves of fossil fuel, particularly coal, exist on Navajo land. Several large power plants located on the reservation serve southern California, and details for additional plants are under discussion. Many Navajo oppose this disruption of their tribal lands; others feel they receive far too little monetary return for the disruption. Another concern some Navajo have is the large amount of water used by power plants. Readily available water is in limited supply on the reservation, and its diversion to power production may restrict the development of irrigated agriculture or other forms of industrialization. Such controversies sometimes split Navajo leadership. Nevertheless, power companies move millions of dollars into the reservation economy each year, and the economic effect is undeniably substantial. Tribal leaders also have expanded the reservation's tourist industry, attracting a number of new industries with its large, available, and now better-educated labor force.

Still, Navajo family income on the reservation remains less than half of that for white rural

On the Navajo Nation reservation, scattered towns like Tuba City serve administrative, social, and provisioning functions for tribal members living in small family compounds or solitary homes miles away in the surrounding desert.

Americans, and the reservation fertility rate is twice the national average. Even so, the general situation is improving.

THE CROSS-BORDER ECONOMY

Perhaps nowhere else in the world do high-income and low-income societies meet in such close geographic proximity as along the U.S.-Mexico border. The economic disparities north-south across the border are striking to even casual observers. It also has long been a relatively open boundary, with movement and exchange easily accomplished and often encouraged.

During World War I and the economic boom times of the early 1920s, large numbers of Mexicans moved across the border to fill labor needs in the United States. They did so again in the 1940s when the United States experienced a labor shortage because of World War II. In 1942, the two countries negotiated the Mexican Labor Program, commonly called the Bracero Program. Under its auspices, Mexican laborers could legally enter the United States and work as seasonal laborers in the agricultural sector. That program lasted until 1964.

In 1965 Mexico started the Border Industrialization Program to attract U.S. labor-intensive manufacturing industries to border communities in northern Mexico. The law allowed foreign companies, called *maquiladoras*, to import equipment and material duty-free into Mexico if the manufactured products were then exported from Mexico. In 1989 the export regulation was eased, and now maquiladoras can sell 50 percent of their total product in Mexico.

At first, maquiladora locations were restricted to a zone within 20 kilometers (8 miles) of the U.S. border to attract foreign investment and employment opportunities to an area far from the center of Mexican economic activity to the south. That restriction, too, has been relaxed, and today maquiladoras can locate almost anywhere in the country outside Mexico City.

For Mexico, the program offered the possibility of jobs for its people. The attraction for U.S. firms in this relationship was the opportunity to use low-cost labor at locations near the U.S. marketplace and sources of supply where transportation costs could be minimized. The result could be a significant reduction in the overall cost of production for many labor-intensive products.

The maquiladoras are *in-bond* plants; they import parts and materials from companies in the United States, and then assemble those parts into a finished product. The finished product is shipped to the United States. American firms pay duty only on the value added to the product in Mexico, not the value of the finished product. And wage rates in Mexico are far lower (perhaps 10 percent of U.S. wages). Mexico benefits primarily through the income earned by the maquiladora workers, who remain in Mexico rather than seeking work north of the border. The maquiladoras are successful because American companies can separate capital-intensive manufacturing and highly skilled work from the labor-intensive, assembly phase of operation, which requires low-skill workers.

Many firms were attracted to the Border Industrialization Program by its cost-saving opportunities during the decade after the program's inception. Little additional expansion took place until the late 1980s. Wage rates in most overseas low-wage areas began increasing, while continued devaluation of the Mexican peso meant that wages there, measured in U.S. dollars, were declining. As a result, each year several hundred new firms were added to the maquiladora list. Currently, about 4000 maquiladoras employ approximately one million Mexican laborers. Maquiladoras lost special tariff status on January 1, 2001, when the North American Free Trade Agreement (NAFTA) was phased in.

A few hints exist to suggest that the U.S.-Mexico border area may be a new region-in-the-making, in spite of the strong cultural and economic differences. Almost paradoxically, the engine driving integration of the American Southwest Border Area and northern Mexico's border region is the border's income differential.

On a daily scale, the differential is shown by the massive movement of people across the border. Millions of American citizens travel to Mexico each year as tourists for recreation, for an easy opportunity to experience a different country and culture, and for purposes of buying goods at lower cost. A few choose to retire there. Mexicans travel north to find products that are not easily available back home, although Mexico's devaluation of the peso in 1994 as part of NAFTA sharply curtailed border area spending by Mexican shoppers on the U.S. side. Mexicans also travel north to visit family and friends on the other side of the border. A significant share of U.S. border state residents in 1990, for example, were born in Mexico, ranging from one in six people in Arizona to almost one in four in Texas.

Almost two dozen pairs of urban places are located along the border, indicating a long period of exchange. U.S. cities and towns paired with Mexican urban places extend from San Diego-Tijuana in the west to Brownsville-Matamoros near the Gulf of Mexico.

Taken together, along with the maquiladoras, these hints of a formative region are only suggestive. For this nascent region to become part of North America's map, the border area and its people must become even more integrated economically and more identifiable by others as distinct from adjacent parts of both countries. The pace at which such integration occurs will be affected by national policies on both sides of the border arising far outside the region itself.

POPULATION GROWTH TODAY

Along with its distinctive cultural diversity, the main identifying features of the Southwest Border Area for most Americans are sunshine and aridity—reasonably accurate images. The Southwest's sunny climate, especially in mild winter areas, is a powerful attraction for many Americans. The spectacular growth of southern California's population during the late nineteenth century and the emergence of its vast metropolitan areas in the twentieth century are a stunning demographic shift.

A new housing development project in Palm Desert, California, sprawls across the landscape. Plans like this one have put several hundred thousand potential second home and retirement lots on the market in rural parts of the state. This real estate boom, reflecting Americans move toward the "Sunbelt," has resulted in a supply of building lots that far exceeds demand.

This burgeoning population and its associated economic growth have spread well beyond California. In the 1990s, Arizona's population grew by almost 40 percent, second only to thinly settled Nevada's 63 percent. Arizona added almost 1,500,000 residents between 1990 and 2000; only four larger states—California, Texas, Florida, and Georgia—added more. All of the region's states during this period grew at a rate well above the national average.

Phoenix led Arizona's population growth. It doubled in size between 1950 and 1960 and added 50 percent more to its population between 1960 and 1970 to reach almost 1 million people. The boom continued, and between 1970 and 2000 the number of people in greater Phoenix grew to over 3.25 million. By 2006, the Phoenix metro area was home to more than 4 million people. It is still a booming urban area, with frequent air pollution to attest to its status. Meanwhile, greater Tucson grew from 266,000 in 1960 to 946,362 in 2006. Both Phoenix and Tucson are spread-out urban areas that roll for miles across large expanses of former desert.

For years the Southwest was attractive due to the healthful effects of the dry climate for people with respiratory ailments. The region's warmer parts today pull in thousands of retired Americans. Large retirement communities have sprouted in the desert. One—Lake Havasu City, Arizona—gained national recognition when its developer purchased London Bridge and carefully transported it to Arizona, where it now arches over a newly created and artificially maintained waterway.

Still, Arizona's growth cannot be attributed solely to retirement migration. Many industries and corporate offices are drawn there. An aircraft industry developed in Phoenix during World War II, taking advantage of its proximity to the large aircraft complex in southern California, plus the promise of good flying weather. Many employers have located in southern Arizona because the environment strongly appeals to its workforce (and to the employers themselves). The state's relative isolation from most major national markets, once a possible drag on Arizona's growth, has lost much of its impact with the emergence of high-value, low-weight manufactured goods, such as electronics.

Other MSAs such as El Paso, Texas, and Albuquerque, New Mexico, roughly doubled in size between 1950 and 1970, and since then have continued to grow rapidly. These cities, along with San Antonio, benefited from the presence of large military bases, although they also share in the diversified growth of light industry common to the region's other cities.

Elsewhere, in New Mexico and in the south Texas border area, population growth has been far more spotty. Many rural counties in the lower Rio Grande valley and most in southern Colorado and eastern New Mexico lost population during the last few decades, sharing the fate of other strongly rural areas in America.

The early Spanish influence on this small town in New Mexico is represented by the church's prominence and architecture.

PERSISTENCE OF A PLURAL SOCIETY

The Hispanic and American Indian cultures give portions of the Southwest Border Area a unique place in the American landscape. The rural highlands of central and northern New Mexico were the principal core of Spanish settlement in the United States; people there continue to live in ways remarkably unaffected by the Anglo tide that engulfed Albuquerque and southern Arizona. Hispanics make up perhaps 70 percent of the highland population of northern New Mexico and comprise the entire population of many small towns. American Indians are a much smaller but highly visible element of the region's rural, non-Anglo culture.

Traveling the back roads north of Santa Fe, one can get the impression of leaving twenty-first century Anglo society behind. Old adobe villages, poverty-stricken agriculture, and public signs in Spanish dominate the cultural landscape. Along the highway near the dominantly Anglo city of Albuquerque, as throughout the north-central part of the state, are centuries-old apartment-like Indian villages called pueblos. Their ancient appearance is in striking contrast to the low, sprawling modern city. Each pueblo controls substantial areas that insulate their population somewhat from the Anglo community.

New Mexico's capital city of Santa Fe, in the heart of Hispanic country, retains a Spanish flavor with its adobe architecture, open central square, and restaurants and stores offering the food and goods of northern Mexico. All this contrives to give Santa Fe a pleasant distinctiveness, despite its rapid growth and increasing Anglo influence. Hispanics, in many cases cut off from the economic advancement offered by the Anglo economy, have become an important political force in New Mexico and frequently hold statewide elective offices.

The lower Rio Grande valley's Winter Garden area in Texas, also overwhelmingly Hispanic, is a major irrigated agricultural area. The average growing season there is longer than 280 days and supports such crops as oranges, grapefruit, winter lettuce, and tomatoes. The Hispanic population has long provided stoop labor for this agriculture. Here, too, politics has become an important activity in the Hispanic community.

In Los Angeles, Hispanic enclaves may contain hundreds of thousands of inhabitants. Despite far more acculturation into Anglo society than has occurred in the upper or lower Rio Grande valley, the Hispanic people in Los Angeles keep their traditions vital. Spanish-language radio stations and newspapers abound, and major Mexican-American festive occasions attract huge crowds.

In the Southwest Border Area, then, the impact of Hispanic and American Indian cultures remains strong. Abundant evidence of a flourishing plural society continues to distinguish this region from others in the western United States.

URBAN CASE STUDY

San Antonio, the Mexican-American Cultural Capital[3]

Springs located at the southern edge of the Texas Hill Country were long popular for American Indian settlement. Spaniards showed an interest in the area for future development in 1691, and in 1718, they established mission San Antonio de Bexar (the Alamo) at the present site of downtown San Antonio. Franciscan friars built four other San Antonio missions along the San Antonio River between 1720 and 1731. The year 1731 also saw the creation of a civilian community at San Antonio, the first real Spanish effort to begin colonization in Texas. The concentration of different institutions at one location made San Antonio, called Bexar during the Spanish period, unique in the Spanish borderlands.

San Antonio remained the chief Spanish, then Mexican, community in Texas until the Texas Revolution. It was overwhelmingly Spanish/Mexican until the 1840s, when German, American, and Irish migrants began entering the city. The beautiful, rolling countryside to the city's north and

[3]Daniel D. Arreola, "The Mexican American Cultural Capital," *Geographical Review*, 77 (1987): 18. Many of the arguments presented here for San Antonio as the Mexican-American cultural capital are Arreola's.

northwest is still called the German Hill Country in recognition of the large number of Germans who settled there in the mid-nineteenth century and whose progeny are still there. By 1850 Mexicans accounted for fewer than half of San Antonio's 3500 residents.

Between 1850 and 1900, San Antonio and Galveston vied for the role of largest city in Texas. During that period, San Antonio established good railroad connections with Mexico, enabling the city to become a major focus for exchange between the United States and Mexico and a center for Mexicans entering the United States. The city attracted tens of thousands of Mexican political refugees in the early part of the twentieth century who maintained ties with their homeland. The Los Angeles Mexican-American population first surpassed San Antonio's population in 1930. Today the California city is home to four times the number of Mexican Americans living in San Antonio.

Still, the "City of Missions" claims the title of Mexican-American cultural capital as the only major U.S. city that is predominantly Mexican-American. San Antonians of Mexican ethnicity cling to the use of Spanish. They are likely to marry within their ethnic group. San Antonians have long played a central role in many Mexican-American political organizations and activities.

San Antonio presents a striking blend of landscapes and influences. Its four U.S. Air Force bases and a major U.S. Army post ensure a strong military presence among the population. The Paseo del Rio, or Riverwalk, offers visitors a pleasant meander for several miles along the San Antonio River through the heart of the city. A German influence, with accordions and a strong polka base to many songs, gives a distinctive sound to local Texas-Mexican music, called *conjunto*. Unlike most American cities, awnings cover many central-city sidewalks, calling some tourists' attention to the hot climate and evoking a frontier community for others. Still, a walk into the west-side barrio reminds visitors of the city's Mexican heritage—signs and conversations are in Spanish—and the annual Fiesta San Antonio in April celebrates that Mexican heritage.

San Antonio's Riverwalk area attracts tourists to the heart of the city by gracefully combining aspects of San Antonio's rich Hispanic heritage in a unique setting.

CULTURAL GEOGRAPHY CASE STUDY

The Hopi Reservation

Land and the attachment to it are of central importance to many people in the world. Some have argued that Americans and Canadians whose forebears willingly relocated and left their ancestral land do not share fully in this attachment. But that argument does not apply to members of American Indian tribes in both countries.

The Hopi, a tribe of about 8500 people, are direct descendants of the Anasazi, or "ancient ones," an agricultural people who built the magnificent cliff dwellings in places like Colorado's Mesa Verde and Utah's Hovenweep. For centuries,

the Anasazi dominated a large area of the Southwest. Then a succession of droughts in the 1200s apparently forced them to abandon their great urban centers and irrigated fields. Some moved southeast to contribute to the emerging pueblo culture along the Rio Grande. But others, after a period of wandering, settled around three mesas in what is now northeastern Arizona. Today these mesas form the core of the Hopi Reservation (Figure 14.5).

The Hopi lived peacefully around the bases of their mesas for several centuries, settling into a series of compact villages (pueblos) in the Anasazi fashion. They continued and still

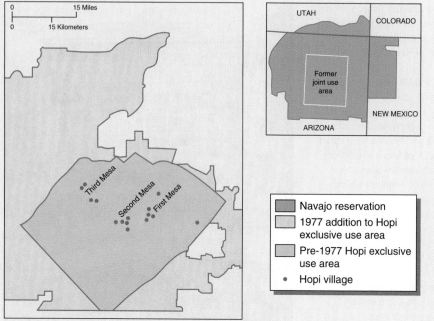

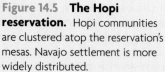

Navajo reservation

1977 addition to Hopi exclusive use area

Pre-1977 Hopi exclusive use area

• Hopi village

Figure 14.5 The Hopi reservation. Hopi communities are clustered atop the reservation's mesas. Navajo settlement is more widely distributed.

practice an agricultural economy focused on crops such as corn and beans. In the late 1600s, they moved their pueblos to the mesa tops to avoid anticipated Spanish reprisals following the Pueblo revolt against European settlers along the Rio Grande. The Hopi also welcomed onto their mesas the Tewa Indians, another pueblo tribe from New Mexico fleeing the Spanish. The Tewa still live in a village on the Hopi Reservation and maintain a separate cultural identity.

At about the same time, the Hopi faced another threat. Navajo people migrated into the area from what is now northwestern New Mexico; they began raiding and then settled on Hopi land. Soon the Navajo surrounded the Hopi, occupying much of the Hopi ancestral land.

The U.S. government ignored this ongoing Hopi-Navajo conflict in the decades after the Mexican War of 1848. It finally granted the Hopi a reservation in 1882 incorporating the area of the three mesas. In doing so, federal officials laid kindling on the embers of Hopi-Navajo conflict with the wording of the order establishing the reservation. It stated the designated land might also be used by "such other Indians as the Secretary of the Interior might see fit to settle thereon."

The Hopi, small in number, initially used only part of their reservation for agriculture. Meanwhile, the numerically dominant Navajo, who had developed a pastoral economy based on sheep, started moving into unused portions. With grazing land scant in northern Arizona's dry environment, the Navajo ultimately moved onto Hopi reservation land with

enough persistence that, before long, only a relatively small portion remained exclusively Hopi.

In 1962, a federal panel proclaimed all of the reservation except the Hopi-dominated section a Joint-Use Area. Because the Navajo were the larger group, however, they occupied virtually the entire Joint-Use Area.

In 1974, the United States Congress addressed the issue by mandating that the Navajo and Hopi divide the Joint-Use Area. However, the tribes failed to negotiate a mutually agreed-upon boundary. Finally, in 1977, a federal district court interceded to divide the area equally between the two tribes. The decision required the relocation of 60 Hopi and more than 3000 Navajo. Despite government promises to provide substantial funding to alleviate costs for moving, many Navajo bitterly resented being forced from their homes.

The Hopi are a conservative people who maintain traditional ways of life. One of those traditions is a strong sense of responsibility to the land. Disturbed at the overgrazing of what they considered Hopi land by the Navajo's sheep herds, they see the 1977 settlement as just. The Navajo, for their part, had lived on what is now Hopi land for a century. While they recognize overgrazing is a problem, they believe the problem results from too little land for their large and rapidly increasingly population. The Navajo believe the Hopi do not use Hopi land effectively. To the Navajo, the federal court's settlement is an unfair limitation of their right to earn a living by traditional pastoral means.

Review Questions

1. What are the major "pull factors" that promote illegal immigration across the southwest border?

2. Discuss the major socioeconomic disparities between the Anglos, Hispanics, and Indians of the Southwest.

3. Explain what is meant by the terminology "cross-border economy."

4. How are *maquiladoras* important to both the U.S. and Mexican economies? Explain the advantages and disadvantages for each country. Are there any unintended consequences?

5. What are some of the attributes of the region that have prompted a significant regional population growth over the past three decades?

6. Describe San Antonio's role as the Mexican-American "cultural capital."

7. What is unique about the location of the Hopi reservation in Arizona? How and why did this situation evolve?

CHAPTER 15

CALIFORNIA

Preview

1. California's physical geography is dominated by north-south oriented mountains/valleys and a Mediterranean climate.

2. The state's location on the Ring of Fire makes it prone to devastating earthquakes.

3. California's agricultural productivity, valued at over $30 billion annually, is greater than any other state in the country.

4. Water is a scarce resource in parts of California and the state has had to undertake large water projects to provide for its growing populations.

5. California's urban megalopolis has been largely shaped by use of the automobile, but automobile-related congestion and pollution continue to be problems.

California's image of a beach-oriented land of surfers shortchanges the great diversity of peoples and environments in the state.

California evokes different reactions in people not living there. Many view California as the desirable ideal of a modern, outdoor-oriented American lifestyle. Others see it as a decadent example of what can go wrong with the country. But whatever the viewpoint, California is a central element in the American cultural fabric (Figure 15.1).

California is part image and part reality. The image is driven by the state's physical environment and the innovative, free-wheeling, and occasionally tumultuous manner Californians use to address their state's problems and opportunities. The reality is guided by the state's physical environment and the fact that California is home to more than 12 percent of all Americans.

By most criteria used in defining regions, California should be treated as more than one unit. One of the state's more important themes, ethnic diversity, is partially addressed in our discussion of the Southwest Border Area (Chapter 14). Culturally, the Imperial Valley's agricultural population in the southeast or the Central Valley's residents differ from San Francisco's or Los Angeles' urban citizenry. Northern and southern Californians have notoriously divergent views, both of each other and of life in general.

Topographically, the striking flatness of the San Joaquin valley contrasts sharply with the conspicuous ruggedness of the Sierra Nevada. Broad deserts cover the state's southern interior while heavily forested slopes drape its coastal north. Lushly verdant, the north coast has an average annual precipitation exceeding 200 centimeters (80 inches). California also boasts the lowest and highest elevations in the coterminous United States. Death Valley and Mount Whitney, respectively, are almost within sight of each other.

California's dramatic and varied physical environment has played a strong role in the state's human settlement. Most of the state's people crowd into a small part of its territory, restricted in where they live by expanses of rugged topography and a widespread lack of water. Californians have invested billions of dollars in tackling problems associated with water. Success is partial, at best, and in many areas precariously fragile.

It is ironic that this mecca for America's worship of the outdoor life, this migration destination for hundreds of thousands seeking to escape from the indoor enslavement of the colder, drabber midsection of the country, should surpass every other state in its level of urbanization. Yet California does. No state has a

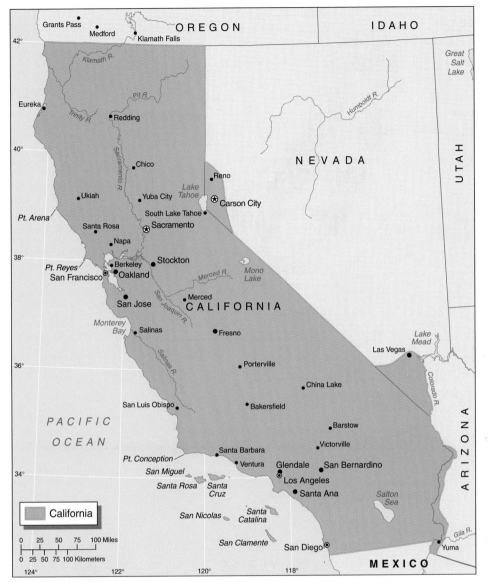

Figure 15.1 **California.**

higher percentage of its people classified as urban by the Bureau of the Census. Several factors account for this status, but restrictive aspects of the physical environment are certainly important.

THE PHYSICAL ENVIRONMENT

California's coast is lined by a series of long, linear mountain ranges that trend in a northwesterly direction. They are collectively called the Coast Ranges. Most are not high—summits are between 1000 and 1600 meters (3000 and 5000 feet)—and they are

heavily folded and faulted because of the pressures from tectonic plate contact just to the west.

The Earthquake Zones

The California faults follow the same northwesterly trend as the Coast Ranges (Figure 15.2). The most famous fault, the San Andreas, extends from the Gulf of California through the Imperial Valley to Point Arena north of San Francisco, where it extends into the Pacific Ocean. Lateral earth movement (i.e., horizontal shifting as opposed to vertical displacement) was as much as 6 meters (20 feet) along the San Andreas

Figure 15.2 **Earthquake fault zones.** The topography of coastal California is linear because of land movement along fault zones generally paralleling the coast.

when the devastating 1906 San Francisco earthquake struck.

Small earthquakes are common across California, especially from the San Francisco Bay Area south toward Bakersfield and from the Los Angeles area southeast through the Imperial Valley. Seismologists (scientists who study earthquake activity) believe places of major continuous stress, like the Coast Ranges straddling the San Andreas, will periodically have large earthquakes because this stress gradually accumulates over time until it releases with violent snaps. Many believe that the San Andreas fault zone, which has had

no major earthquake since 1906, is due for "the big one." The 1989 San Francisco earthquake, while destructive to property and life, was not it. The Los Angeles vicinity is thought to be especially vulnerable to a very large quake, because the San Andreas fault has been relatively sleepy in the south for 120 years.

California's two major foci of population growth, the San Francisco Bay Area and the Los Angeles Basin, are principal centers of seismic activity. Thus, the majority of Californians live with the threat of an earthquake. Major highways, commercial and industrial concerns, and thousands of homes sit atop or near known fault zones. Enterprise and individual earthquake insurance is prohibitively expensive. Insurers fear catastrophic financial losses in the event of a major earthquake.

One of the most visible impacts of the earthquake threat is on the skylines of California's larger cities. Until recently, local "earthquake laws" limited how high buildings could be to protect against great loss of life during an earthquake. For many years, this moratorium on construction mandated a low, even skyline in San Francisco, and in Los Angeles only City Hall rose more than a few stories.

These homogeneous skylines (especially San Francisco's) visually set these cities apart from their eastern counterparts. Innovations in structural materials and construction techniques have since led to modification of many of these laws. Now tall buildings punctuate the skylines in both cities. Many native Californians find these substantial changes in the state's urban landscapes jarring, especially in San Francisco.

Valleys and Mountains

California's Central Valley lies east of the Coast Ranges. Extremely flat, the valley reclines adjacent to much of the mountain string, extending 650 kilometers (400 miles) north to south and nearly 150 kilometers (95 miles) wide in places. In ancient geologic times, the Central Valley was a massive extension of the Pacific Ocean. But over time, erosion material carried off the Sierra Nevada's western slope filled it in with sediments. The contemporary benefit is a low-relief landscape richly fertile for large-scale agriculture.

East of the Central Valley the western face of the Sierra Nevada rises gradually and is heavily eroded. In contrast, the mountain's eastern face offers a dramatic change in elevation. This different aspect comes from the fact that these are *fault-block mountains*. Such mountains form when large rock masses uplift as a unit. In the Sierra Nevada, the eastern side was lifted far more sharply than the western. Because they reach high elevations and contain few passes, the Sierras have been a major barrier to movement between middle and northern California and places east. The well-known tragedy of the Donner Party or the exploits of Chinese laborers who worked building the first transcontinental railroad are testimony to the problems of overcoming this barrier.

Two other landscapes, less associated with California by outsiders, complete the state's topography. In the north, the alternating mountain-valley-mountain pattern breaks down to become more consistently mountainous. The central plateau directly north of the Central Valley has two major volcanic peaks, Mount Lassen and Mount Shasta, indicating that California's northern tier of mountains is a southern extension of the Cascades in Oregon and Washington. The other notable landscape lies southeast of the Central Valley.

San Francisco's modern skyline sprang up after the repeal of "earthquake laws" that had limited building height. Contemporary structures tower above older, more squat buildings that extend out toward the Golden Gate Bridge.

There, the Basin and Range area of the interior extends into California, creating a terrain of low-lying mountains interspersed with large areas of fairly flat land.

Climate and Vegetation

California's weather begins offshore, along the eastern margins of the Pacific Ocean (Figure 15.3). When moisture-laden, maritime air moves across the state from the northeast Pacific, abundant precipitation may result. Which part of the state the storm systems reach is affected by the strength of a stable atmospheric high-pressure cell, usually found off the west coast of Mexico. Such a stationary center blocks storms emerging from the maritime air mass, forcing them eastward onto northerly shores. This blocking high drifts north during summer and south during winter in conjunction with the latitudinal movements of most climatic zones during the year.

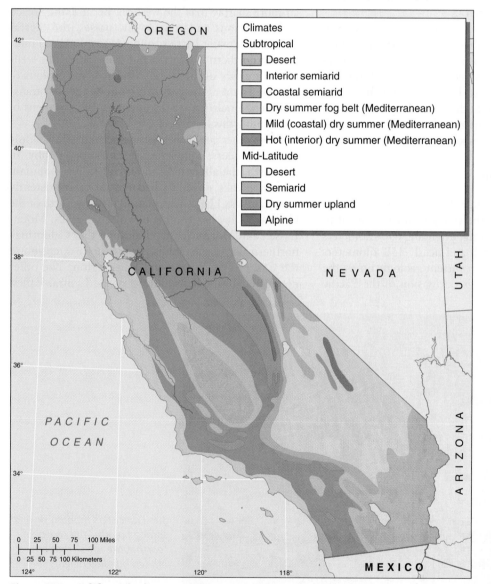

Figure 15.3   California's climates. (Adapted from *California: The Pacific Connection* by David W. Lantis, Rodney Steiner, and Arthur E. Karinen. Chico, CA: Creekside Press, 1989.) Topography plays a substantial role in establishing the climatic pattern of the state.

A definite north-south gradient in average annual precipitation occurs in California due to the interaction of these two major air masses. The north is much moister on the average than the south (see Figure 2.3). Furthermore, summers are characteristically drier than winters, especially in the south. In summer, southern California frequently experiences long periods during which there is no rain. Consequently, wooded mountain slopes grow dry as tinder. Forest fires, another of the region's recurring environmental problems, are most frequent in late summer and fall, toward the end of the long dry period.

A climate with moderate precipitation concentrated almost entirely in the winter months, and one that has mild winters and hot summers, is called a *Mediterranean climate*. It is so named because such a climate is found in much of the Mediterranean Basin lands of Europe, Africa, and Asia. In California, the entire coast from San Diego northward past San Francisco Bay, all of the northern Central Valley, and the western margins of the southern valley represent the only zone of Mediterranean climate in North America.

Along the coast north from San Francisco, average annual precipitation increases greatly, and its seasonality partially disappears. California's northern coasts have a climate similar to that of the Pacific Northwest—mild temperatures with relatively little seasonal variation, plentiful year-round precipitation, and frequent periods of overcast skies. Such a climate is called Marine West Coast, an indication of its customary location along the western margins of continents in the middle latitudes (see Chapter 16).

The Central Valley is much drier than the state's coastal margins. As air masses pass over the Coast Ranges, they rise, cool, and lose much of their moisture. When those air masses descend into the valley, they warm. Little moisture is added to the air, and its relative humidity (the percentage of the absolute humidity represented by the moisture actually in the air) declines. The result is a much lower average annual precipitation.

Annual precipitation in the Central Valley is usually less than half that found at a similar latitude on the western slopes of the Coast Ranges. This dryness is especially pronounced in the rain shadow along the valley's west side. For example, Mendocino, on the coast north of San Francisco, averages 92 centimeters (37 inches) of precipitation annually, while Yuba City directly east of Mendocino in the heart of the Sacramento valley averages only 52 centimeters (21 inches). In the state's southern half, coastal San Luis Obispo averages 52 centimeters (21 inches), while inland Bakersfield makes do with only 15 centimeters (6 inches) annually.

Summer temperature differences between coastal and inland points at similar latitudes are equally dramatic. San Luis Obispo's average July temperature is 18°C (64°F); Bakersfield's average is nearly 12°C (20°F) higher. Daytime high temperatures in San Francisco in late summer are usually under 27°C (80°F), while Stockton, 100 kilometers (60 miles) east, bakes in

Fires, such as these that devastated the Santa Clarita area in October of 2007, are a common environmental threat in California when warm, dry winds (known as Santa Ana winds) blow from the interior of the state.

temperatures greater than 38°C (100°F). Much of this difference is based on the moderating effect of the cold ocean current offshore and the usual pattern of afternoon and evening fog in summer along the coast from Point Conception northward (Figure 15.3).

To the interior of the Coast Ranges and the Sierra Nevada, in California's southeast, is a broad section of arid steppe or desert environment. The entire area frequently comes under the influence of dry air generated from the south and east. During the late summer months, desiccating winds occasionally press west to the coast, bringing extremely low humidity and temperatures that can register more than 40°C (104°F). California's southeastern interior receives on the average less than 20 centimeters (8 inches) of precipitation annually.

Vegetation patterns closely parallel variations in climate. Nearly all of lowland southern California and the area east of the Sierra Nevada-Cascade Ranges is covered with sage, creosote bush, chaparral, and other characteristic desert and semidesert flora. The Central Valley and the valleys of the southern Coast Range are somewhat better watered than are areas farther south; they are steppe grasslands. These grasses are green during the moist period from early winter through spring and turn light brown during the dry, summer season. Wrapping around the Central Valley and following the coast from Santa Barbara to Monterey Bay are mixed open forests of live oaks and pines. The coast from Monterey Bay north is home to coastal redwoods, the world's tallest trees. At higher elevations in the Coast Ranges and Sierra Nevada, mixed forests of pine and fir exist, and high in the Sierra Nevada, subalpine hemlock-fir forests, including those of the sequoia, dominate.

CALIFORNIA'S CLIMATE AS CATALYST AND LIMIT

California's climate, which provides much of the state's unquestionable attraction, is also the basis of some of its greatest problems. Viewed from a purely economic standpoint, it is almost inconceivable that one out of every eight Americans live in California. Geographers and other scholars have long viewed a place's population growth to be a result of several factors; two of the most important are the natural resources of that place and the strategic advantages of its location. (For a discussion of relative location, or situation, see Chapters 3 and 4.)

California's greatest disadvantage, at least until recently, is its location at the far western periphery of the country. The United States was settled primarily from east to west. Still today, its major population concentrations and most areas of principal economic importance are near or east of the Mississippi River. This puts California as much as 3500 kilometers (2200 miles) from the country's most vital areas of economic demand and supply. This relative isolation is compounded by the nature of most of the land between the Sierra Nevada Mountains and the South and Midwest regions; except at a few sites, this broad section generates little local freight. Therefore, for years freight moving between California and cities east of the Mississippi River absorbed the high shipping costs demanded by the long-distance exchange. Normal economic considerations suggest that these high transportation costs would limit California's economic growth. But obviously, the text-book formula did not hold in light of the state's boom.

California is not without geographic advantages; San Francisco Bay, for example, is one of the world's premier harbors. Still, these are not sufficient to account for the state's population size. More than anything else, climate is the key in overcoming the state's ostensible locational disadvantage; climate has nourished both the state's settlement history and its agricultural development.

Settlement History

Perhaps 10 percent of Native Americans living in what was to become the United States and Canada dwelt in California when Europeans first arrived in the Americas. These first inhabitants of California were hunters and gatherers. For food, they depended on seafood and game, or on wild grains and nuts they collected and ground into flour. Few large tribes existed. Instead, countless groups organized themselves into small units of perhaps 10 to 20 families. They usually spent their lives in small territories, never venturing far beyond. One result of their geographic isolation was substantial cultural variation among California's Indian groups.

Spanish explorers brushed the edge of California in the mid-1500s, claiming it as a part of Spain's large North American holdings, but they did little to buttress their claim for the next two centuries. The region was at the extremity of Spain's North American empire (see Chapter 14). Not until concern arose over expansion by other colonial powers in western North America, just before the American Revolution, did Spanish missionaries establish a string of missions from San

Diego to Sonoma. These mission settlements, now some of the most sentimentally revered vestiges of the state's past, were joined by presidios (forts) and a few pueblos (towns). It was during this late-eighteenth-century period of Spanish settlement that most of the Native American population disappeared, victims of mistreatment and European disease.

The Spanish and Mexican governments granted a series of large landholdings (*ranchos*) to encourage immigrants. Still, the area remained peripheral. Towns were small and ramshackle, and hides and tallow were the *ranchos'* most important exports. This backwater effect lingered because Spaniards (and later, Mexicans) had no interest in California other than using it as a buffer against the Russians, who at the time were planting settlements down the coast from Alaska.

The United States seized California in 1846. Soon after came the great 1848 gold strike in the foothills of the central Sierra Nevada, an event that ushered in the first significant change in the region's settlement fortunes. Within a year, 40,000 people had come to the gold fields by sea, passing through San Francisco harbor. Roughly the same number came overland. By 1850, California was a state. The frantic gold rush period lasted only a few years, but it succeeded in breaking the state's isolation from the rest of the country. It also stimulated the growth of San Francisco, growth that easily weathered the end of the gold boom. San Francisco remained the largest United States city on the West Coast until World War I.

Mission Santa Barbara is one of several missions originally established by the Spanish in their attempt to secure a territorial toehold on the Pacific coast. The building pictured was completed in 1820 and replaced earlier structures on the site.

Southern California, the center of prior Spanish occupation in the state, did not share in the early population expansion. But its quiet existence ended abruptly when the Southern Pacific Railroad connected to Los Angeles in 1876, soon followed by the Atchison, Topeka, and Santa Fe Railroad. To create demand for their facilities, the railroads advertised widely for settlers; they lowered fares and aided new arrivals in finding housing and jobs. For one day in 1887, rail fares on the Southern Pacific from Kansas City to Los Angeles were $1. During the first southern California land boom, between 1881 and 1887, Los Angeles' population grew from 10,000 to 70,000. Land speculators offered literally hundreds of thousands of town lots for sale in communities throughout the Los Angeles Basin, and many paper millionaires were created overnight.

Although this first land investment bubble burst in 1888 due to wildly excessive development schemes fueled by a land speculation frenzy, the growth of Los Angeles, like that of San Francisco after the gold rush, continued. Promotional literature emphasized its delightful and healthful climate, appealing to those suffering from tuberculosis and asthma. As this propaganda spread, thousands of ailing people flowed into the area.

A large number of crops were also introduced into southern California during the 30 years after 1881; the navel orange (1873), the lemon (1874), the Valencia orange (1880), the avocado (1910), and the date (1912). These fruits were in demand in eastern markets, and at that time in the United States, only southern California could provide them in large quantities. Several technical improvements in food handling, notably artificial dehydration (1870) and refrigerated freight cars (1880), meant that these foods could reach eastern markets with only minimal loss owing to spoilage. Agriculture was to remain the backbone of southern California's economy until after World War I.

California's Agriculture Today

California, by some measures the country's most urbanized state, is also its most agricultural in terms of total farm income. In 2005, the total market value of agricultural products sold in California was $31.7 billion, more than the combined total of its closest competitors, Texas and Iowa. California's agriculture, though typified by specialty products like artichokes and kiwi fruit, is, in fact, broadly based (Table 15.1).

TABLE 15.1

California Agricultural Production as a Percent of United States Production	
Crop	Percent of 2002 U.S. Production
Alfalfa & Hay	6
Almonds	99
Apricots	93
Artichokes	99
Asparagus	55
Avocados	90
Broccoli	92
Carrots	76
Cauliflower	86
Celery	95
Corn, Fresh Sweet	15
Dates	99
Figs	99
Garlic	84
Grapes	91
Honeydew Melon	75
Kiwifruit	99
Lemons	87
Lettuce	74
Onions	32
Peaches	75
Pears	30
Plums	99
Spinach	67
Peppers, Bell	43
Strawberries	85
Tomatoes	86
Walnuts	99

Source: California Agriculture Statistical Review (Sacramento, CA: California Agricultural Statistics Service, 2002).

The variety of climatic regions and the market demand of California's own bulging population are the principal contributors to the state's crop diversity. But the importance of its many specialty crops does much to explain why its farmers have had such success in penetrating the markets of the distant East. These are crops that can be grown, or at least grown on a large scale, in few parts of the country. Most require long growing seasons. Thus, there is no local competition in the demand areas. Cooperatives such as Sunkist have been organized to facilitate marketing and to guarantee low shipping costs. Southern California, especially the Imperial Valley, is also able to provide vegetables, such as tomatoes or lettuce, during winter when competition is minimal and sale prices are at a maximum.

Although California as a whole produces many agricultural products, local areas within it tend to specialize in one or a few crops. There are several reasons for this pattern of local specialization. One is the general trend toward agricultural specialization that exists across the country. The image of a typical farm as one growing a variety of crops and being self-sufficient in terms of food is no longer accurate in American agriculture. Often, local markets are structured to handle only the few products local farmers emphasize. This ties to economies of scale in handling and marketing, just as farm size growth is linked to equipment and operational scale economies (see Chapter 11).

A second factor is the major role played by large agricultural operations. Some specialty crops produced in California are grown by only a handful of farmers. In the San Joaquin valley (the southern half of the Central Valley), some landholdings extend over thousands of acres. Such large-scale growers have been leaders in the move toward further specialization.

Finally, the diversity of climatic and physiographic niches influences regional specialization. Many specialty crops are particularly sensitive to what may seem minor variations in climate or soil type. The Coastal Range valleys opening onto the Pacific are frequently foggy with moderate temperatures. Vegetables, such as artichokes, lettuce, broccoli, or Brussels sprouts, grow well under such conditions. The Salinas valley "Salad Bowl" is a prime example of such a locational advantage. Varietal grapes for some of America's best wines need a mild, sunny climate like that found in the inland Coastal Range valleys around San Francisco Bay. The grapes of the San Joaquin valley or of southern California, where summer temperatures are much higher, are used for table grapes, raisins, or for what most (though by no means all) wine drinkers consider less distinguished wines. Most flowers grown for seed in the country are planted in the Lompoc valley west of Santa Barbara. Navel oranges and particularly lemons are grown almost exclusively along the coast and the interior surrounding the Los Angeles Basin. It is unlikely that early spring frosts would occur there to damage the emerging

California's diverse agricultural landscapes produce much of the American food supply. Here, lettuce is grown on irrigated land near Blythe.

flowers or that heavy winter frosts would take place to kill the trees themselves.

These local crop concentrations, though economically advantageous, mean that moderate environmental changes in a relatively small area of California may impact national production of specialty crops. Because of this, California's expanding urban sprawl is an acute problem for many agricultural areas in the state. For example, the Napa valley, with many of the state's finest vineyards, is being inundated by waves of residential construction. Although most vintners are relocating elsewhere (total wine grape acreage in the state has increased in recent years), their new locations are less suitable for growing quality wine grapes. South of San Francisco, coastal valleys such as the Santa Clara valley are also urbanizing rapidly and threatening production of crops like apricots, prune plums, and pears. Almost all of California's orange and lemon groves, concentrated in the Los Angeles Basin through the 1940s, have been forced out of the Basin by urbanization and smog damage (as well as competition from the Florida citrus industry; see Chapter 10).

Whenever land-use competition develops between urban and agricultural users, it is the agricultural users who lose. Urban users, both commercial and residential, are willing and able to pay far more for land than are their agricultural counterparts. In addition, rural land has traditionally been taxed in the United States on the basis of its appraised sale value and not on its value as a strictly agricultural unit. Thus, the twin pressures of increasing

taxes and high offers from people who wish to buy land—both normal consequences of urban expansion—may convince even the most dedicated farmer to sell. (See the discussion of this point in Chapters 4 and 6.)

The consequences of this changing land use, estimated to result in an annual changeover of 800,000 hectares (2 million acres) from rural to urban uses in the United States, have been of little national concern in the past. It appeared to many that enough rural land remained for farming and to offer the aesthetic reward inherent in a trip to the country. But this lack of concern is changing.

Many cities are located in the midst of good farmland, and urban land users prefer the same flat land that is most valuable to farmers. High-quality farmland is lost to urban expansion at a better than average rate near cities. In an increasing number of places, residents have come to feel that in losing local farmland they are losing a valued part of the landscape.

Attempts to preserve farmland in an urbanizing environment have followed several tactics. Most common is to grant special tax breaks to farmers, usually accomplished by assessing their land as farmland for tax purposes rather than as potential urban land.

One problem, however, is that urban expansion, already quite jumbled in its spatial pattern, may become even more so as some farmers choose not to sell their land until surrounding land is urbanized. This kind of urban spread would make even less orderly than at present the provision of urban services (water, sewer, streets, etc.); public service cost increases are the result.

Within the Los Angeles region, farmers in several dairying areas tried a different approach. They attempted to protect themselves by incorporating, thereby creating their own tax structure. This system broke down when individual farmers within the "town" broke ranks and sold to residential developers. Soon new residents outnumbered the farmers, and the urbanites' demands for improved services pushed taxes up.

Other areas have anticipated urban sprawl by enacting zoning restrictions, such as 2- or 4-hectare (5- or 10-acre) minimums for residential plots. This may slow the transfer of land away from farming, although farmers are often ambivalent about the approach. While affording them some protection from urban tax rates, zoning limits farmers' options. In any case, the future of such restrictions is in doubt because some courts have found this type of zoning to be illegal.

Today nearly every state has laws that provide some property tax relief to farmers. In order to be eligible for reduced property taxes, California's landowners must enter into a long-term agreement not to develop their land. Some 300 counties nationally have some type of agricultural zoning to preserve farmland. A few places have programs to purchase development rights to ensure that the land remains in farming.

Water Supply

California's agriculture, more than its manufacturing or urbanization, has created a massive demand for water. California has more irrigated farmland than any other state, about 3.5 million hectares (8.6 million acres). Only Texas has as much as half as many irrigated acres as California. Nearly all of California's cotton, sugar beets, vegetables, rice, fruits, flowers, and nuts are grown on irrigated farmland. The state's farmers use more than one-fourth of all irrigation water used in the United States. On the average, irrigated land in California receives about 1 meter (40 inches) of "artificial" water annually.

Crop selection at a site depends on water availability as well as other factors such as soils, drainage, terrain, and growing season. But the potential for irrigation is usually critical in California. A transect across the San Joaquin valley shows livestock grazing in the Sierra foothills; dry farming for grains in the flatter land below where land is still too high for irrigation; irrigated fruit trees and vine crops in the better

drained soils near the valley floor; and irrigated field crops such as cotton, vegetables, and sugar beets on the flat valley floor.

About 70 percent of California's precipitation falls in the northern mountains and valleys and in the Sierra Nevada. About 80 percent of the water that falls as precipitation in the state is consumed on farms and in cities in the drier south. Of California's major farming regions, only areas north of San Francisco and a few coastal valleys to the south receive as much as 50 centimeters (20 inches) of precipitation in an average year. Two of the most important farming areas—the southern end of the Central Valley and the Imperial-Coachella area in the southeast—average annually less than 25 centimeters (10 inches) of precipitation.

Farms are the principal users of water, but city officials have been the ones initiating development of California's tremendous water movement complex. At the beginning of the twentieth century, Los Angeles outgrew its local groundwater supplies and identified a supplementary source in the Owens valley, east of the Sierra Nevada and about 300 kilometers (185 miles) north of the city. By 1913, the Los Angeles Aqueduct carried water to Los Angeles, much to the dismay of Owens valley farmers who lost virtually their entire water supply (Figure 15.4). This aqueduct still provides half of the city's needs.

In 1928, Los Angeles and 10 other southern California cities formed the Metropolitan Water District to develop an adequate water supply for their entire

California's orange groves, squeezed out of the Los Angeles Basin by urban expansion, are concentrated in irrigated areas in the southern San Joaquin valley. Wind machines and smudge pots protect trees from frost damage, especially during the early spring blossoming period.

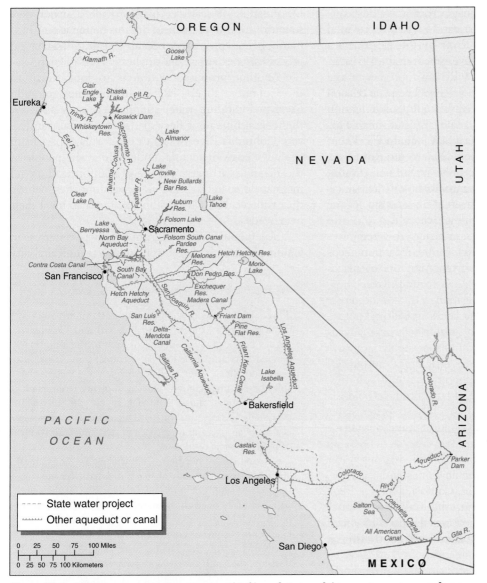

Figure 15.4 Water movement systems. Chief beneficiaries of this massive system are farms in the San Joaquin and Imperial valleys, and the cities of the Los Angeles Basin.

area. By 1939, the group completed the Colorado River Aqueduct, which carried water from Parker Dam on the Colorado to coastal cities from San Diego to the Los Angeles Basin. Today the Metropolitan Water District, which serves 6 counties, more than 130 cities, and nearly 18 million people, is one of the state's most powerful political bodies. It estimates that in its jurisdiction alone over 200,000 additional people per year will need water in the coming decades.

Farmers were by no means ignored, however. Perhaps the most spectacular episode in the state's water history occurred in 1905 in the Imperial Valley. In 1901, private groups constructed canals to carry water from the Colorado River into the Imperial Valley; the result was an immediate agricultural land boom. Then, in February 1905, the Colorado River flooded, breaking out of its channels and flowing into the irrigation ditches. Before a massive effort returned the river to its channel in the fall of 1906, 1100 square kilometers (400 square miles) of the Imperial Valley had been flooded, creating a new lake in an ancient, dry lake bed. Called the Salton Sea, it is still maintained by discharging water and drainage from irrigated farmland. Most of the valley's irrigation water is provided today from the Colorado River by the All-America Canal, built by the federal government in the 1930s.

The federal Bureau of Reclamation began the Central Valley Project in the 1940s. The project's aim was simple—improve local availability of irrigation waters in the Central Valley. The best land suitable for irrigation was in the southern San Joaquin valley, but most water was in the northern Sacramento River.

Today the project is fully functional. Water, removed from the Sacramento River by the Delta-Mendota Canal, flows south along the west side of the San Joaquin valley to Mendota, where it is put into the San Joaquin River. This transferred water raises the river's volume enough to meet the irrigation needs of the San Joaquin valley from Mendota to places north. What constitutes the San Joaquin River's normal flow can be used for irrigation in the southern part of the valley, opening up that land's agricultural potential. Today the southern San Joaquin valley is the leading agricultural region in California, supplying nearly 40 percent of the state's total agricultural output. Of the United States' ten leading agricultural counties, measured in terms of value of farm products sold, six are in the San Joaquin valley. Only six other states exceed the combined value of agricultural output for these six counties.

In 1957, California brought all existing water projects and all schemes for new ones under a state-administered plan called the California Water Plan. Under the plan's direction, the California Water Project emerged. This project was designed to obtain water where available in California and transfer it based on need.

Easily the largest water movement program in American history, the project focuses on a massive system to move water from north to south in the state. The California Aqueduct, starting from the same Sacramento River delta area as the Delta-Mendota Canal, carries water south partly to be used in the western San Joaquin valley and partly to supply the ever-growing water needs of southern coastal urban centers. Most of the cost of the southern extension is borne by users in the Metropolitan Water District. The system's total cost continues to increase, with estimates for improvements now exceeding $20 billion.

Even this monumental system will satisfy California's increased water demands for only a few decades, with critical water shortages likely by the year 2020.

Desalination of seawater may provide a major new source of freshwater by then. But this option is only hesitantly being considered, because existing desalination processes are expensive and are heavy users of energy.

Another proposed option is to look north to the Columbia River—its flow dwarfs the Colorado River's—for future water supplies. As early as 1949, the federal Bureau of Reclamation suggested transferring Columbia River water to California. Apart from the huge costs of such an undertaking, the people in the Pacific Northwest outside of California show no inclination to accept such a suggestion; they believe they will eventually need all of the water to meet their own needs.

Several points emerge from a study of water use in California. One is that a tremendous technical ability

Water Problems: They Won't Go Away

Three problems consistently complicate water issues in California. First, much of the Imperial and Coachella valleys face a gradual buildup of salts in the soil. Suspended salts are carried in large concentrations in the Colorado River water used for irrigation. Once in the soil, they can be removed only by applying large quantities of water to flush them out. If left, the salts will eventually destroy the soil's productivity. This damage is already apparent in Mexico's irrigated lands just south of the California border.

Second, many northern Californians are not happy about the transfer south of much of what they view as their water supply. They feel that their part of the state will eventually need all the water it can get. So far, their political clout has not kept the water from flowing south. California's center of population is in the mountains just north of Los Angeles. Hence, in any statewide political decision, the south has far more potential voting power.

Third, many Californians question the economic and environmental costs of water transfer projects. In 1980, the state legislature proposed construction of a peripheral canal east of the delta to carry water from the Sacramento River south to increase the volume of flow to the arid south. The state estimated that the project would cost $5.4 billion; opponents argued that $20 billion was a more accurate figure. Environmentalists were concerned that the loss of freshwater would damage the rich aquatic and waterfowl populations located in the intricate lacework of waterways of the delta. Delta farmers feared loss of their rich vegetable-producing lands. San Franciscans again worried over a water loss to the south. A coalition of these diverse interest groups defeated the canal in a 1982 referendum.

exists to move water from one place to another, given adequate financing. Another is that when more water is available, demand rapidly increases. Still another is that a clear tendency exists to reach across ever-increasing distances to meet mounting needs.

In recent decades, California's water consumption increased even more rapidly than its population. Suburban swimming pools dot southern California and growing industrial demand has been important, but farm irrigation has been the major user. Between 1964 and 1987, California's total farm acreage dropped by more than 20 percent, yet both the number of acres irrigated and the volume of water used per acre increased. Water use increased because irrigated land in this generally dry environment produces very high yields. California's cotton production per acre, for example, is double the national average. And because government agencies absorb most costs involved in irrigation projects, water is provided to farmers at an artificially low price, and this encourages them to expand consumption.

As more people choose to live in California and its neighboring states, water demands and the cost of new projects designed to meet those demands rise. But California's water supply is limited, as is that of the entire Southwest. The region is quite close to using its entire supply.

URBAN CALIFORNIA

Despite the national importance of California's agriculture, agriculture's unique character, and the fact that its influence on state affairs outstrips the number of people employed in it, California's population is overwhelmingly, increasingly urban. More than 90 percent of California's residents live in one or another of the state's 25 Metropolitan Statistical Areas (MSAs). Between 1980 and 1990 Los Angeles passed Chicago to become the nation's second largest city. Most of the state's people live in one of its two major urban regions, one centered on Los Angeles, the other on San Francisco.

The Southern Metropolis

"A city in search of a center!" This is what many call the loose assemblage of cities normally lumped under the general name of Los Angeles (Figure 15.5). The land boom of the 1880s led to the establishment of several score of cities scattered across the Los Angeles Basin and southern California coastlands. As their populations increased, these communities squeezed out the intervening rural lands that initially separated them; yet, despite the spatial merging, they maintained their jurisdictional independence. The result is a jumbled complex of politically independent but economically intertwined units. The cities have surrendered some of their individual decision making and taxing jurisdiction in recognition of common regional needs that can only be met cooperatively. The Metropolitan Water District is the best example of such a regional agreement. Still, the political complexity results in many problems in taxation and planning (see Chapter 4).

Most of the 300-kilometer (185-mile) stretch of coastline from Santa Barbara through Los Angeles to San Diego is now occupied by one long megalopolis. Home to about 19.2 million Californians in 2000, this

The California Aqueduct carries water south to irrigate the San Joaquin valley and to provide water for metropolitan Los Angeles.

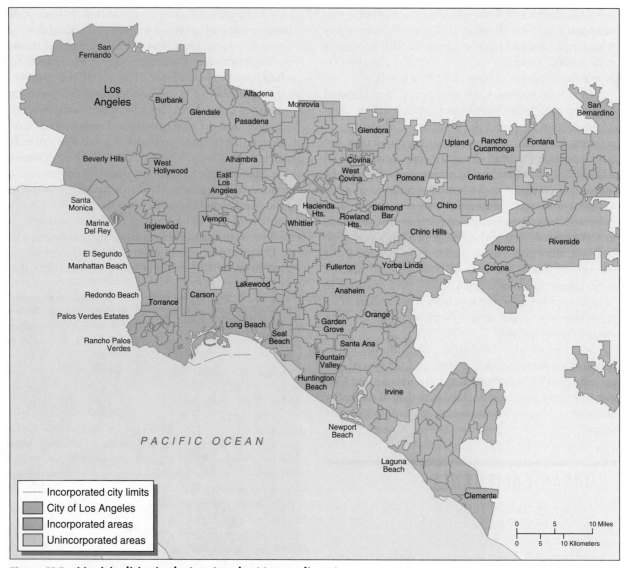

Figure 15.5 Municipalities in the Los Angeles Metropolitan Area. Los Angeles annexed large portions of the San Fernando valley northwest of the city's early center to avoid being entirely surrounded by other independent urban places.

vast spread of urban landscape reflects a growth process shared by other North American cities and, at the same time, a strikingly unique metropolis.

Basically, the entire complex is a creation of the twentieth century. In 1900, its population was less than one-tenth its present size, and it was not until after World War I that the area began to shift from an essentially agricultural to an urban region. Thus, many elements typical of eastern cities, elements placed on the landscape largely during the nineteenth century and the first two decades of the twentieth century, are

not present in and around Los Angeles. There are no four- or five-story walk-up apartment buildings, comparably high warehouses, fixed-rail elevated public transportation lines, or a strong nodality centered on the central business district. Even the subway is a relatively recent addition. Knowledgeable Angelenos could undoubtedly expand this list, but it is representative.

The Automobile as Landscape Definition

True to its twentieth-century character, California's southern metropolis was shaped most importantly by

Follett

Follett Corporation

Packing List Report

ACKING SLIP

Return Date : 5/11/2017

Returned By : 8622236 patron ID

Guinevere Henrikson
3415 20th ave. S apt 412
Grand Forks ND,58201
US

Ship-To : Bemidji State University Bookstore
Hobson Memorial Student Union,
1500 Birchmont Drive NE
Bemidji MN,56601-2600

Email : gpanda.henrikson14@gmail.com
phone 218-452-1290

ental Check-In

Item ID	ISBN Due Date	Author	Title	Edition	Publisher	Qty
34168136	2900205990794 5/11/2017	Spradley	Conformity & Conflict			1
34168135	2900470098263 5/11/2017	Birdsall	Regional Landscapes of US & Ca			1

Mailing Instructions

Suspended

ase use a laser or laser-quality printer.
lude the Packing Slip in the box.
pe the Shipping Label on the box. NOTE: Make sure the box has no other labels, barcodes or addresses on it.
ace label so that it does not wrap around edge of postage.
ch shipping label number is unique and can be used only once - DO NOT PHOTOCOPY OR FAX. Only the original label can be accepted.
ckages may not weigh more than 150 pounds. Shipping boxes may not exceed 108 inches in length or 165 inches in length plus girth(L+2W+2H). For more information go to www.fedex.com
op off package at an authorized FedEx pick up location.
ase write your return address on the shipping label.

***** End Of Report *****

1. P
2. I
3. T
4. F
5. E
6. P
7. D
8. F

the family automobile. Los Angeles—and some other western cities, like Phoenix—grew as automobiles became widely used. In these new cities, unlike cities in eastern North America's older urban areas, land-use patterns were not historically entrenched. Places to live, work, shop, and recreate sprang up in and around the city in locations that took advantage of the new spatial flexibility made possible by automobiles. Similarly, public transport's linear inflexibility was no longer important; people in Los Angeles did not have to locate homes and jobs along streetcar lines.

Today fully half of the central portion of Los Angeles has been surrendered to the automobile, either for roads or parking. A dense system of freeways encourages high-speed movement across the metropolitan region. The Los Angeles area has more cars per capita than any other part of the country; only a minimal public transportation system exists. Neither fact should be surprising given the diffuse pattern of housing, employment, and commercial facilities, but it means private cars are the means of choice for an overwhelming percentage of each day's travelers.

Increasingly heavy auto use has resulted in congestion on major arteries. Los Angeles County has four of North America's five busiest freeways. On most mornings during rush hour, 40 percent of the region's freeways are clogged with commuter traffic moving no more than 25 kilometers (15 miles) per hour.

Freeway construction did not keep pace with growing congestion. Until a 1990 vote endorsed a major highway funding program, California spent less per capita on highway construction than any of the other 49 states. The rapid geographic expansion of greater Los Angeles, especially east toward San Bernadino and the Mojave Desert, will only increase commuting distances and make congestion worse.

Famous as an ode to L.A. lifestyle, automobiles are also partly responsible for one of Los Angeles's most infamous features—smog. Ringed by mountains to the north and east with hot dry deserts beyond, and bordered by the cool waters of the Pacific to the south and west, the Los Angeles Basin frequently is subjected to temperature inversions. Temperature inversions occur when a body of warm air lies above cooler air. This warmer air forms an atmospheric lid, and pollutants in the cooler air are unable to rise and dissipate into the higher, warmer air. In Los Angeles, onshore winds push pollutants inland, but the combination of the inversion lid and the surrounding mountains keeps much of the pollution in the Basin (Figure 15.6). Palm Springs, just east of the Basin, has developed as a high-status community because it is beyond the smog's reach. In reality, Los Angeles produces only moderate amounts of pollution; little heavy industry exists, burning is carefully controlled, and strict controls govern automobile emissions. As for the automobile's culpability, total emissions

The Los Angeles River, usually a trickle but a major stream after winter rains, is paved for much of its length across the Los Angeles Basin, giving the river the appearance of an unused freeway.

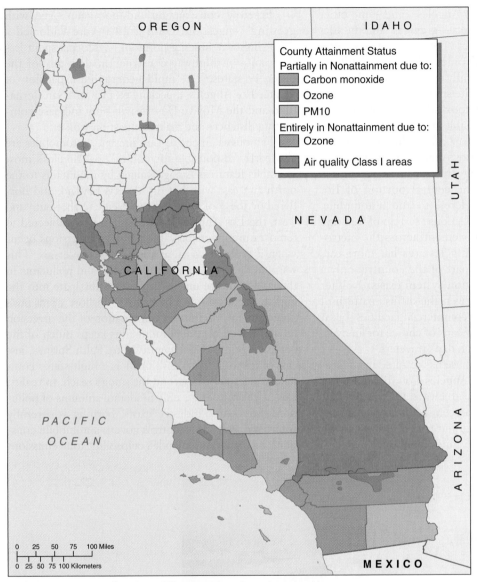

Figure 15.6 **Air quality in California.** Natural conditions and human activities continue to pose considerable challenges to attaining acceptable air quality standards, especially in Southern California. Ozone is the major contributor to nonattainment in urban areas, while excessive particulate matter results in partial nonattainment levels in rural areas. *Source*: Environmental Protection Agency.

have actually been reduced since the mid-1950s, even though the Basin's automobile population has more than doubled.

The city has made a continuous effort to lessen smog problems since World War II. Overall smog levels in the Basin today are lower than they were 25 years ago. Responsibility for developing far-reaching plans to deal with the smog problems was assigned to the South Coast Air Quality Management District. The area's peculiar physical geography, a pattern that led early Spanish explorers to comment on the buildup of smoke from Indian fires, is the sole nagging and challenging problem. A semiserious suggestion arises from time to time to carve a large hole through the Basin's eastern mountains and set up huge fans to blow the smog out. People living in areas beyond the mountains, already hit by some effects of the smog that drifts over the mountains, are understandably opposed to such an idea.

Much about southern California's urban landscape reflects the impact of the automobile. Population density in Los Angeles is 3040 per square kilometer (7876 per square mile); in San Diego it is about half that. By comparison, Philadelphia has an average density of 4336 per square kilometer (11,232 per square mile). Housing units throughout southern California are predominantly of the detached, single-family type. Even low-income areas tend to have such housing, a feature often leading easterners to conclude mistakenly that slum or near-slum conditions do not exist in Los Angeles.

Its automobile-molded urban landscape has yet another facet: Los Angeles is a city without a center. The singular central business district (CBD), geographically defined as the focus of urban activity, barely exists there. Many of the communities comprising the greater Los Angeles area have their own independent centers. These days, the metropolitan area seems to be coalescing into about 18 "urban village" cores, which are business and retail focal points within the generally low-density urban landscape. Los Angeles itself is just the largest. Some cores, such as Pasadena or Westwood, are older communities that have grown into regional hubs. Others, like Ontario and Costa Mesa-Irvine-Newport Beach, seem to have simply emerged out of the expanding urban sprawl. The terms city center and suburb, commonly used elsewhere to describe the geography of an urban area, have little meaning in Los Angeles.

Resource Contributions

Local minerals and the local climate have been important in Los Angeles' economic and urban growth. Three of the United States' major petroleum fields are located in southern California. Offshore development began in 1965. Long Beach has reaped a large financial harvest from the sale of leases and from a tariff on oil pumped. Visitors to the area find the offshore pumping stations, built as islands complete with trees and waterfalls, visually less offensive than they had expected. The heavy demand for petroleum products, especially gasoline, results in the local consumption of virtually all of southern California's production.

More important economic benefits are attributable to the region's climate. Southern California is famous worldwide as the location of Hollywood, long the center of the American motion picture industry. In the early days of filmmaking, outdoor settings and natural light were the norm. The area's cloudless skies and short cold-temperature periods made its streets and fields a fine setting for countless motion pictures. Los Angeles remains one of the centers of American filmmaking and television, both of which have been invaluable in advertising the state nationally and internationally.

Early on, silent movies introduced California's palm-lined streets to the rest of the world, and now contemporary television glamorizes the region. Today, the industry plays a relatively small role in the Los Angeles economy. Only about 75,000 people in Los Angeles County are employed in motion picture and television production, which is less than 2 percent of the metropolitan area's total employment. But the economic benefits accruing to the region through its advertised image are immeasurable.

Southern California's Mediterranean climate and varied coastal scenery also supported growth in the region's recreation industry—so much so that the area is classed as one of the country's centers of outdoor recreation. As early as the late 1800s, towns and railroads extolled the virtues of the area's fine climate and scenery for recreation. Pasadena's Rose Parade was developed partly to advertise the area's mild winters. Today, these natural advantages have been supplemented by some of the country's largest and best developed recreation facilities. San Diego's Balboa Park, with its excellent zoo, Knott's Berry Farm, and Marineland are major southern California attractions, and Disneyland is the main destination of countless tourists.

A Postindustrial Place

Still, all of this does not add up to support the area's 19.2 million people. It does not fit the normal logic of location theory. Heavy resources, such as coal or iron ore, are virtually nonexistent. San Diego has a good natural harbor, but Los Angeles's harbor, entirely human-made, is only average. So to understand this population concentration better, additional elements must be considered: general national affluence during the growth of a postindustrial economy and the role of federal government spending.

In the United States and Canada, the declining share of total employment held by manufacturing and the corresponding growth of the tertiary and quaternary sectors freed American workers from the locational constraints that bind manufacturing to its physical resource base. This employment shift is essentially a transition into what is called a postindustrial economy. Because so many people are employed in the service and information economic sectors, places without mineral resources can be home to businesses and industries that employ millions of people.

The nature of manufacturing has also shifted; increasing shares of workers are in light industries, where the value added by manufacture is high and worker skills are valuable. Electronics is such an industry, and it is important in southern California manufacturing.

In this postindustrial period, new factors, such as the availability of adequate administrative support staff or the presence of major universities with their diverse intellectual and research resources, become more meaningful locational considerations. The question of where workers and, more particularly, management want to live also becomes highly significant (see Chapter 10). Southern California, with its mountains, seashore, mild climate, and outdoor image, offers the kind of natural environment many Americans find especially desirable.

Southern California also profited from government spending far more than most other areas of the country. California now receives about 20 percent of all Department of Defense spending and nearly half that of the National Aeronautics and Space Administration. San Diego is the West Coast home of the U.S. Navy, and the Navy is easily the city's principal employer. San Diego has relatively little manufacturing employment for an urbanized area of more than 2.9 million, lending credence to its claim of being one of the country's most livable cities. Southern California has been the country's leading aircraft-manufacturing center for nearly 40 years, with much of the business generated by orders for military aircraft. This means that a portion of the regional economy is cyclical, with the fluctuations dependent on government defense spending.

Ethnic Diversity

Close to Central and South America and Asia, southern California is also the major destination for migrants entering the United States from these parts of the world. So enduring have been the migration flows that today Anglos are a minority of Los Angeles County's 9.9 million people. Greater Los Angeles may be the largest Mexican metropolitan area outside Mexico; the second largest Chinese metropolitan area outside China; the second largest Japanese metropolitan area outside Japan; and the largest Korean, Filipino, and Vietnamese metropolitan areas outside those countries. With such numbers, its diverse ethnic population offers some security to new arrivals. It is estimated that by 2010 the metropolitan area will be 40 percent Anglo, 40 percent Hispanic, 10 percent Asian, and 10 percent non-Hispanic black.

This cultural diversity will challenge certain basic institutions, such as schools. For example, one school-child in four in the Los Angeles school district speaks one of 104 different languages better than English. Los Angeles' diversity also renews America's challenge to itself that its people of many backgrounds can live together with tolerance.

The Los Angeles cityscape reflects its ethnic diversity. Recent migrants often settle in ethnic neighborhoods. Little Tokyo, long a part of the city, has a renewed vibrance. Monterey Park in the San Gabriel valley is 50 percent Asian, making it the most heavily Asian city in the mainland United States. A rich diversity of ethnic restaurants, many found within ethnic enclaves, dot the city.

In 1950, seven of the world's twelve largest cities were European or North American. By 2000, only two—New York City and Los Angeles—still fit this category, and of the two, only Los Angeles is still growing rapidly. Los Angeles has a higher dollar volume of retail sales than the New York metropolitan area does, and the value of its manufactured goods is higher. A decade ago, Los Angeles passed San Francisco as the West Coast's financial center; in 1986 it surpassed Chicago in banking deposits; and it now ranks second nationally in population after New York City. The twin ports of Los Angeles and Long Beach together form the fastest-growing major cargo center in the world. The dollar value of their import-export oceanborne cargo easily surpasses that of the Port of New York and New Jersey. U.S. trade with Pacific Rim countries is 30 percent greater than U.S. trade with Atlantic Basin countries, and 60 percent of that trade passes through Los Angeles's two ports and its international airport.

San Francisco Bay

San Franciscans view their city as old, cosmopolitan, and civilized. There is something to be said for this attitude.

San Francisco was the northern core of Spanish and Mexican interest in California, and it served as the supply center for the California gold rush. By 1850, it was the largest city on the West Coast, a ranking it maintained until 1920. When the first transcontinental railroad joining San Francisco to eastern cities was

California's diversity can be seen in its numerous ethnic festivals, such as this Chinese New Year parade in Los Angeles.

In San Francisco, packed housing stacks up on one of the city's famous hills.

completed in 1869, the city's size and its excellent harbor made it not only the focus of western U.S. growth, but also the key location for U.S. commerce with the Pacific. Large numbers of Asian immigrants, especially Chinese, came into the city, as did substantial numbers of other foreigners. Together they created a cosmopolitan, ethnic mixture that today is a vital aspect of the city's character.

The romantic flair of San Francisco's early history is a piece of the mosaic that makes it one of the most popular American cities. Its physical geography provides a splendid setting. Steep slopes offer dramatic views of the Pacific Ocean and the San Francisco Bay. Its mild climate allows escape from the sometimes staggering summer heat of southern California. Quantitative evaluations of the quality of American cities, which tend to emphasize readily available socioeconomic statistics, generally give San Francisco good marks. Whatever the meaning of such evaluations, they are another component of the pattern that places San Francisco at

or near the top of a list of desirable places in studies of urban perception. A less publicized consequence of this perception has been unemployment levels generally above the national average as newcomers without jobs are attracted to the city. Events such as the "dot.com bust" during the 2000–2001 further exacerbated the unemployment problem.

San Francisco's image is so powerful that its cachet spreads to the entire Bay Area. In fact, the city of San Francisco is currently home for fewer than one-ninth of the Bay Area's 7 million people. Hemmed into a small peninsula, San Francisco is actually losing population while the rest of the Bay Area is growing (Figure 15.7).

The Bay Area today is composed of several different areas, each with its own special character. The East Bay is the most varied, with a mix of blacks, Berkeley students, large tracts occupied by middle-class residents, and most of the region's port facilities and heavy industry. The San Jose-South Bay area is upper middle class, white, with new houses and fine yards, interrupted by

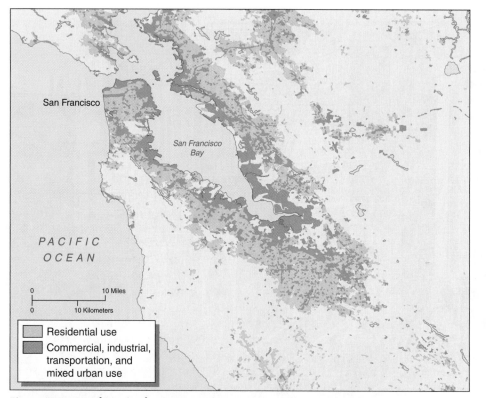

Figure 15.7   Land use in the San Francisco Bay Area. Industrial and transportation activities hug the margins of the bay. Accessibility provided by this superb site has been critical to the area's economic growth.

major regional shopping centers. Along the Bay north of San Jose is Silicon Valley, so named because of its concentration of businesses engaged in chemical and electronics research associated with production of computer components. North of the Golden Gate Bridge, cities are smaller, manufacturing is sparse, and the conflict between agricultural and urban land uses is actively waged. Here, well-to-do urbanites search for a place in the country.

The city of San Francisco itself, with its grid pattern of streets incongruously placed on hilly terrain, its closely spaced late nineteenth- or early twentieth-century houses, and its ethnic neighborhoods, has a special appeal, despite the recent appearance of skyscrapers along the skyline. San Francisco adopted zoning ordinances to help maintain the city's character. Buildings taller than 18 meters (60 feet) are banned from many areas; architectural appropriateness is encouraged in public housing; and private redevelopment plans are encouraged to mimic the city's tone.

San Francisco is not the focus of the Bay Area it once was. Oakland is the principal center of the East Bay; San Jose is the hub of Santa Clara County and is now more populous than San Francisco. The bay, so important to the quality of San Francisco's site, in some ways is also a problem. Movement between different sections of the area has been difficult and expensive, with few bridges crossing the bay. As the urban area grew, water barriers made continuing the regional focus on San Francisco difficult. The Bay Area Rapid Transit (BART) system, one of the few new mass transit systems in the country, was designed as an innovative attempt to meet the difficulties of urban movement. But only three of the area's nine counties chose to join the system. So BART serves only the East Bay and San Francisco, doing little to ease circulation in the entire region.

Unlike Los Angeles's improbable setting, the existence of a major urban center in the Bay Area is not surprising. Its excellent harbor and good climate are key site factors. Access to the bay is so important and sizable tracts of flat land for construction are so limited that large portions of the bay were filled in to create new land. Fearing they might suffer economic and ecological losses if the bay were filled excessively, communities around the bay created a regional government authority to restrict such landfill development. The authority has allowed little reshaping of the bay in recent years.

San Francisco Bay's relative location, or situation, was and is excellent. By volume, it is the major Pacific port in North America. Its rail and highway ties to the East are the equal of those for any other West Coast city. Just as the East's Megalopolis is America's hinge with Europe, San Francisco is its hinge with Asia.

Cities of the Valley

Outside the southern coastline and the San Francisco Bay region, the largest cities in California are in the Central Valley. Sacramento, the state capital, is the most populous with a 2000 population of 1.8 million in its metropolitan area. Like all Central Valley cities, it is a major agricultural processing center, processing fruit and vegetables for canned or frozen products from nearby sources. Sacramento also has a substantial aerospace industry. Two large air force bases not far from the city contribute to the regional economy. Fresno (2006 MSA population, 1,002,284) and Stockton (2006 MSA population, 673,170) are regional service and agricultural processing points for the central and northern San Joaquin valley. Bakersfield (2006 MSA population 780,711) serves a similar role at the San Joaquin's southern end. Also influenced by local petroleum extraction, Bakersfield is home, too, of country-western music in California.

A STATE OF IMMIGRANTS

A native Californian is a rarity.
California is a state of immigrants,
mostly from the Middle West.

These are commonly held beliefs about Californians. As with most such statements, they contain a kernel of truth, but they are also misleading. Although more than two-thirds of native-born Americans live in their state of birth, fewer than half of Californians were born in the state. California has been a major destination for U.S. internal migration in nearly every decade since 1850 and has become the residence for people born in an area extending from California east and northeast through the Midwest. The political climate, especially in San Diego and Los Angeles, often reflects the conservatism of the area's southern and midwestern immigrants. Between 1940 and 1990, California's share of the total U.S. population grew from 5 percent to more than 10 percent. Today, the state accounts for more than 12 percent of the U.S.

Off-road vehicles can do significant damage to the environment, especially where natural recovery from a mere tire imprint takes decades.

population. For decades, California attracted more migrants than any other state until Florida gained that distinction in the 1970s. California regained the number-one position during the 1980s and has held onto that ranking since then.

California's Future

California has established itself as a dominant political and economic force in American life. In national elections, a presidential candidate traditionally has been well advised to focus on winning California; the state has nearly 20 percent of the electoral votes needed for victory.

In economic matters, Los Angeles alone produces enough goods and services each year to rank the metropolitan area among the world's 12 wealthiest countries. As the Pacific Rim countries continue to develop economically and as U.S. trade with Rim countries increases, California benefits.

California vies for the role of the country's cultural core, challenging New England and the Middle Atlantic states. It has taken a leadership role in varied efforts such as local tax reduction, environmental legislation, term limits for elected officials, and gay and lesbian rights. Certainly, it represents what many Americans view as the good life. It is a new kind of America, one that is copied, if not totally admired.

California's problems, however, are as great as its possibilities, with growth linked to both. The state's physical environment, so admired and desired, is fragile; the desert and alpine ecosystems are easily damaged. The outdoor recreation demands placed on the environment by over 36 million Californians are immense and can easily outstrip the state's resources. Only so much seacoast exists, and much of it is already unavailable for public use. Although the state has some of the country's most rigorous, well-enforced environmental laws, the magnitude of the task tests the adequacy of these laws.

Perhaps California's two greatest problems are related to urban expansion and water supplies. Although cities are expanding across North America, farmland lost here poses a serious threat to the domestic production of certain crops. The greatest losses of land, centered around Los Angeles and San Francisco, have thus far been offset largely by expansion elsewhere, particularly in the San Joaquin valley. But most of the easily available new land is now in use. Additional expansion will require irrigation on a massive scale, and this will exacerbate conflict between the need to maintain agricultural production levels and other demands for the state's limited water supplies. If present trends continue, water demands could outstrip California's available supply early in the next two decades.

It may well be that, in the long run, irrigation is not the most economical use for the state's water. The monetary return for water invested in crop growth is relatively small when compared with alternative industrial uses. Development of new sources of supply has allowed California to sidestep this difficult question. Soon it may no longer be able to do so.

The problem of water sufficiency makes the future of California's agriculture difficult to predict. Although California leads the nation in the value of its agricultural products, the relative importance of agriculture within the state's total economy is much less than it is for states like Iowa or Nebraska. Also, many of the crops grown, like avocado, kiwi, pistachios, and pomegranates, are not basic parts of the American diet. Demand for specialty crops may fall if their prices are pushed up too greatly by more expensive irrigation water. Finally, the state's expensive irrigation projects help relatively few farm operators, who are often corporations rather than individual farmers.

Despite the political clout these few operators have, a decision might eventually be made to shift water from agricultural land to urban areas where the vast majority of Californians live and where most of the state's income is generated—no matter the loss it would represent to corporate farmers, to the state, and to the nation.

PHYSICAL CASE STUDY

Plate Tectonics and the Ring of Fire

California is associated with what geologists have called the "Ring of Fire," a belt of intense earthquake and volcanic activity that encircles much of the Pacific Ocean. In what amounts to a landslide of intellectual activity during the last 35 years, earth scientists have improved their grasp of the reasons for such a pattern of concentrated, violent earth movement. Their findings are important to understanding both the physical and human geography of California and, indeed, most of the West Coast of North America.

Geologists now agree that the crust of the earth, the *lithosphere*, is composed of large, solid pieces called plates that move slowly on the molten rock which lies beneath them. In general, oceanic plates are rather thin, averaging about 50 kilometers (32 miles) in thickness, but are dense and heavy; continental plates are thicker, averaging 150 kilometers (95 miles) thick, but are brittle and more buoyant.

In midocean areas, these plates may pull apart, forming a deep trench bordered by ridges. A mid-Atlantic ridge, stretching nearly the entire north-south extent of that ocean, is a prominent example of this feature.

In other places, plates may be pushing together and colliding. Where oceanic and continental plates collide, the denser oceanic plate usually plunges deep beneath the continental plate in a process called *subduction*. Deep ocean trenches may be created along the zone of surface contact between the two plates. The crust of the oceanic plate may melt as it is subducted. The resulting *magma*, or molten rock, can move to the surface, creating a line of volcanoes well away from the surface contact zone.

A third possibility is that the two plates are in contact but are merely sliding alongside one another, with no movement toward or away from each other. Faults usually run along the line where the two plates contact. Faults are fractures in the earth's crust where relative crustal displacement has occurred, whether mountain building or lateral displacement. The general theory given to these associated explanations is called *plate tectonics*.

The entire West Coast of the United States and Canada is a zone of active plate contact. Most of the continent is dominated by a single plate, called the North American plate. An even larger Pacific plate borders it along the West Coast. In California, from the Imperial Valley northwest to Cape

Mendocino, the Pacific plate extends inland to meet the North American plate roughly along the line of the San Andreas fault (see Figure 15.2). Coastal California north of the Transverse Ranges, which form the northern edge of the Los Angeles Basin, is sliding northwest along the North America plate at a steady rate of 2 centimeters (1 inch) or more a year. The two plates are colliding along the Transverse Ranges. This entire zone of contact is an area of frequent earthquake activity.

From northern California north to British Columbia a much smaller oceanic plate, the Juan de Fuca plate, is being subducted by the North American plate as the latter follows a generally westward drift. One dramatic consequence is the line of volcanic peaks from northern California to Washington. This line marks the zone of crustal melt for the subducted oceanic plate. This is a zone of active volcanic activity, as the 1980 eruption of

For most of its length, California's San Andreas Fault lies invisibly beneath the earth's skin. But in the Carrizo Plain between Bakersfield and Santa Barbara, this famous strike-slip fault shows its immense power in the feature called the Dragon's Back pressure ridge.

Mount St. Helens in southwest Oregon reminded us (see Chapter 16).

When comparing maps of the coastal margins of the East and West coasts of the United States (see, for example, Figure 2.1), the Atlantic Coast looks indented and irregular. This is a classic example of a coastline of submergence, where the land is sinking relative to sea level. The coastal plain itself is broad and flat, especially in the South. Mountain building associated with plate contact has generated a very different appearance along the Pacific Coast. The coastline is one of

emergence. It is very regular, with few embayments and no estuaries of significance. Offshore, the broad continental shelf that marks the Atlantic Coast is replaced on the Pacific Coast by deep trenches near the coast. There is no coastal plain. Instead the land is generally hilly or mountainous to the water's edge. Rivers seem to just suddenly meet the ocean, with none of the gradual transition from river to estuary to ocean so common on the East and Gulf coasts. Few large West Coast cities, and none north of San Francisco, are at the ocean's edge.

POLITICAL CASE STUDY

Dreaming of a Divided California[1-5]

It's been called the modern-day equivalent of the Boston Tea Party. Disgusted by the way state government represented their interests, hefty numbers of northern Californians muttered secession in 1992 and backed a proposal by Shasta County Assemblyman Stan Statham to sheer off into a separate, rural state. This wasn't California's first secessionist rumbling. Since the Golden State achieved statehood in 1850, there have been 27 propositions to split California. The 1992 proposal was the first time the possibility had a serious chance of success.

Statham, whose home county encloses the northern end of the Sacramento valley, initially suggested in 1991 that California subdivide into two states—a rural north and an urban south. As with so many previous proposals, his proposal reflected a conflict between rural and urban interests and lifestyles. North of the San Francisco Bay Area, small towns, farms, forests, and mountains spill across land occupied mostly by white, conservative people whose ethos is more akin to those in eastern Oregon, Idaho, Montana, and Wyoming, than to people living in the bulging, ethnically and culturally mixed urbanscapes elsewhere in the state.

Rural northern Californians contend they get economic short shrift because the massive needs of the urban regions overshadow theirs. Proposition 13 is one such example. Passed in 1978, this anti-tax measure limits property taxes and, consequently, hobbles a county's ability to raise revenues. By

1991, California's sparsely populated rural counties were teetering on bankruptcy as they tried to find money for services they needed such as welfare and criminal courts that were no longer financed by the state. On top of this, the state legislature often passed unfunded mandates—new, required, and expensive programs with no state money provided. Rural Californians also contend their political power was sapped by earlier redistricting, so their representatives no longer could protect rural interests when urban legislators pass laws suited to urban, but not rural, areas. Top off this accumulated dissatisfaction with California's economic woes in the early 1990s, and people were grasping for solutions.

Nourished by this environment of discontent, Statham devised a plan whereby counties north of a line running from San Francisco to Sacramento to Yosemite National Park would split off from California and form a separate state. Faced with the precedent of 26 prior secessionist tries, Statham's plan was unique in a key way. This was the first time the issue was put to a vote. In June 1992, supervisors in 31 counties put the measure on the ballot for a nonbinding, advisory opinion from the voters. By huge majorities, 27 counties voted yes. Of those that rejected the plan, the two in the San Francisco Bay Area indicated their objection was not to secession but that the proposed division would place them in the same state as Los Angeles.

Using this feedback, Statham modified his proposal to suggest division into three, not two, states (Figure 15.8). The rural northern state originally proposed would remain the same. The San Francisco Bay Area and the Central Valley would make up a second state, and southern California would comprise the third state.

The blueprint Staham and his supporters presented in hearings across California addressed not just feelings but practical concerns as well. Each proposed state had the option of rewriting its constitution, electing a new legislature, and starting fresh. Citizens' concerns about possible tax increases

[1] Katherine Bishop, "California Dreaming, 1991 Version: North Secedes and Forms 51st State," *New York Times National*, November 30, 1991.

[2] Eric Brasil, "Up Country: It's Divide and Rule," *San Francisco Examiner*, February 9, 1992.

[3] S. J. Diamond, "Split Personality," *Los Angeles Times*, July 25, 1993.

[4] Ed Fishben, "Just One California Not Enough for Many Residents of That State." *Houston Post*, December 4, 1992.

[5] Stan Statham, personal communication, 1998.

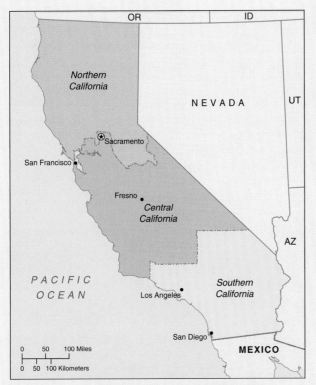

Figure 15.8 **The 1992 proposed subdivision of California.** California is so diverse in so many ways that proposals are made from time to time to subdivide the state into two or more new states.

were eased when calculations of each region's contribution to the state General Fund and the subsidies and services used suggested that a deeper reach into taxpayer pockets was not necessarily likely. Parched southern Californians were reassured the water-blessed north would continue to help quench the drier region's thirst. Also, the three states would share a host of nondivisible resources. For example, each would recognize professional credentials and licenses, guarantee the integrity of public retirement programs, and provide each state's students equal access to universities and public colleges throughout the three new states. This kind of thoughtful planning provided the ammunition to shoot down enough obstacles so that public support for the proposal grew throughout California.

A proposal to form a new state from the division of an existing one must be approved by the parent state legislature and the United States Congress. Statham's three-state plan was the first in 134 years to move past one house of the state legislature, passing the California Assembly on a vote of 46 to 34. After the proposal went to the state Senate, however, it died in the Rules Committee by a single vote.

This will not be the last serious proposal to divide California into two or more states. As with other political units in North America, California is a unit of many parts. Over time, the balance shifts between what holds California together and what seeks to separate it.

Review Questions

1. What are the challenges inherent in designating California as a distinct region?

2. What was the long-term impact of the gold strike in the foothills of the central Sierra Nevada in 1848?

3. How can Los Angeles be described as a city without a center?

4. What is the origin of the infamous Los Angeles smog?

5. What are the advantages of San Francisco's site and situation? Have these remained constant over the past 150 years?

6. What are some of the major problems concerning California's water supply?

7. How is California associated with the geological term "ring of fire"? Why is much of the region prone to earthquakes?

8. What would be key issues to address if California were to be divided politically into two or more states?

CHAPTER 16

THE NORTH PACIFIC COAST

Preview

1. The physical geography of the North Pacific Coast is defined by rugged terrain and a wet climate with moderate seasonal temperatures.

2. Because of its remote location, the region was explored and settled by Europeans and Americans later than other areas of North America.

3. Except for Vancouver, Seattle, and Portland, there are relatively few large metropolitan areas in the region.

4. Agriculture and logging are important elements of the region's economy, but both sectors are faced with relatively long distances to other North American markets.

5. Seattle's economy is dominated by healthcare and biotechnology sectors and by two commercial firms: Boeing and Microsoft.

landscapes. Another component is embedded in the region's situation, or relative location. Both aspects are represented in an attitude held by many North Pacific Coast residents, one that flies in the face of traditional definitions of progress and success.

By most widely accepted socioeconomic markers, regional progress is measured by a region's ability to attract people to fill its empty land. However, people in the North Pacific Coast scuttled this belief in the 1960s. A change of attitude surfaced in the southern half of the region as residents of Oregon and Washington realized they had something good and the influx of newcomers threatened to destroy it. Their fear was given nationwide attention in 1976 by a plea made on national television by Oregon's governor, Tom McCall. He said: "Come to visit us again and again.... But for Heaven's sake, don't come here to live." That inherently exclusive attitude is now ingrained in the region's image, even though not all of the region's residents share or approve of it.

Clear mountain streams tumble down rock-strewn courses to a rugged coastline where precipitous, fog-enshrouded cliffs rise out of pounding surf. Lofty, snow-capped mountains loom in the distance. Tall pines, spruces, and firs drape a mantle of green over the land between. Where they exist, cities look new, blending into the landscape with a crisp, clean look that makes them outshine most other urban places. The people are satisfied and friendly. They bask in the beauty of the land and a life filled with the North American virtues of independence and self-reliance. This is North America's North Pacific Coast, or more popularly called the Pacific Northwest (Figure 16.1).

Though laced with hyperbole, such a description matches how many Canadians and Americans visualize the coastal zone stretching from northern California to southern Alaska. The conjured images have induced many to seek their fortunes there and have kept countless of them settled in the region when their monetary aspirations evaporated. The inescapable appeal of the North Pacific Coast lies in its natural environment and the variety of outdoor activities it supports, leading one writer to label this region Ecotopia.[1]

Much of the North Pacific Coast's regional character is tied up in the beauty and bounty of its

[1] Joel Garreau, *The Nine Nations of North America* (Boston: Houghton Mifflin, 1981).

Heceta Lighthouse stands dramatically in the midst of Oregon's rocky coastline. Natural landscapes in the North Pacific Coast have a spectacular beauty that encourages the region's residents to enjoy outdoor recreation as nowhere else in North America.

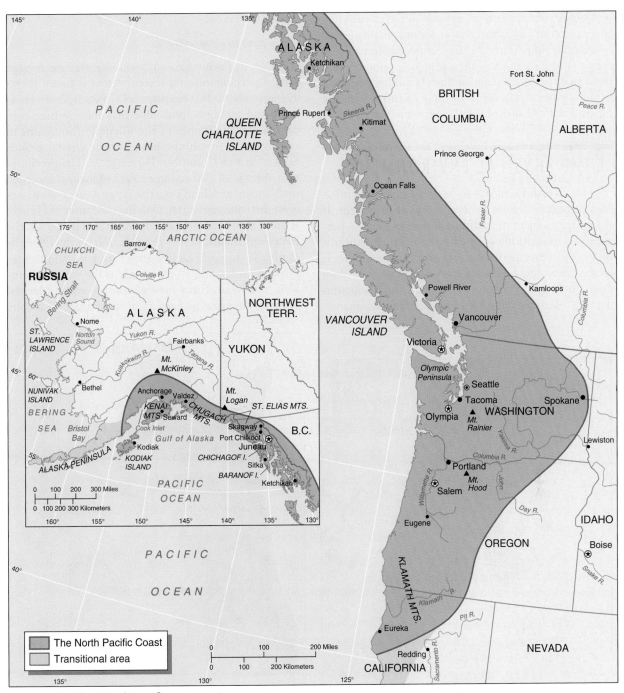

Figure 16.1 The North Pacific Coast.

Another important element of the North Pacific Coast's regional character is its location relative to the rest of North America. Substantial distances of arid or mountainous terrain separate the region's populated sections from other principal centers in the United States and Canada. Residents tend to view this separation as positive because it provides a geographic buffer against what they consider the less desirable rest of the world. Economically, however, it is a hindrance. High transportation costs inflate the price of North Pacific Coast products in distant eastern markets, and they also discourage some manufacturers from locating in the region.

But as important as the economic impacts of regional separation are, the sense of regional independence that the separation fosters among residents is often expressed in terms of their relations with the North Pacific Coast's physical environment.

THE PHYSICAL ENVIRONMENT

The North Pacific Coast is, with one or two notable exceptions, defined by its physical environment. It is strongly subject to maritime influences and is pervaded by dramatically rugged terrain. Precipitation is high. Lush vegetation associated with heavy moisture is located near the coast, but marked variation in flora exists over short distances because microclimates develop due to the influence of surrounding mountains.

Climate

The mountain is out today.

The North Pacific Coast is wet. The greatest average annual precipitation on the continent is found there; averages above 190 centimeters (75 inches) are common, and averages are double that amount measured on the western slopes of the Olympic Mountains in northwestern Washington and in the coastal mountains of British Columbia (Figure 16.2). The precipitation average of 600 centimeters (230 inches) recorded on Vancouver Island's northwest end easily surpasses even these measures and, with the exception of Hawaii, is more than twice that of any other area in the United States or Canada. Heavy precipitation on the Olympic Peninsula supports rainforests where ferns and mosses grow in profusion and trees such as the western hemlock, red cedar, Sitka spruce, and the world's largest Douglas fir grow to heights of more than 60 meters

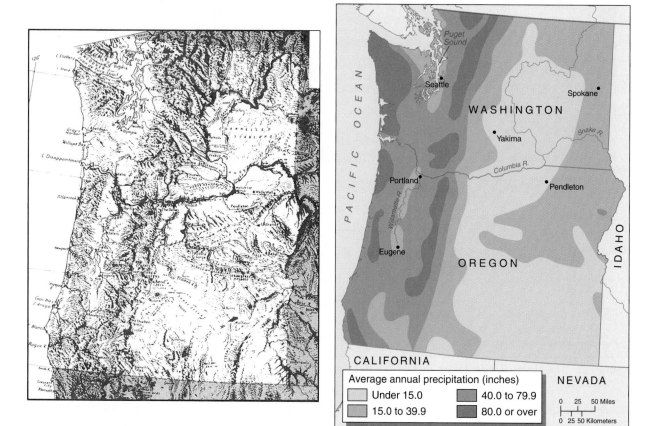

Figure 16.2 **Precipitation and topography.** (Topographic map from Raisz, *Physiographic Diagram of the United States*.) Notice the association between areas of rugged topography (left map) and higher precipitation (right map).

(200 feet). Naturalist Roger Torey Peterson suggests that the Olympic rainforest has more weight of living matter per unit of land than anywhere else in the world.

The northern Pacific Ocean is a spawning ground for great masses of moisture-laden air. As these air masses move, they are pushed south and east by prevailing winds onto North America's Pacific shores. A high-pressure system located off the coast of California in summer and off northwestern Mexico in winter prevents many of these maritime air masses from drifting farther south and thus ensures that most of their moisture falls over the North Pacific Coast.

The seasonal movement of weather belts has a noticeable impact on precipitation in the North Pacific Coast. Everywhere in the region, winter precipitation amounts are higher than summer levels, but the seasonal difference is more marked on the region's southern margin. For example, in southern Oregon and northern California, coastal areas receive less than 10 centimeters (4 inches) of rain during July and August, only a tenth of what falls there between December and February.

The pattern of high precipitation does not hold for the entire North Pacific Coast. Considerable portions are semiarid. Parts of the borderlands of Washington's Puget Sound receive only about 60 centimeters (25 inches) annually, no more than the amount that falls on Wichita, Kansas, in a typical year. Portland's average, like Seattle's, matches that for Boston or New York and is far less than the typical annual precipitation of a southeastern U.S. city like Atlanta.

To a casual eye, areas in the Northwest seem wetter than they are. This climatic illusion has several sources. Cloud cover is frequent from midfall to midwinter. During winter, for example, cloud cover is so constant in Seattle that residents tell each other when nearby Mount Rainier is visible. Furthermore, when precipitation falls, it seldom takes the form of heavy thundershowers; more typical is a gentle, light, frequent rain that feels like a heavy mist. This form of precipitation lessens runoff, normal in heavy rains, and enables vegetation to make maximum use of the moisture.

Mountains are the main reason for both the high precipitation along the Pacific Northwest's coast and the substantial climatic variations that exist inland in close proximity. As a Pacific air mass passes over land, moving east and southeast, it strikes the mountain ranges lining the North Pacific Coast and is forced to rise. As the air rises, it cools, and its moisture-carrying capacity is reduced, resulting in precipitation. Such precipitation is called *orographic* (mountain-induced) precipitation, and it is common on the windward side of most extensive mountain chains.

The same air masses that provide abundant coastal moisture and heavy precipitation on the western slopes of coastal ranges produce substantially less moisture along the lowland trough east of the mountains. Along a belt extending from south-central Oregon to southwestern British Columbia, the Coast Ranges in the United States and the Insular Mountains in Canada are backed by a trough of low-lying land, including the Willamette Valley in Oregon, the Puget Sound lowland in Washington, and the Fraser River lowland in British Columbia. As the east-moving air descends into these lowlands it warms, and its moisture-carrying capacity increases. Thus, because additional moisture is not introduced into the air, little precipitation occurs.

This high-moisture, low-moisture sequence repeats itself as the air mass continues to negotiate the topography. A second north-south trending range of mountains lies east of the first lowland. As the moving air hits it, it must rise again, and again orographic precipitation results, although not as heavy as along the coast. These more inland mountains, called the Cascades in the United States and the Coast Mountains in British Columbia, are tall; Mount Rainier in Washington is 4390 meters (14,410 feet) high, and many peaks in the range reach 2750 to 3650 meters (9000 to 12,000 feet). Winter precipitation there falls in the form of snow, making the Cascades-Coast Mountains the snowiest portion of North America.

Finally, the air masses warm as they descend into the eastern reach of the region beyond the Cascades in interior Washington. What little moisture remains in the air is retained, so most of eastern Washington as it merges into the high plateau areas of the Empty Interior averages less than 30 centimeters (12 inches) of precipitation each year. Only by irrigating and using dry-farming techniques can most of the area be cropped.

South and north of this mountain-valley-mountain system, mountain ranges merge and the separating valley disappears. Heaviest precipitation amounts concentrate in a single band along the coast, especially the central and northern coast of British Columbia and Alaska's adjacent panhandle. This area receives substantial moisture throughout the year, and average cloud cover percentages are among the continent's highest. The climate of Alaska's panhandle in particular is dominated by moisture and cloudiness.

North and west of the panhandle, average precipitation levels drop sharply along the coast. Most of central Alaska's southward-facing coast averages 100 to 200 centimeters (40 to 80 inches) annually. Since many of the weather systems created off southern Alaska's shores move south and east—often just brushing the shoreline of central Alaska—Anchorage residents tend to be disdainful of the southern peninsula's damp climate, favoring their own clearer, colder weather.

While the rain is a prominent feature of North Pacific Coast the region's maritime location provides moderate temperatures. Summers are cool. Winters are mild, although dampness renders the air raw and makes it less comfortable outdoors than the thermometer suggests. Still, Seattle is warmer in winter than St. Louis, and Juneau, Alaska, is only as cold as Washington, D.C. Snow is uncommon along the coast south of Vancouver, but thick accumulations can smother the coast of northern British Columbia and the Alaskan panhandle.

Another consequence of the seasonal movement of air masses is frequent periods of high winds along the coastal margin, especially in Oregon, northern California, and southwestern Washington. Large storms build up during passage across the North Pacific and can strike the coast with tremendous ferocity. While not hurricanes, they can pack winds exceeding 125 kilometers (80 miles) per hour at their winter worst. Although the coastal mountains provide some protection and winds are generally lower in the summer, even in the summertime eastern portions of the region sometimes deal with the consequences of high winds. When they do, the danger of fire in this arid area is aggravated.

Topography

The North Pacific Coast is a land of mountains; few places in the world offer more breathtaking views when the weather is clear. In the region's northern extremity, Mount McKinley, at 6200 meters (20,300 feet), is the highest peak in North America. Canada's St. Elias Mountains are the world's highest coastal mountains, with Mount Logan looming at 6000 meters (19,700 feet).

Elevations are generally lower in the North Pacific Coast's south, but it, too, is a land of high mountains and rugged terrain. The peaks of the Coast Ranges of Oregon and Washington are the lowest among the region's mountains. The ranges are fairly continuous in Oregon, with elevations reaching about 1200 meters (4000 feet). In Washington they are discontinuous; several rivers, notably the Columbia and the Chehalis, cut pathways across them. Coast Range elevations in Washington are seldom above 300 meters (1000 feet).

Mount St. Helens, one of a series of volcanic peaks that punctuate the southern Cascades, erupted violently in May 1980.

In northern California and southern Oregon, the Klamath Mountains offer a jumbled topography that is a product of Pleistocene glaciation and stream erosion. Little pattern is apparent in its terrain, in marked contrast to the parallel ridge and valley structure of the Coast Ranges. The Klamath Mountains are a rugged, empty area.

The lowlands of western British Columbia and Oregon are part of a structural trough created when the area sank at the time the Cascades were elevated. This trough extends north in the form of straits separating Vancouver Island from the rest of British Columbia, then passes through the complex of islands that line the Alaska panhandle and provides the Inside Passage north as far as Juneau.

Inland beyond the trough, the Cascades run northward from the Klamath Mountains into southern British Columbia. The Cascades' southern section is a high, eroded plateau topped by a line of volcanic peaks. Splendid in their isolation above the plateau, the peaks' linear march marks the zone along which the Pacific plate is passing beneath the North American plate (see Chapter 15). Tectonic pressure keeps the zone alive, though on a geologic timescale. Until 1980, Mount Lassen in California was the only volcano in the United States outside Hawaii to be active in historic times. Then the eruption of Mount St. Helens redefined the cone's meaning for the region's residents. Dramatic in a different way, the northern Cascades are ruggedly complex. But there, too, extinct volcanoes like Mount Rainier provide high elevations and well-defined peaks.

In British Columbia, the Cascades—known there as the Coast Mountains—merge with the Insular Mountains and push against the ocean. This is a land of dramatic coastal mountains cut by glacially eroded fiords and islands. Land transportation along the coast is nearly impossible, and coastal shipping provides the principal transportation connections. Between Vancouver and the Alaska panhandle, overland transportation lines connect the coast with the interior at only three places—at Prince Rupert in British Columbia and at Port Chilkoot and Skagway at the northern end of the Inside Passage in Alaska.

Beyond the Alaska panhandle and the massive, glacier-covered St. Elias Mountains, the mountains again divide in southern Alaska. The Coast Ranges, notably the Chugach and Kenai Mountains, decline in elevation from east to west. The interior mountains, the Alaska Range, are taller and more continuous. A

large lowland at the head of Cook Inlet is south of a gap through the Alaska Range, and here Anchorage, easily Alaska's largest city (metropolitan population 359,180 in 2006), sits with its good harbor and easy connection to the interior.

Juneau, Alaska's capital, is located on a narrow coastal lowland on the panhandle. Juneau's only transportation connections to the rest of the state are by air or water; the farthest one can drive from town is 15 kilometers (10 miles). When Alaska's wealth was in the panhandle's forests and salmon fisheries and when access to the Yukon gold fields through Skagway was a consideration, this location for the capital was reasonable. But as Alaska's economy changed and other resources became more important, the panhandle languished. Fairbanks (population 31,324 in 2006) in central Alaska grew to match Juneau's population, and Anchorage, which is accessible to the state's southern coast, easily surpassed both cities in population. The capital city's population was only 30,987 in 2006.

Alaska's voters and political leaders, dissatisfied with Juneau's inaccessibility, voted in 1974 to move the capital to a point about 80 kilometers (50 miles) north of Anchorage. The cost of constructing a new capital city, now estimated at more than $3 billion, led many Alaskans to reconsider that decision. In 1978 and again in 1982, voters refused to allocate money for the move, so despite its location, Juneau remains the state's capital.

Trees and More Trees

There are magnificent redwood stands in the Klamath Mountains; Douglas fir, hemlock, and red cedar in Washington and Oregon; red cedar, hemlock, and Douglas fir in British Columbia; and Sitka spruce on the Alaska Peninsula. This is a land not just of forest, but of exquisite expanses of tall trees reaching straight for the sky, trees that are among the largest on earth, trees that encourage people to stand and stare in awe or admiration—and to cut for profit.

Except for the tundra above the treeline and the drier lowlands, all of the North Pacific Coast is, or was, covered by forest. The region's plentiful moisture and moderate winter temperatures encourage tree growth. The trees' size and abundance encouraged some to think of exploiting this resource, and forest products have long been an economic mainstay of the region. Even today, although the southeastern United

States produces more wood for pulp and paper products, no other part of Canada or the United States provides as much lumber as the North Pacific Coast.

PATTERNS OF HUMAN OCCUPATION

Except for the polar areas, no other coastline was explored by Europeans as late as the North Pacific Coast. Vitus Bering had claimed the Alaska coast for Russia by 1740, but it was not until 1778 that Captain Cook sailed the coast from Oregon to southeastern Alaska. By then, several million colonists on the other side of the continent were fighting a war for independence from the very country for which Cook was sailing. By the time Lewis and Clark crossed the Cascades to the mouth of the Columbia River in 1805, Philadelphia and New York City, each with about 75,000 people, were vying for the title of the country's largest city. By the mid-1840s, when substantial numbers of American settlers began traveling the Oregon Trail to the Willamette Valley, New York's population was approaching 500,000, and its city administration was wrestling with problems of urban transportation and waste disposal.

Since European influence was related to a place's distance from Europe, this late exploration and settlement of the North Pacific Coast is understandable. In the days of sail and slow overland transportation, no nonpolar coast in the world was as far from Europe as this one.

American Indians

The region's pre-European population was relatively large. The moderate environment provided plentiful food throughout the year. Deer, berries, roots, shellfish, and especially salmon were a natural bonanza that seemed without limit. The American Indians responded to this by maintaining a hunting and gathering economy; corn or other crop cultivation was not necessary to support their numbers.

Concentrated along the coast, the native population was divided into many distinct ethnic groups, each occupying its own, often small, valley. Typical houses were large and impressive, built of red cedar planks; people went to sea in dugout canoes made of the same

wood. Despite intergroup diversity, the North Pacific Coast's Indian people shared many cultural features. Notable were the *potlatch* (a ritual during which property is given away in a large redistribution celebration) and the carving of totem poles (the recording on an upright log of the key incidents in an individual's life).

Along most of the Pacific Northwest's coast, Indian people seemed to melt away when Europeans arrived. Because their pattern of living in places separate from each other made organized opposition difficult, the small tribes tended to succumb quietly to the intrusion and impact of European settlement. Their numbers quickly dwindled as they left or died from European diseases.

Today, few Indian tribes remain in the North Pacific Coast's southern portion. Of those who do, some are reasserting ancestral rights and seeking clarification of old treaty terms. The Puyallups, for instance, won in the courts the right for all Indians to half of Washington's total salmon catch. The court's decision, based on interpretation of an early agreement in which Indian residents surrendered their land in return for vaguely stated rights, resulted in restrictions on salmon fishing among the state's many non-Indian sport fishermen. Farther north in northern British Columbia and the panhandle of Alaska, the Indian population remains a substantially larger presence.

Early Canadian and American Settlement

The early European settlement history of the North Pacific Coast reads like a collection of pulp novels, with prospectors, lumbermen, land barons, fur trappers, settlers, sober-minded preachers, railroad magnates, and imported brides trooping across the land in the best Bret Harte fashion. It is the stuff of good historical fiction, and it is a wonder that the region's writers haven't done better with it. Several events in this history are so important to the North Pacific Coast's development.

Russians were the first Europeans to establish permanent settlements along the coast. They came late in the eighteenth century, motivated by the predictable European search for easily extracted riches. In the North Pacific Coast, these riches proved to be furs, and the Russians established a series of trading posts and missions that were concentrated in southeastern Alaska and extended as far south as northern California. The outposts never became self-sufficient in foodstuffs, and the cost of maintaining these scattered sites usually exceeded the income from fur sales.

So the Russians wanted out. Following several attempts to negotiate the sale of what became Alaska, Russia and the United States agreed on a $7.2 million sale price in 1867. Many Americans considered the price too high, and the U.S. government was roundly criticized for the purchase. Evidence of Russian occupation still exists in parts of Alaska, visible in the architecture of wooden, onion-domed Eastern Orthodox churches and in cemetery headstones.

When the Hudson's Bay Company moved its fur-trading operations into the Columbia River Basin early in the nineteenth century, it set the stage for another consequential episode in North Pacific Coast history. The Company was the dominant influence between northern Oregon and British Columbia until the late 1830s. Then American missionaries and settlers began the long journey across the Oregon Trail which started in Missouri and ended in the Willamette valley. Most of the new American settlers moved into the valley, but others pushed north. Soon their numbers outweighed the Pacific Northwest's small British population.

In their quest for land and filled with a sense of their "manifest destiny" (see Chapter 3), the cry of these American settlers was "Fifty-four forty or fight." Also significant in the U.S. presidential election of 1846, the rhetoric laid claim to all land as far north as 54°40′ N. The threat was deemed real enough that, in that year, the British and American governments agreed to establish the Canadian-United States boundary as it exists today at 49° N. But however acceptable the boundary decision may have been politically, it disrupted the normal north-south movement patterns along the Puget Sound and Columbia River transportation corridors.

When the international boundary was drawn in 1846, Victoria, on Vancouver Island, was the main urban focus for the British-held portion of the region. Established in 1843, it was the first permanent European settlement in the North Pacific Coast, and it remained the largest town in British Columbia until 1886. But in that year, the consequences of thwarted north-south movement settled the city's future. Vancouver, located on the mainland adjacent to the mouth of the Fraser River, became the western terminus of Canada's first transcontinental railroad. In subsequent years, while Victoria continued to be the province's administrative and commercial center, Vancouver was on its way to becoming Canada's main port of entry on the Pacific.

Railroads were key to the eventual growth of Oregon and Washington as well. In 1870, these two states combined had fewer people than San Francisco. In 1883, the Northern Pacific Railroad was completed to Seattle, and a decade later, the Great Northern Railroad linked to the city. This ended the region's overwhelming dependence on oceanic shipment, with ships sailing via the southern tip of South America to eastern United States and European markets. By rail, the region's bulky resources could be transported more quickly and in volume.

For a period after completion of the railroads, twice as many people moved to Washington each year as moved to California. The immigration that flowed into Washington and Oregon between the 1880s and the beginning of World War I created an interesting difference in the two states' ethnic composition. In 1880, Oregon had 175,000 people, and Washington had only 75,000. But by 1910, Washington had outpaced Oregon's growth so that its burgeoning population of 1,140,000 was nearly double the southern state's 672,000 people. Whereas most of Oregon's residents were immigrants from other parts of the United States, booming Washington attracted hefty numbers of European immigrants as well, especially Scandinavians. Today ties to that ethnic heritage can be important in Washington, and many of the state's politicians have been of Scandinavian descent. Oregon, by comparison, exudes a strong New England heritage, the result of early overland settlement.

Population Distribution Today

More than 10 million people live in the North Pacific Coast. Their numbers are increasing at a more rapid rate than the national averages for either Canada or the United States. Most live in a long cluster of cities and towns along the lowland extending from the Fraser River in southwestern British Columbia through the Willamette valley of Oregon (Figure 16.3). Three of these cities, Vancouver, Seattle, and Portland, have metropolitan area populations of more than 1.9 million people each. Ironically, this land of the great outdoors, like nearly every other part of the two countries, is peopled by an urban population. At the same time, the cities represent the region's character in urban form.

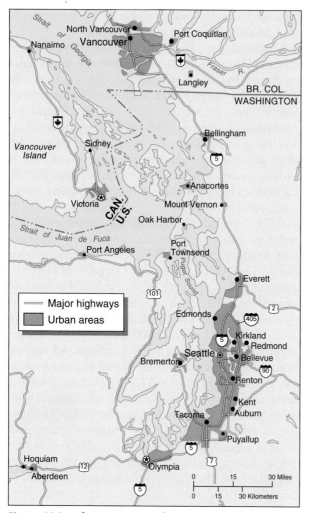

Figure 16.3   The Puget Sound–Fraser River lowland.
This narrow corridor, home to a majority of the North Pacific Coast's residents, benefits from excellent access to oceanic shipping and a desirable recreational environment.

Vancouver

Canada's third largest city, this bustling metropolitan area of nearly 2 million people has grown greatly in size and cosmopolitan character since World War II. Its urban area now extends south almost to the U.S. border, and local planners expect the city's population to double during the next 30 years.

This growth is attributable to the fact that Vancouver has become the western headquarters for nearly all of Canada's major businesses. Most of British Columbia's manufacturing, with wood processing most important, is in or near the city. Only the fact that the provincial capital, Victoria, is across the Strait of

Georgia provides a modicum of balance to Vancouver's dominance in British Columbia.

Also in support of its growth, Vancouver's transportation function has become more substantial as the western outlet for interior Canada. Most of the Prairie Provinces' products, formerly shipped to and through the cities of eastern Canada, now funnel west to Vancouver. The development of Asian markets for western Canada's wood products and wheat has encouraged expansion of the city's port facilities. With an annual cargo volume of more than 62 million metric tons (68 million tons), Vancouver is easily Canada's busiest port. So crowded are Vancouver's port facilities and the eastern rail lines serving it that Canadian officials have encouraged use of the alternative port at Prince Rupert, where a new major grain transshipment facility has been built.

Seattle

Seattle has been the largest city along the North Pacific Coast since its boom era of the late nineteenth century. Founded as a logging center, Seattle achieved regional dominance when it was linked to the northern U.S. transcontinental railroads.

Sometimes dubbed the world's largest company town, Seattle was home to Boeing Aircraft after World War I. A mainstay of the construction of bomber aircraft in World War II, Boeing Aircraft's fame subsequently spread when it introduced a series of highly successful passenger jets. For a time, Seattle employment opportunities flourished along with Boeing. But employment at the company then fluctuated wildly as airplane orders responded to changes in market demand and the general condition of the world economy. From a high of 103,000 in the late 1960s, employment at Boeing fell to about 50,000 during the early 1970s. The economy of the city, and to an extent that of Washington, suffered as a consequence.

Seattle's population declined following the heady growth of the 1960s, and only nine states grew less rapidly than Washington during the economic slowdown. But by the late 1980s and on into the 1990s, Boeing's employment levels recovered and steadied at more than 106,000. This, combined with local efforts at economic diversification, enabled the area's economy to rebound. One of the great successes of the Seattle area is, of course, Microsoft. Spin-off research, along with the University of Washington's own work, has expanded the information technology, biotechnology,

At Vancouver, British Columbia, excellent harbor facilities and easy access into the interior up the Fraser River valley help make this handsome city Canada's major western urban center.

and healthcare sectors of Seattle's economy. In King County alone, Microsoft employs about 16,000 people. The attraction of job opportunities at Microsoft and other high tech firms fueled in-migration, as did the growth of Seattle's cultural appeal as the center of the "grunge rock" movement of the early 1990s. Today, the Seattle metro area has grown to 3,263,497 (2006) and continues to experience substantial immigration.

But the future is unsettled for Seattle. In 2001, Boeing decided to move its corporate headquarters out of the region to Chicago. This move shocked the Puget Sound area and signaled that even long relationships

Much of the vehicular traffic entering and leaving Seattle is squeezed by hilly terrain onto I-5. This feature of urban life can make rush hour, even in Seattle's pleasant setting, difficult.

between cities and industry can change dramatically and suddenly. Between December of 2001 and late 2003, thousands of Boeing jobs were cut in the Seattle area. It is still uncertain how many jobs might be created in the region from the production of Boeing's next generation aircraft, the 787 Dreamliner, announced in December of 2003 and scheduled to enter service in 2008.

Seattle's urban core is tucked onto a narrow isthmus bordered by Puget Sound on the west and Lake Washington on the east. Like Vancouver, it has a beautiful site, and views of mountains and water offer panoramic scenes to residents living on its many hills. The city is largely middle-class, with scattered neighborhoods of tree-lined streets. Evaluations of the quality of life in the largest United States cities during the mid-1970s and again in 1989 concluded that Seattle was the best city in the country in which to live, and throughout the 1990s and into the twenty-first century it continues to score high. But, ignored in these evaluations are urban realities such as regular traffic congestion on the cities' arteries, constricted by its scenic setting; as Seattle's population rises, these issues are increasingly problematic.

Portland

Two studies of urban quality conducted about the same time as those lauding Seattle as the country's most livable city disagreed with its conclusion. Instead, a study by the federal Environmental Protection Agency declared, it was Portland that earned the honor. Its "most livable city" label gained support from yet another study, produced by Midwest Research Institute in Kansas City. Evaluating more than 100 variables, it agreed that Portland was indeed the best of the country's 67 largest cities. (Eugene, Oregon, in the Willamette valley, was judged the best medium-sized city in both studies.)

Portland's residents greet such laudatory studies with mixed emotions. They are inclined to agree, but they fear such publicity will bring waves of newcomers whose numbers would destroy the qualities of the city that attracted them. Considered old by regional standards, Portland's economy is more diversified than Seattle's, and its relations with the region's interior are closer because the Columbia River valley provides a lowland route east. This accessibility facilitates Portland's role as a major transshipment point for grain from eastern Washington. And because the lower Columbia River is navigable, Portland rivals Seattle as an ocean port, even

though it lies inland about 160 kilometers (100 miles) from the coast. In 2006, the Portland metropolitan area included 2,337,565 residents.

THE REGIONAL ECONOMY

The economic structure of the North Pacific Coast is dominated by the production of staple products (see Chapter 6) and by the region's distance from major markets in the United States and Canada. This latter factor is especially applicable for the region's United States portion.

Fewer than 3 percent of Americans live in the Pacific Northwest, although about 10 percent of Canadians live in British Columbia. While the region has many engaging features, its residents must deal with a tough economic reality that affects their mix of work opportunities: Most of the major markets of both countries are more than 3500 kilometers (2200 miles) away. British Columbia is better situated because it is the exclusive western terminus for national economic activity in Canada. But goods and people passing in or out of the United States through Pacific Coast ports can opt to go through California, bypassing Oregon and Washington completely.

Freight-rate structures also place many of the region's products at a competitive disadvantage in major markets. Although the North Pacific Coast has long supplied North America with high-demand products like lumber and foodstuffs, its producers bear high movement costs that reduce their ability to get products to market. In economic geography terms, this means that great distance limits the *transferability* of products. The effect distance has on market areas is that buyers turn to sources of supply that are nearer, and goods are less expensive to obtain. This theoretical scenario has played out for North Pacific Coast businesses. The higher freight rate charged for finished wood products in Canada, for example, encourages manufacturing in British Columbia to remain heavily involved in the initial wood-processing stages only.

Agriculture

Agriculture in the North Pacific Coast is a good example of the impact that transferability problems can have on a regional economy. The crops grown in the North Pacific Coast are similar for the most part to those grown near areas of major consumption in the East. Washington apples, for example, are in direct competition in Eastern markets with apples from Wisconsin, Pennsylvania, and Virginia. The same is true for a variety of other fruits and vegetables grown in the region. This competitive aspect does not apply to central and southern California, where a Mediterranean climate allows growers to produce crops that are not viable in the colder north and to provide, as well, a greater volume of common fruits and fresh vegetables during winter. Thus, for decades California has cornered the market for many of the North Pacific Coast's agricultural products (see Chapter 15). Distance to market has little negative impact on California's agricultural economy because intervening opportunities (other, closer sources of supply) are not available.

The result is that much of the North Pacific Coast's agricultural produce is for local consumption, not export. Aggressive advertising campaigns and cost advantages accompanying innovations in air cargo transport increased the East Coast market share for certain fruits like Washington apples and pears and Oregon strawberries, but the relative impact of food exports on the region's economy is modest.

Oregon's broad Willamette valley is easily the largest agricultural area near the Pacific Northwest's coast (Figure 16.4). Land there has been cultivated for more than a century, and its farms are prosperous and well established. Much of the valley's farmland is in forage crops, and many farmers follow the practice of burning these fields in the fall—resulting in several weeks each autumn when parts of the valley are covered by a layer of smoke.

Dairy products, generated mostly for local markets, are of greatest agricultural importance in the Willamette valley. Strawberries are perhaps the most valuable specialty crop, though a variety of others like hops, grass for turf seed, cherries, and spearmint thrive in the valley's climate. Even grape production, supporting a local wine industry, has increased in recent years. The total value of these crops, however, does not match the area's income from dairy production.

The Puget Sound lowland in Washington is another significant dairy area, as is the Fraser River delta in British Columbia. In these places, too, farmers grow a variety of specialty crops, with peas leading the pack. Quick frozen and then shipped to markets throughout North America, this cool weather crop is particularly well adapted to the local climate.

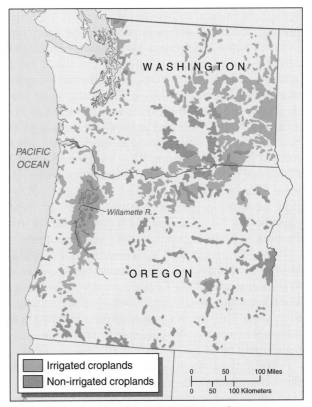

Figure 16.4   Cropland in Washington and Oregon. (Reproduced with permission from *Atlas of the Pacific Northwest*, 8th edition, edited by Philip L. Jackson and A. Jon Kimerling, © 1993 Oregon State University Press.) Oregon's Willamette valley and the Inland Empire of eastern Washington are prominent among agricultural areas in the North Pacific Coast.

The area east of the Cascades in Washington presents a very different kind of agricultural landscape. Most of this area is semiarid, and grasses and desert shrubs replace the majestic evergreens of the coast and mountains. Although called the Columbia Plateau, the area has little of the characteristic flatness one expects from the term *plateau*; rolling hills dominate it. Elsewhere in central Washington, the landscape has a series of steepsided dry canyons called *coulees*. In ancient geologic times, the land the coulees cut was covered by lava flows; floods from glacial melt during the Pleistocene ice retreat subsequently eroded the coulees' canyons. The visual effect today is of a surface dotted with lava pockets that resemble scablike knolls. The area, echoing this effect, is called the "channeled scablands."

Most of the Columbia Plateau falls within the western Empty Interior because of its aridity, nodal settlement pattern, and economic focus on minerals, grazing, and transportation services. But an exception is found along the Oregon-Washington border and across much of eastern Washington. In this zone, a substantial farming area exists, easily the most important in the Pacific Northwest. Physically, even aesthetically, separated from the coast, residents of the area—who call it the Inland Empire—nevertheless feel just as associated with the coast as they do with other parts of the Empty Interior. They sell crops down the Columbia River; they share a common state government with the coast; and they look to Seattle and Portland for many goods and services. For all of these reasons, the Inland Empire is a part of the North Pacific Coast, although it is certainly an atypical part.

The Inland Empire's agriculture is quite varied. Within its domain, the hilly country of east-central Washington, called the Palouse, averages between 35 and 65 centimeters (15 and 25 inches) of precipitation each year. This amount is marginal for farming, yet somewhat more abundant than other parts of the adjacent Empty Interior.

Today the Palouse is one of the most important dry-farming areas in the country. Dry farming is usually confined to subhumid areas and emphasizes water conservation more than other types of farming. Wheat is the Palouse's primary crop, with farmers growing both spring and winter varieties. In the Palouse, farmers normally plant wheat on a given field every other year. In alternate years, they let the land lie fallow; that is, it is plowed, but nothing is planted. This practice retards evapotranspiration and allows soil moisture to increase.

In other areas within the Inland Empire, irrigation has played a major role in agriculture in recent decades (Figure 16.4). Two places especially have benefited. Water from a number of streams flowing eastward out of the Cascades is used to irrigate the mountains' relatively narrow valleys like the Yakima and Wenatchee. The result is the country's premier apple-producing area that effective national advertising has helped create a market for by touting their apples' crisp superiority. Subsequently, pears from the Yakima, Wenatchee, and neighboring valleys have received equivalent marketing attention and distribution.

Another place irrigation economically transformed is the area around Grand Coulee northwest of Spokane. At Grand Coulee, the Columbia River was dammed primarily to provide hydroelectric power. But it also made available volumes of irrigation water to the Big Bend area of south-central Washington. When these waters became available in the late 1950s, crop acreage in the Big Bend expanded considerably. Its current major agricultural products include sugar beets, potatoes, alfalfa, and dry beans.

Washington give the area an appearance that is more like parts of the Great Plains. Here, a combine harvests wheat near Palouse, Washington.

In contrast to the typical image of a verdant, mountainous Pacific Northwest, the rolling wheat fields of eastern

Economics and Ecosystems: The Struggle for Balance

Changes in the practices of the forest products industry are visible in the North Pacific Coast's landscape. Formerly, loggers cut everything in their path and moved on, always sure more trees could be found elsewhere. Today, they plant a new tree for every one cut, although wide swaths are still clear-cut during the logging process. Governments of both the United States and Canada, as well as private companies, carry on extensive research programs to develop stronger, straighter, healthier, faster-growing trees; and young trees are now being fertilized from airplanes and helicopters. Computers control the process by which trees are cut into lumber, with each tree cut to obtain the maximum amount of lumber. But research and technology aside, new stands still take 40 to 75 years to reach maturity, and it is arguable whether or not yields of most tree varieties can be sustained indefinitely under these methods of operation.

Because lumber and pulp production are vital to the region's economy, people feel strongly about and widely discuss any issue that affects the availability of trees for cutting. Two major controversies have been subject to debate in recent years, especially in the United States. The first involves northern California's redwoods, the world's largest tree and among its oldest. Most redwoods—the chief timber sought by California loggers—are on private land. In the late 1960s, a major environmental drive developed to keep these trees from being cut and led to the creation of Redwoods National Park. But conservationists contend that the park is too small and that some lumber companies are hastily cutting trees on land that may and should be set aside as additional parkland. Loggers respond that expanding parks to protect the redwoods would ruin the lumber industry, cost people jobs, and cause large increases in forest product prices.

The second controversy involves *clear-cutting*, a practice by which all trees in a given area are cut at one time. Loggers argue that clear-cutting is the most efficient way to remove trees and ease reforestation. Many conservationist groups, however, respond that clear-cutting destroys forest ecosystems, contributes to soil erosion and stream pollution, and is an ugly blotch on the landscape.

The issue was addressed head-on in 1973 when the Izaak Walton League sued to halt clear-cutting in the Monongahela National Forest in West Virginia. In their argument before the court, the League contended that the Organic Act of 1897 provided only for the cutting of "dead, matured, or large-growth" trees. They won, and the ruling was applied elsewhere. In 1975, cutting was halted in Alaska's Tongas National Forest on the same basis. A wave of concern swept the lumber industry of the North Pacific Coast, although clear-cutting continues to be practiced there. The solution to the "clear-cut" versus "no clear-cut" issue may eventually come from the U.S. Congress, but the conflict is indicative of continuing disagreements between conservation groups and the forest products industry.

Forest Products

British Columbia is easily the largest lumber producer in Canada, supplying about 54 percent of all timber harvested in that country. In the United States, Washington, California, and Oregon together provide more than half of all timber cut, and Washington and Georgia vie for the lead in pulp and paper production. Although forestry was the region's first major industry, the North Pacific Coast's rich forest did not become nationally important until late in the twentieth century when improved transportation facilities, coupled with the destructive overcutting of most eastern forests, opened the region's woods to lumbering.

Douglas fir—of prime importance as structural supports for houses and for flooring, doors, and plywood—is the region's major lumber tree. Each section of the region has its own mix of trees to harvest. In British Columbia, for example, spruce, hemlock, and balsam are most important. In northern California, redwoods are locally important, although their numbers are dwindling. The western red cedar is widely cut in the region from Oregon northward.

The large size of many of the trees, as well as the distances to market, tends to encourage large scale logging operations. Effective marketing has enabled the region's lumber products to penetrate all markets of both countries, even the southeastern United States, America's other major forestry region. The Japanese have also become major buyers, especially for lumber and lumber products from British Columbia and Alaska; most of Alaska's lumber now goes to Japan.

Power and Dams

Precipitation and rugged topography provide the North Pacific Coast with a hydroelectric potential for power unmatched anywhere else on the continent.

Clear, or block, cutting, seen here in Washington's Mount Baker-Snoqualmie National Park, is a controversial approach. Loggers argue that it is the most efficient method, but some environmentalist groups counter that clear cutting increases ecological damage.

Fully 40 percent of the United States' potential is contained in Oregon and Washington alone. The Columbia River, in particular, has a flow volume larger than the Mississippi River's and a drop of 300 meters (1000 feet) during its 1200-kilometer (750 mile) course from the Canadian-U.S. border to the sea; it is a power developer's delight.

Begun in 1933, the Grand Coulee Dam was the first of many dams constructed on the Columbia River, and it remains the region's largest. Its construction was followed by no fewer than ten dams downstream. The Dalles, between Oregon and Washington, where Lewis and Clark portaged around dangerous falls, is now a lake. British Columbia and the United States agreed to construct in Canada three additional dams that would store water during periods of heavy flow and then release the water back into the river when flow was low to guarantee consistent power generation. British Columbia's government was paid $273 million as part of the agreement. This guaranteed flow instigated the addition of new generators at Grand Coulee Dam to triple its capacity, making it the world's largest single power producer. In British Columbia, power production has been expanded for provincial use, most notably in the Peace River area.

These developments have provided inexpensive electricity for the North Pacific Coast. Inexpensive electrical power, in turn, has attracted manufacturers that are heavy power consumers, such as the aluminum-smelting industry. Kitimat, isolated on the northern coast of British Columbia, is now the site of one of the world's largest aluminum refineries. Kitimat's coastal location permits cheap transportation of bauxite, a bulky ore, and a nearby dam provides the very large quantities of hydroelectric power needed for aluminum production.

Fishing

Forestry and fishing once formed the backbone of the North Pacific Coast's economy. The North Pacific's cold waters were, and to an extent still are, fertile fishing grounds. Large numbers of whaling vessels were attracted to these waters during the late eighteenth century and first half of the nineteenth century. The vessels overharvested the North Pacific's whale population, reducing it to a small fraction of former levels. Currently, most nations have agreed to an international ban on whale hunting, and the few remaining whalers have turned to waters off Antarctica for most of their catch.

Salmon has been the fish of greatest import in the North Pacific Coast for a long time. It was a major food and economic mainstay of coastal tribes before Europeans arrived and is still the principal fish caught in the region. Salmon are anadromous; that is, they migrate upstream from the ocean to spawn in freshwater. Years ago their five to seven annual spawning runs choked the rivers, and unbelievably massive catches were obtained by people on the banks. The Salmon Chief of the Colvilles, near Grand Coulee, reportedly could once catch 400 fish a day, each weighing about 14 kilograms (30 pounds), simply by scooping them up and throwing them onto shore.

Today the size of the salmon catch is greatly diminished; fish experts put the yield at less than half its level 5 decades ago. Most salmon are now caught off the Alaskan and British Columbian coasts—in their ocean-dwelling, not spawning stage. The decline of (nonhatchery) salmon is irrevocably linked to modern economics.

When the region's streams were dammed, access to many of the salmon's traditional spawning grounds was blocked, especially on the upper Columbia and its tributaries. Solutions have been implemented, but with mixed results. *Fish ladders*—a series of gradual water-carrying steps that allow fish to jump from level to level and thus bypass a dam—were built around some of the dams. Salmon could negotiate the ladders at smaller dams, but the larger ones remain a barrier. As a consequence, nearly all of the Snake River and its tributaries as well as all branches of the Columbia above Grand Coulee are closed to salmon, and the once flush rivers are all but empty of these fish. This has kept controversy between hydroelectric power developers and both working and sport fishermen heated. The salmon and those who depend on them have been the chief losers in dam construction.

Salmon are not the only fish in the region that have commercial importance. Runs of herring and smelt provide significant catches; so does halibut, hooked in ocean depths. But overfishing has had a substantial impact. As they fish the North Pacific, Canadian and American commercial fishermen are joined by cohorts of other nationalities, most notably the Japanese. Because of this foreign competition, many U.S. fishermen in the Northwest supported the federal government's 1976 decision to extend exclusive offshore fishing rights to 320 kilometers (200 miles) for a wide variety of fish, including the salmon.

West Coast fishermen still vigorously debate the wisdom of extending the offshore fishing limit. While the extension cuts foreign fish harvest and provides larger catches for domestic fishermen, it encourages other countries to enforce their own established offshore limits or to enact comparable 320-kilometer limits that U.S. fishermen must observe. This kind of reciprocity is especially bothersome to tuna fishermen, based primarily in California. Much of their catch originates off Ecuador and Peru. Until the 1976 federal decision, U.S. fishermen largely ignored the Ecuadorian and Peruvian limits, claiming they were illegal. After the United States imposed its own 320-kilometer offshore fishing limit, the claim of illegality was hard to make.

Juneau, the capital of Alaska, is cut off from land access with the rest of the state by the sea and encircling mountains.

Another source of contention over marine resources lies in differences between the United States and Canada. Controls on fishing exerted by Canadian governmental bodies have not always been matched by equivalent conservation measures by the United States. One result has been occasional disputes between American and British Columbian fishing interests, similar in nature to disagreements between American and Atlantic Provinces fishermen in Atlantic fishing areas.

ALASKA—A POLITICAL ISLAND

Coastal southern Alaska is part of the North Pacific Coast but must be viewed as separate from the rest of the region. No railroad connects Alaska with North America's population centers, and only one partly paved highway connects it through interior Canada to the rest of the United States. Crowded by coastal mountains onto a shoreline rarely more than a few hundred meters wide, people in southeastern Alaska's panhandle look to air and sea transportation for connection with the rest of the world. This physical isolation leads to a greater sense of detachment than is typical of the rest of the region. It leads, too, to a greater sense of separation from activities in the "lower 48" and saddles this part of the United States with an economy of high prices due to scarcity and high transportation costs.

A general impression of the Alaskan economy is that it is based heavily on minerals, lumbering, and fishing. But it is the federal government, primarily the Department of Defense, that is the primary employer in the state. Even the petroleum development boom on Alaska's North Slope modified, but did not eliminate, this employment pattern. Anchorage, home of one in three Alaskans and the most expensive city in the United States in which to live, depends on the military as the cornerstone of its economy.

Alaska is almost as close to Japan as to the coterminous United States, and Japan is the major market for many Alaskan products. Most lumber cut in Alaska, for example, is destined for Japan. Energy-hungry Japan would also be a ready market for North Slope petroleum, should the United States decide to export the oil rather than ship it south to meet domestic energy demands.

LAND OF CONFLICTING OPPORTUNITIES

The Pacific Northwest, where the unspoiled is not yet spoiled, attracts a different kind of person. The weather is too cool, or at times too uncomfortable, to attract the sun-seeking elderly. Economic opportunities rarely provide a quick kill. The Northwest is for people more interested in being than achieving.... And in the Northwest it is not necessary to own a great deal to share in a great wealth, the outdoors all around.[2]

This statement, written by a lifelong resident of the Northwest more than three decades ago and still pertinent, says much about its people and environment. It is a place where many want to live, and most who are there seem relatively satisfied, but it is not the home of a booming economy. Unemployment is frequently high, particularly when demand for one of the region's few staple products declines, as happened with Boeing Aircraft in Seattle in the early 1970s. Although the growth of a diversified manufacturing economy in the northern Willamette Valley is diminishing the lumber industry's role, unemployment in Oregon was 13 percent in early 1982 as a national decline in home building suppressed the local lumber market. In 1998, unemployment hovered around 5 percent, but climbed back to between 7 and 8 percent in 2003. Limited and volatile economic opportunity is a major reason why more people have not moved there.

To those who are there, this is a land of the great outdoors. The environment seems to call people to go out and enjoy it. Most do. That so much of the land is publicly owned simplifies their search. Much of British Columbia remains true mountain wilderness in spite of development projects undertaken in the last several decades across the province. The province's tourist-oriented slogan—"Super, natural British Columbia"—speaks to the land's attributes. Literally dozens of magnificent state parks and campgrounds dot Oregon's majestic 650-kilometer (400 mile) coast. To reduce litter, Oregon was the first state to pass a law requiring that beer and soda be sold only in returnable containers. In other parts of the region, nearly 200,000 pleasure boats crowd Puget Sound every year. Washington has more than 500,000 licensed fishermen. As Thomas Griffith puts it, "Northwesterners are a people by nature possessed."[3]

[2] Thomas Griffith, "The Pacific Northwest," *The Atlantic Monthly*, April 1976, p. 51.

[3] Ibid., p. 47.

Alaskan Land: Who Is It For?

When the United States acquired Alaska in 1867, nearly all of the territory's 150 million hectares (375 million acres) fell under the control of the federal government. The Alaska Organic Act of 1884 recognized vaguely that Indians and Eskimos in the territory "shall not be disturbed in possession of lands actually in their use or occupation or now claimed by them." Otherwise, most of Alaska remained in the public domain.

Historically, all public land initially controlled by the federal government in a newly acquired area was opened to private ownership through one of a series of land transfer laws, such as the Homestead Act of 1863 or the Mining Law of 1872. In Alaska, however, the Department of the Interior imposed a ban on private preemption or settlement that remained in force for nearly a century. Thus began a spirited struggle for Alaskan land that continues to the present, a struggle that has involved argument between the U.S. federal government and Alaska state government, between native groups and each government, between settler and native groups, and between developers and environmentalists. The opponents in these

arguments have occasionally shifted in their alliance with others involved in the issue, but the scope of the struggle has grown dramatically and regularly.

The federal government chose to maintain its hold on Alaskan land until 1958, the year of statehood. On the eve of statehood, only 200,000 hectares (500,000 acres) were owned privately, with most private land located around Anchorage and Fairbanks—and most of it patented recently under the Homestead Act. The federal government had reserved a number of plots for military purposes and a large area north of the Yukon River for a Naval Petroleum Reserve.

The Alaska Statehood Act of 1958 ignored Indian and Eskimo claims and granted the state the right to select 42 million hectares (104 million acres) for state use by 1983. Most of what has been selected is in the Alaska Range and the Yukon valley, although the state also controls a large area on the North Slope.

In 1968, petroleum was discovered at Prudhoe Bay on the North Slope. At least partly to dispose of native claims to Prudhoe Bay and to accelerate petroleum development generally, Congress passed the Alaska Native Land Claims Settlement Act in 1971. This gave the state's Eskimos, Aleuts,

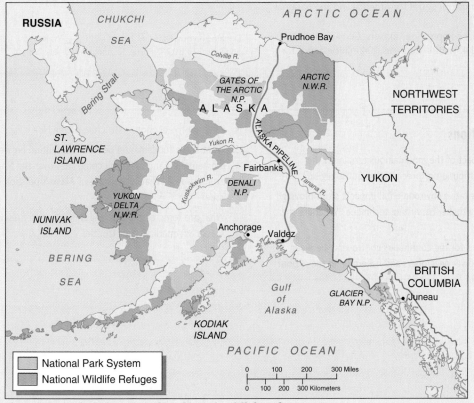

Figure 16.5   Alaska's national parks and wildlife refuges. The state has nearly one-third of the United States' entire national park and refuge acreage.

and Indians nearly $1 billion and the rights to 18 million hectares (44 million acres) of land. Most of this land was given to profit-making regional development corporations, although about 40 percent went to serve 200 villages that maintain control of surface rights.

The transfer of land to the state and to native groups proceeded slowly amid a tangle of conflicting claims. It was during this period that environmentalists became concerned about what they saw as the potential loss of a rich natural resource, the vast wilderness areas of Alaska. In 1978, the U.S. House of Representatives passed the Alaska National Interest Lands Conservation Act, which withdrew 58 million hectares (144 million acres)—more than a third of the state—primarily for national parks and wildlife refuges. A threatened filibuster by Alaska Senator Mike Gravel kept the Senate from acting on the bill, but President Jimmy Carter used a little-known 1906 Antiquities Act to create 22 million hectares (54 million acres) of Alaskan national monuments. Furthermore, he proposed another 22 million hectares for wildlife refuges and 4.5 million hectares (11 million acres) to be exempted from mining.

Outraged at these moves to set aside this land from possible development, many Alaskans argued that land withdrawal at the scale proposed would strangle the state's economy. The governor at the time, Jay Hammond, spoke for this view when he complained that "Alaska is being called upon at once to be the oil barrel for America and the national park for the world."

Congress passed a compromise Conservation Act in 1980. This act designated 42 million hectares (104 million acres) for parks and refuges, including 23 million hectares (57 million acres) specified as wilderness (Figure 16.5). It also expedited the conveyance of land to native groups and to the state, and it extended the transfer deadline to 1994. Less than 20 percent of the state's total land to be transferred from the federal government had actually been shifted when transfers were halted because of conflicting claims. The 1980 commitment became further clouded a decade later as discussions within the federal government raised the possibility of opening parts of the wildlife refuge along the Arctic coast to oil exploration.

The struggle for control of Alaskan lands is probably not over, because conflicting goals remain:

"Fundamentally, the issues have been ones of priority. The environmental movement argues that Alaska's lands belong primarily to all the people of the United States and require legal protection to prevent undue development, particularly of mineral resources. The native corporations established by the Alaskan Native Claims Settlement Act of 1971 argue that their rights to land should take first priority because they represent aboriginal rights recognized in the Alaska Organic Act of 1884 and implicitly confirmed in the Alaska Statehood Act of 1958. The state, while recognizing the paramount legal rights of the federal government and the rights of the natives, argues that the recent massive federal withdrawals of land from individuals have seriously impinged upon the rights confirmed to Alaska by the Statehood Act."[4]

[4] Donald F. Lynch, David W. Lantis, and Roger W. Pearson, "Alaska: Land and Resource Issues," *Focus*, 31 (January–February 1981): 2.

Review Questions

1. What is the impact of the mountainous terrain on the amount and distribution of precipitation in the region?

2. What are the negative environmental impacts of employing clearcutting as a timber harvesting technique? What are the advantages?

3. What is the basis for the continuous controversy between hydroelectric power developers and fishermen in the region?

4. Describe the economic structure of the North Pacific Coast.

5. What is the fundamental point of contention between environmentalists and major oil companies over the North Slope explorations? What is the state of Alaska's position on this issue?

6. What are some of the challenges in trying to preserve Alaska as America's "last frontier"?

CHAPTER 17

THE NORTHLANDS

Climate

If North Americans were asked to give a one-word definition of the nature of the Northlands, "cold" would probably be the most commonly used adjective. It is a valid judgment. Average January temperatures there range from a high of about −7°C (20°F) along its southern Great Lakes margin to a full −40°C (−39°F) in parts of arctic Canada, where temperatures can be as low as −60°C (−76°F).

Not only are winter temperatures frigid across most of the region, but winters are long. The frost-free period—or average time between the last frost in the spring and the first in the fall—is roughly 135 days at the region's southern margins but only about 14 days along parts of the Arctic Ocean borders of the two countries. Most of the region has a frost-free period of

The American historian Frederick Jackson Turner said that because the frontier experience is still recent, it continues to have a noticeable impact on our society. The high value Americans and Canadians attach to land ownership, by comparison with most Europeans, stems in part from the widespread availability of land and the drive to possess it that helped push settlement westward. Ignoring or ignorant of American Indian territories, they believed more land was always within reach over the next hill or beyond the next river.

The Canadian-American frontier is largely gone today. But one vast region in North America remains sparsely settled, with tens of thousands of square miles totally unoccupied. This is the North American Northlands (Figure 17.1).

Extending as far south as the northern Great Lakes states and including most of interior and northern Canada and Alaska, it is easily the largest North American region. If it were a separate country, its area would make it the world's sixth largest. Hudson Bay is larger than the Sea of Japan; Canada's Arctic archipelago, with 20 major islands and 18,000 smaller ones, is the largest in the world. The inhospitable nature of the Northland's physical environment as well as the consequent thinness of settlement gives the region its special character.

A HARSH ENVIRONMENT

Cold temperatures, long winters, thin soils, poor drainage, and low precipitation are key features of the Northlands environment. In combination, they limit broad human use of this region.

In Canada's far northern region of Nunavut, an Inuk hunter wearing caribou and polar bear fur jumps an opening, or "lead," in the sea ice. Inuk is the singular form of the word Inuit.

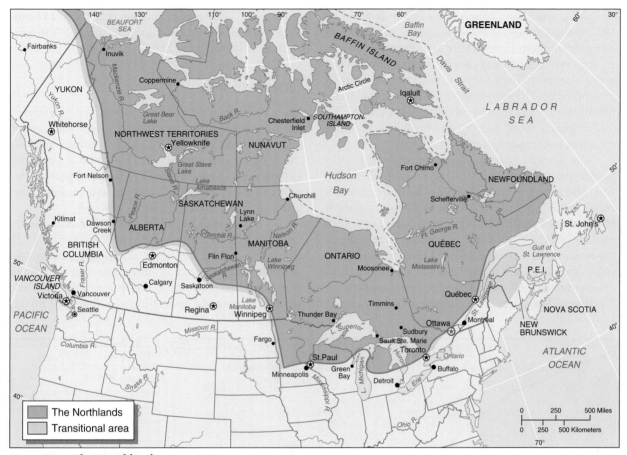

Figure 17.1 The Northlands.

less than 90 days. Because virtually all major food crops need a growing season longer than that, crops are raised in just a few small areas along the Northland's southern margins.

The Northland's climate is largely continental. Maritime moderation is significant only along the peripheries, mainly in the east and west. Thus, along with low winter temperatures come brief but warm summers. When asked by visitors what they do in summer, residents of the Yukon and the Northwest Territories jokingly respond, "Well, if it falls on a weekend, we go on a picnic."

Dramatic seasonal variation in temperature results from great shifts in length of day and the angle of incidence of the sun's rays. The earth's axis—that straight line connecting the two poles—tilts at an angle of 23°30' from the plane of revolution around the sun. As the earth follows its annual path around the sun, the north pole is tilted toward the sun in North America's summer and away from it in the winter. Thus everywhere north of the Arctic Circle, 66°30'N latitude, is in darkness for at least one day at midwinter and has at least one 24-hour period without the sun setting at midsummer. Moreover, during winter, the sun remains low on the horizon when it rises. This means the sun's rays strike the earth at a low angle and, thus, cover a relatively large area. As a result, much less heating occurs in winter than in summer, when the sun's rays are more direct.

In midwinter, Fairbanks, Alaska, makes do with light from a sun hanging just above the horizon for perhaps 4 hours per day. In midsummer, the city relishes a brief sunset around midnight. Even southern sections of the Northlands receive only 6 or 7 daylight hours in winter. For people accustomed to sleeping in darkness and being active during daylight, this seasonal fluctuation demands substantial psychological adjustment.

Precipitation amounts vary widely across the Northlands. Highest levels occur in the far southeast where both winter and summer storm systems move along the coast and combine to dump more than 100 centimeters (40 inches) of annual precipitation along the southern shore of Labrador. Precipitation levels drop markedly toward the interior and north. Most of the Northwest Territories in Canada receive less than 25 centimeters (10 inches) annually; parts of the Arctic Islands average less than 15 centimeters (6 inches). The cold air usually found over this area holds little moisture and, therefore, produces little precipitation.

The Arctic Ocean is also culpable for the scant precipitation. Covered by ice for all or much of the year, the sea supplies little or no moisture through evaporation. The northern Arctic Islands are the most arid parts of Canada and the United States. Snowfall amounts in the Northlands are consequently much lower than might be expected. Whereas parts of Labrador and interior Quebec can average more than 150 centimeters (60 inches) of snow a year, most of the Northwest Territories receive less than 60 centimeters

(24 inches). The Arctic Islands get only about 25 centimeters (10 inches) of snow annually, roughly the same amount as Washington, D.C.

Terrain and Soils

Despite the paucity of precipitation, the Northlands does not look like a dry environment. In summer, much of the region is covered with standing water, and the mosquito stories people tell match those generated anywhere on the continent. The standing water is due partly to the low levels of evaporation and evapotranspiration typical of cold climates.

In the region's northern portions, standing water is also caused by widespread *permafrost* (Figure 17.2), which is a subsurface layer of permanently frozen ground. Permafrost can be only a few meters thick but commonly has a depth of 100 meters (330 feet) and sometimes extends down for more than 300 meters

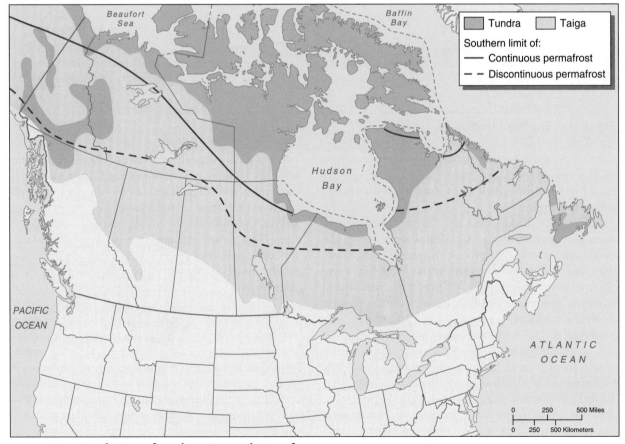

Figure 17.2 Distribution of tundra, taiga, and permafrost. A harsh winter and a short growing season dominate much of interior and northern Canada. Most Canadians live south of the taiga.

(1000 feet). In warmer areas, permafrost is discontinuous, meaning areas of frozen ground are interspersed with unfrozen soil. As the surface layer thaws during the short summer to a depth of perhaps 1 meter (3 feet), water is held on the surface by the frozen layer underneath instead of penetrating the soil, creating an extremely boggy, shifting surface.

Construction in permafrost is difficult. Buildings must be placed on piles sunk deeply into the permafrost for stability, and roads must be repaired extensively each year to maintain any resemblance of an even roadbed. Heat given off by buildings or absorbed from the sun's rays by road surfaces causes additional melt of the permafrost under the construction. Unless care is taken, a building constructed on permafrost can sink gradually into the ground, plunging deeper and deeper as heat given off by the building continues to melt the permafrost directly under it. Roughly half of Canada and most of Alaska are underlain by continuous or discontinuous permafrost.

Although considerable local terrain variation is found, much of the Northlands' topography is either flat or gently rolling. Alaska's north slope and much of the central Canadian Arctic is a broad, flat coastal plain, as are the broad valleys of the Mackenzie River in northwestern Canada and the Yukon River in central Alaska.

Almost all of the area to the south and east of the Arctic coastal plains is part of the Canadian Shield, North America's largest physiographic province. The Shield is underlain by ancient, heavily metamorphosed rocks. Substantial local relief is found mainly along the Shield's southeastern edges, especially just north of the St. Lawrence valley. Nearly all of this 3-million-square-kilometer (1.1-million-square-mile) area was scoured and contoured repeatedly during the past million or so years of the Pleistocene (Ice Age), removing much of the Shield's soil and depositing it in valleys to the south. This process greatly disrupted drainage patterns. Today, the landscape is dotted with thousands of lakes and bogs that mark former lakes and is drained by rivers following tortured, twisted paths to their destinations.

Northland soils are varied but are generally acidic, poorly drained, and of low agricultural quality. Soils in the region's southern portion are mostly boralfs or spodosols, soils of a cool needleleaf forest environment. To the north, water-saturated and often frozen tundra soils dominate. Fertile soil is confined to some of the old river valleys and to those lakes that have been filled by sedimentation and decayed vegetation.

The rate of soil formation in the Northland's cold climate is slow. Relatively little organic material deposits on the surface, a result of slow growth rates for most vegetation. In addition, the long cold season means that decomposition and humus formation, accomplished during warm periods, is hindered.

Vegetation

Most of the Northlands can be placed in two large, distinct vegetation areas. Stretching across the entire southern arc of the region is a coniferous forest called the *boreal forest*, or *taiga*. Covering hundreds of thousands of square miles, the taiga's closely ranked spruces, firs, pines, and tamaracks blanket the landscape in a dark, almost black, mass when seen from the air. Slow growing and never really tall, these trees decrease in number and height from south to north across the taiga. Around the Great Lakes, a mixture of pines and hardwoods predominates.

Passing just south of Hudson Bay, then angling northwest to the mouth of the Mackenzie River and across the northern edge of Alaska, is the treeline. The treeline, which is a transition zone rather than an actual line, identifies the boundary between the taiga and tundra in the Northlands (see Figure 17.2). Beyond it, climatic conditions are too harsh for treelike vegetation, and a broad tundra of lichens, grasses, mosses, and shrubs rolls north. At first glance, tundra vegetation appears homogeneous because it is all of the same low height. But in fact, considerable variation exists based on temperature, precipitation, and soil differences.

Arctic Ice

The great Arctic ice pack covers 4.8 million square kilometers (1.8 million square miles). Only 3 to 6 meters (10 to 20 feet) thick, it is a thin but rugged sheet of nearly salt-free ice floating on the Arctic Ocean. In winter, the Arctic ice pack extends south to enshroud all of the Arctic coast of Canada and northern Alaska. The ice minimizes any moderating impact by the Arctic Ocean on the Northlands' climate. Summer melt briefly frees all but the northern islands of Canada's Arctic archipelago from its grasp.

The ice limits ocean transport in the Arctic to a fleeting, often hectic period each summer. An unusually early return of the ice can block ships and whales that migrate into the Arctic Ocean each summer from escaping to open water to the south.

A caribou drinks at a pond amidst the autumn tundra in Canada's Northwest Territory.

The Arctic ice pack holds as much water as all the freshwater lakes in the world. It is slowly melting, apparently in response to gradual global warming, and has lost about 1 percent of its bulk in the last 100 years.

Rising summer temperatures, attributed to global warming, have produced dramatic changes around the margins of the Arctic Ocean. By 2007, the Arctic ice pack was sharply diminished to an area smaller than any on record. Coastal areas of the Northlands' islands were completely ice-free. This climate change promises to impact human lifestyles and local wildlife, if it continues as is expected. But winters remain very long, dark, and cold.

HUMAN OCCUPATION

Nearly all of the Northlands is sparsely populated, with highest densities found along its southern margins. American Indians, Métis, and Inuit (Eskimos) are numerically dominant over much of the region north of the United States. Inuit are the predominant population in most of the Arctic (Figure 17.3). American Indians live mainly in the boreal forest area (Figure 17.4). In pre-European days, the Indians were widespread and survived by hunting and fishing. The Métis are the result of intermarriage between Indian women and white men during the early fur trading period of European settlement in the taiga. Métis now outnumber American Indians in the boreal zone.

The Inuit culture is uniform across its broad distribution, which extends around the Arctic margins of the continent from Siberia to Greenland. An Inuit in Greenland is understood by one in northern Alaska, on the other side of the Arctic Ocean. The most recent arrivals of America's pre-European population, the Inuit crossed the land bridge from Asia about 4500 years ago and settled largely along the Arctic's water margins. They lived by fishing and hunting and developed an effective, self-sufficient lifestyle in that harsh environment. Today, Canada's Inuit population totals about 45,000.

The arrival of Europeans in the Northlands brought an end to many of the traditional lifeways of

In the Inuit village of Pond Inlet on the north coast of Baffin Island, prefabricated housing, electricity, and motorized transportation reflect changes from traditional, pre-European ways of life.

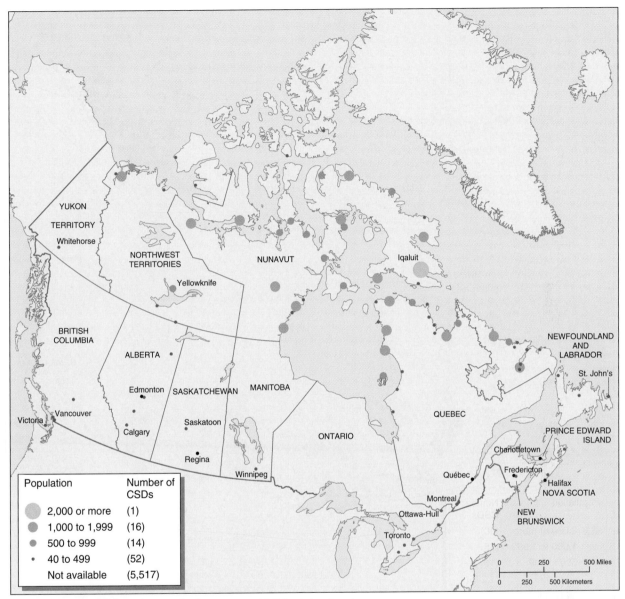

Figure 17.3   Canada's Inuit population. *Source:* 2001 Statistics Canada.

the American Indians and Inuit. Early fur traders acquired pelts from the Indians of the taiga, and bartered European goods entered the Indian economy as a result. Along with these trade goods came European diseases, liquor, and Christianity, all of which worked against the structure of Indian culture. Although these changes did not reach many Inuit until comparatively recently, the results have been much the same. Where Inuit continue hunting and fishing, the motorboat, rifle, and snowmobile have usually replaced the kayak, bow, and dogsled.

Most Northlands Indians and Inuit no longer exist by hunting and fishing. They have moved in substantial numbers into towns, or the economy of their villages has changed. Many urban places in the Northlands now have large native populations. Generally occupying the bottom rungs of the social and economic ladder, they endure unemployment, disease, and poverty. Most are poorly educated and trained and are not able to compete for the better paying urban jobs.

Because the Northlands was the last part of Canada and the United States incorporated by the

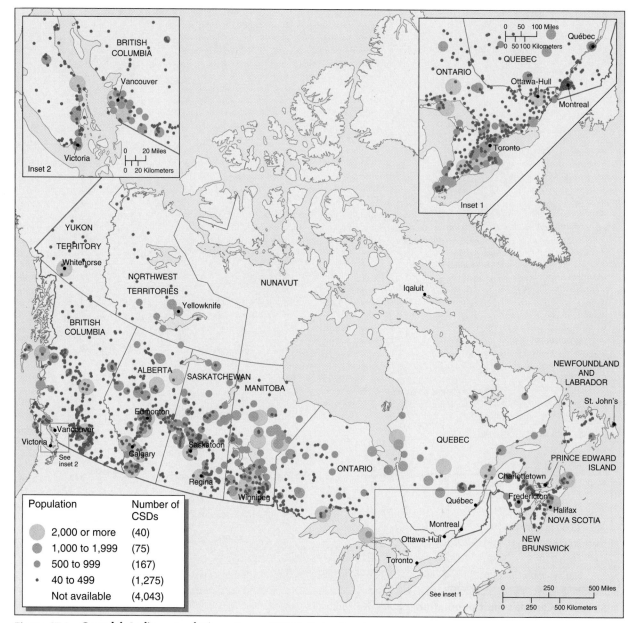

Figure 17.4   Canada's Indian population. *Source:* 2001 Statistics Canada.

countries' governments, the governments' attitude toward the region's native peoples was not as paternal or destructive as it had been initially elsewhere on the continent. The Canadian government had never adopted the "remove them" approach that characterized American attitudes toward Indians. Even in the United States, contact with Alaskan natives came sufficiently late to allow for some softening in these harsh attitudes. The native population remains scattered

across Alaska, and the American government has recently approved the return of large blocks of state land to Indian and Inuit control (see Chapter 16).

Canadian Inuit and Indian groups in the Northwest Territories and parts of Quebec claim aboriginal rights to the land, rights not surrendered by treaty as elsewhere in Canada. Canada established a Native Claims procedure in 1974 in which the government agreed to negotiate settlements with native groups whose rights had not been surrendered by treaty. In 1975 Quebec reached an accord with local American Indians on the control and use of land around the massive James Bay hydroelectric project. In 1983 the Indians of the Yukon Territory agreed to surrender most of their land claims in return for a $183 million settlement.

A dramatic change occurred on April 1, 1999. On that date, Canada's Northwest Territories was divided into two parts: The western 45 percent of the land (1.53 million square kilometers [589,559 square miles]) remained the Northwest Territories, but the eastern 55 percent of the land became a new Canadian territory. The new territory's name, Nunavut, means "our land" in Inuktitut, the Inuit language. Nunavut is governed and administered within Canada by its residents, approximately 80 percent of whom are Inuit.

Many of Nunavut's administrative challenges are embedded in its Northlands location. The bulk of the territory's 31,113 people (2007) live in 28 settlements scattered across its 1.9-million-square-kilometer (733,000-square-mile) area, an area larger than the combined states of North and South Dakota, Nebraska, Kansas, Oklahoma, Texas, Arkansas, and Louisiana. There are no railroads or roads in Nunavut. Travel is by dogsled, by snowmobile, or, for longer distances, by village-hopping small plane. There are few doctors or other residents with professional education. Still, Inuit have long sought regional self-governance, and Nunavut will provide that for the first time in Canada.

Early European Settlement and the Northern Economy

The Northlands offered little of interest to most Europeans who came to North America. Early settlers generally selected the greater agricultural promise of warmer climates and better soils to the south. Where Northlands' settlement did occur, its focus was usually either extractive or military.

French voyageurs, fur trappers, and traders in search of animal pelts were pushing their canoes beyond agricultural settlements along the lower St. Lawrence by the middle of the seventeenth century. They extended French political control across the Great Lakes and north into the headwaters of streams draining into Hudson Bay. The Hudson's Bay Company, a British fur-trading company, established itself on the margins of Hudson Bay and then pushed south and west, effectively blocking French expansion to the west. By the mid-eighteenth century, the Hudson's Bay Company, having been granted a trade monopoly to the area by the British government, controlled the entire boreal forest reaching from Hudson Bay west to the Rocky Mountains, extending influence into the Arctic as well. This vast extractive empire brought with it only a minimal number of small and widely scattered settlements.

The voyageurs—woodsmen and boatmen in Canada's frontier north—and the Hudson's Bay Company relied on the Northlands' numerous lakes and streams for transportation, and they located small forts at control points along the water routes. At places where major streams met lakes, where streams ended and an overland portage began, or where rapids or falls were encountered, vessels had to be unloaded, and their goods moved and reloaded onto other boats. These break-in-bulk points provided effective control of the entire water system. Today, cities are at the sites of many early French forts, continuing the break-in-bulk function. Ottawa and Sault Ste. Marie in Canada, and Chicago, Detroit, and Pittsburgh in the United States can claim this as a reason for their existence (Figure 17.5).

Logging

The boreal forest of the Northlands' southern half has the largest area of uncut forest remaining in North America. Until recently, lumbering and pulp and paper industries only nibbled at the edges of this vast reserve. The upper Great Lakes area was logged on a massive, devastating scale during the late 1800s and the early twentieth century. Tree removal was so complete there it became known as the Cutover Region. Because reforestation was seldom done and the cold climate slows regrowth, much of the Cutover Region is only now recovering its previous appearance and logging has still not reasserted itself as economically important. Repeated fires, fueled by wood debris left from wasteful logging, and a hardpan soil that developed after the vegetation cover was removed, further slowed the reforestation process.

Canada is the world's leading exporter of forest products. Lumber and, especially, pulp and paper operations dot the southern margin of the boreal forest from Quebec to Manitoba, and cutting is gradually moving north. The rich spruce forests south and southeast of Hudson Bay are the prime source of raw material for paper mills. The mills are located on some of the plentiful water power sites along the southern margins of the Canadian Shield, many of which are in

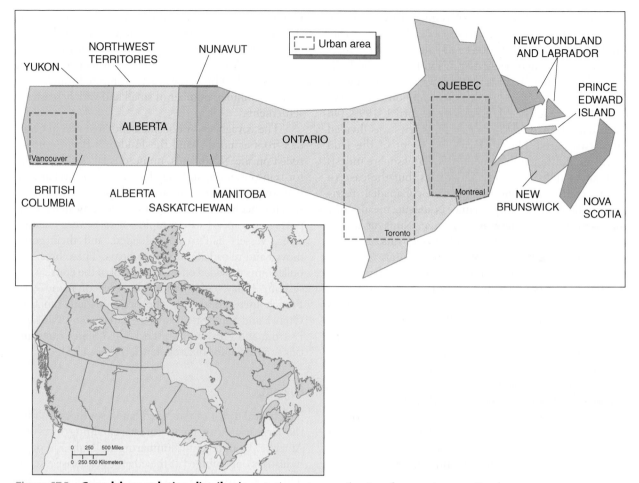

Figure 17.5 Canada's population distribution. In the cartogram, the size of an area is proportional to its population. Most Canadians are urbanites and about 30 percent live in one of the country's three largest metropolitan areas (Toronto, Montreal, and Vancouver). So few live in the Northlands that the region is hardly visible here.

the St. Lawrence lowland. Quebec leads the region in Canadian pulp and paper production. Canada used to produce about 40 percent of the world's newsprint, but drops in demand in the late 1990s and early 2000s has lowered production. After motor vehicles, pulp and paper products remain Canada's most important export in value, with lumber also ranking high.

Capital for the Canadian lumbering and pulp and paper industries, and most other Canadian extractive industries, has come primarily from outside Canada. Foreign investors control nearly a quarter of Canada's nonfinancial industries, including 40 percent of all manufacturing. Until the end of World War I, the United Kingdom was the principal supplier of investment capital, but since then an increasing majority has come from the United States (see Chapter 6).

Today about 70 percent of all foreign direct investment in Canada comes from the United States. Because adequate development capital for large-scale growth was unavailable locally, foreign capital has aided Canada in positive ways. However, outside ownership frequently has also meant outside control. This rankles Canadians, who feel their country is restricted in its ability to govern its own extraction policies.

Setting these concerns aside, the governments of the two countries signed the United States-Canada Free Trade Agreement (CFTA) in 1988, the goal of which was to make trade and investment across the border less cumbersome and more secure. Under the pact, all goods produced in both countries would be allowed to cross the border duty free by 1998. But, CFTA was suspended in 1994 with the implementation

Logging and related industries remain important in Canada. Here, smoke rises from a pulp and paper mill in Espanola, Ontario.

of the North American Free Trade Agreement, or NAFTA. An agreement among the United States, Canada, and Mexico, NAFTA has brought about an elimination of almost all tariffs among the three countries.

The total value of Canadian exports in 2006 was US$407 billion; about 82 percent of that export was sent to the United States. Imports totaled just under US$357 billion, with the United States claiming a 55 percent share. This bilateral trade flow is the largest between any two countries in the world.

The composition of Canadian exports to the United States has changed markedly. In 1960, more than 50 percent of all exports by value were fabricated raw materials (lumber, newsprint, refined metals, and the like); 20 percent was crude materials like mineral ores; 10 percent was foodstuffs. Today, manufactured products account for about half of the total, and fabricated materials make up 25 percent.

Mining and Petroleum

Canada produces a wide range of metals and other minerals. It is the world's leading producer of nickel, zinc, and asbestos, and it is a major supplier of potash, copper, lead, and iron ore. The majority of all production comes from the Northlands (see Figures 2.8 and 2.9). The upper Great Lakes area is the United States' leading source of iron ore and also has a substantial supply of copper. Alaskan North Slope petroleum has substantially added to America's energy supply, presently pumping about 25 percent of the country's total domestic production. In short, mining and petroleum extraction together with lumbering comprise the bulk of the Northlands' economic activity.

As with logging, the Northlands' accessible peripheries became the region's first important mining districts. The Mesabi Range in northern Minnesota, an area of gently rolling, elongated hills, along with neighboring areas in Minnesota, Wisconsin, and Michigan, developed into America's chief source of iron ore late in the nineteenth century. The Great Lakes waterway was an easy route for moving those ores to industrial centers in the manufacturing core (see Chapter 5). To date, billions of tons of high-grade ore have been moved by rail to Lake Superior ports, then loaded onto large specialized lake ships for a trip to ports in northeastern Ohio. As it was off-loaded, the ore was transferred to railroads again, this time headed to iron- and steel-processing plants in the Pittsburgh-Youngstown area. Today, most ore goes to newer integrated iron and steel facilities at the southern end of Lake Michigan. Locks at Sault Ste. Marie that connect Lake Superior with the rest of the Great Lakes are the busiest in the world, largely because of this ore traffic.

Most of the high-quality iron ores are now gone from the Lake Superior mining district. Attention has turned to a lower-grade ore called taconite, also found in huge quantities in the district. The iron content of taconite, roughly 30 percent compared to double that for the richer ores, is so low that shipping the ore to the lower Great Lakes for processing is far too expensive. Thus, the bulk of taconite is reduced (and its iron content increased) through a process called beneficiation. In beneficiation, the ore is ground into a fine powder, much of the rock removed, and the resulting material pressed into small pellets having a higher iron content. This greatly lowers the cost of shipping taconite, thereby increasing its transferability. That is, taconite's movement costs are sufficiently low relative to the value of the improved ore and this makes it

Canada's second largest port is Thunder Bay. Situated in Ontario at the western end of Lake Superior, Thunder Bay is a major transshipment center for Canadian grain, ore, and coal. A similar facility for American ore is found a short distance southwest at Duluth-Superior.

profitable to ship. Without reducing its bulk, taconite from the Lake Superior mining district might be ignored in favor of sources that, though more distant, offer an ore that is cheaper to move to market because of its higher ore content.

The cost of shipping low-grade ores is the major factor in choosing to locate many smelting operations near the source of supply instead of near the market. For example, copper, which seldom represents as much as 5 percent of its ore and often less than 1 percent, is nearly always refined near the mine. The smelting and refining of ores is the major form of manufacturing employment in the Northlands, and the large smoke-stacks of the refineries are the central element of the skyline of some of the region's largest cities.

These smokestacks also reflect a key problem with this and almost any mining company—environmental disruption. Lake Superior iron ore is mined by the open-pit method. Until recently, the land was simply left in its disturbed condition to recover as it might, leaving abandoned mines as ugly scars on the landscape. With the topsoil gone, sterile materials exposed on the surface, and normal drainage patterns destroyed, natu-ral recovery is extremely slow. But in places today, attempts are being made to aid recovery by recontour-ing land, planting, and fertilizing. Some successes exist, but such efforts tend to be difficult and expensive.

Other environmental problems develop at proc-essing plants as well. A vast quantity of waste material, or slag, is created during ore processing. The topogra-phy around many refining centers gradually changes as the waste is piled into great hills. Also, many kinds of refineries produce gaseous and liquid wastes that can be harmful. If one had visited Sudbury, Ontario, dur-ing the mid-1960s, for example, the first indication that this mining and smelting center was nearby would have been summertime's trees covered with brown leaves. Substantial improvement is apparent now, but treatment of waste is expensive and often only partially successful.

The mining frontier pushed northward from the Great Lakes early in the 1900s. At first, gold and silver were the principal minerals removed; then attention turned to metals like copper, nickel, lead, and zinc. Sudbury, at the southern end of this great mining con-centration along the Ontario-Quebec border, was the first mining town. With a 2006 urban area population of 157,857 people, it is the largest urban center in the Northlands today. North of Sudbury, the Clay Belt district crosses the provincial border from Timmins to

Val d'Or, with copper the leading mineral. The Chi-bougamau area, 350 kilometers (220 miles) farther northeast, is a newer mining region that produces a variety of metals, including lead, copper, and zinc. Together, these several mining districts comprise the major mining region in Canada.

Of the other metallic mining areas scattered across the Shield, the most notable is the Labrador Trough along the northern border between Labrador and Quebec. The existence of large bodies of high-grade iron ore in this locale was first confirmed early in the twentieth century. But its inaccessibility coupled with adequate supplies from the Lake Superior district slowed development. Then the decline in the quality of the ores in the Lake Superior district, together with growing demand in Canada and the United States encouraged exploitation of the Labrador Trough. A railroad, built at a cost of nearly $400 million, opened in 1954 connecting Schefferville, the early center of production, with Sept Iles, some 550 kilometers (340 miles) south on the St. Lawrence River. Today, Schefferville's mines are closed, and the community has been abandoned. However, other mines in the Labrador Trough, such as those at Wabush and Carol Lake, Newfoundland, and Gagnon-Fermont, Quebec, maintain the two provinces' leadership in Canadian iron-ore mining. Recent interest in the Labrador Trough's uranium deposits, among the world's largest, may bring mining activity to the area once again.

Smaller concentrations of mining activity are found in other locations in Canada's Northlands, including mines north of Lake Superior in western Ontario (iron ore); Flin Flon on the Manitoba-Saskatchewan border (copper and zinc); Thompson, some 320 kilometers (200 miles) north of Flin Flon (a leader in nickel production); and Yellowknife on Great Slave Lake (a center of gold mining).

Canada is the world's third leading metal pro-ducer, trailing Russia and the United States. Because much of that product is exported, the Canadian mining industry is susceptible to international economic shifts, especially those felt by the U.S. metals industry. For example, a recession in the U.S. metals industry between 1980 and 1983 depressed the sale of Canadian copper by one-third, iron ore by more than one-third, and nickel by nearly one-half. The economic results were devastating to many of the single-industry min-ing communities of the Northlands.

Accessibility remains a major problem for North-lands mineral production; construction of railroads was

Thompson, Manitoba, like many towns of the Northlands, is a mining center. Its skyline is dominated by the chimney of the refinery that processes the bulky ore and reduces the metals' shipment costs.

necessary before large-scale mining could begin in many of these areas. Many other sites, particularly in the far north, still have not been fully explored geologically.

Until recently, little petroleum was produced in the Northlands. Canada's major proven resources of liquid petroleum are on the prairies of Alberta; American demands were met by domestic production or by imported supplies. Recent declines in developed domestic reserves encouraged further exploration and development in both countries. In Alberta, a major initiative financed by both public and private funds is underway to expand development of the Athabasca tar sand petroleum reserves (see Chapter 12). Elsewhere, geologic conditions suggest that the best prospects for petroleum and natural gas discoveries are found along the west's mountain margins to the Mackenzie River delta and in a broad band extending from Alaska north of the Brooks Range (the North Slope) across Canada's

northern Arctic Islands. Oil deposits have definitely been indicated in the Mackenzie delta, and gas has been found on the Arctic Islands. Exploitation of these northern fields will be gradual, however, slowed by the high cost of developing a resource that is cheaper at other locales and by Canada's growing desire to husband its natural resources.

The United States, in contrast, moved rapidly to develop its North Slope petroleum fields. Some oil producers paid well over $1 billion just for the right to search for oil in the area. America's fear of an inadequate energy supply resulted in the country's most dramatic technological feat since the boom days of space exploration in the 1960s.

Transporting the crude petroleum was the principal problem involved in opening the North Slope fields. At first, many assumed that large tankers could carry the petroleum through waterways between

A pipeline snakes through the town of Inuvik in northern Canada. Elevated above ground, the pipeline is designed to help ensure that the heated petroleum traveling through it does not melt the permafrost.

Canada's Arctic Islands to the Atlantic Ocean, and one specially reinforced ship did make the voyage. Canadian concern for the long-term ecological damage that an oil spill could cause in cold Arctic waters, coupled with the near impossibility of making the trip during any period other than the few warm months, forced abandonment of this plan.

The only real alternative was to construct a pipeline. For a time a joint American-Canadian project was considered in which a pipeline would be built extending from the North Slope perhaps as far as the lower Great Lakes. However, differences over management and construction timetables and the huge cost of construction led to its abandonment as well. A pipeline costing $8 billion and crossing central Alaska to the port of Valdez on the Pacific was finally built and opened in 1977.

Hydroelectricity

About 70 percent of Canada's electrical power comes from hydroelectric facilities, located mostly along the Shield's southern margins where streams crossing the Shield's hard rock drop onto the lowlands of southern Ontario and Quebec. Until Quebec finished an even larger project on the margin of James Bay, the Churchill Falls project in Labrador claimed the position of largest in the country. These projects, like many other developments, have extended north onto the Shield. The electricity generated by them is both cheap and abundant. As a result, Quebec is a major center for aluminum smelting, a process that requires large amounts of electricity. The province has also contracted to sell surplus electricity to New York, Ontario, and several power companies in northern New England. Canada now exports to the United States about 10 percent of its generated electric power.

Transportation

In a region where isolation is pervasive, transportation is key. Developing transportation in the Northlands involves more than solving environmental difficulties, however important that factor might be. The Northlands' sparse population and especially its lack of cities means that even if transportation routes could be constructed cheaply, they would be used relatively little. The large nodal growth necessary to support a well-integrated transport network is unlikely to develop in the region. Hence, it is difficult to justify high transportation expenditures unless a large deposit of a particularly needed mineral is available.

Today only the Mackenzie River is used to any extent for river transport to the north. Diesel fuel is the most carried freight. Few railroads penetrate into the Northlands. They include one to the Labrador Trough from Sept Iles; one that carries mostly export wheat from the Prairie Provinces to Churchill on Hudson Bay; one to the mining developments on Great Slave Lake; and one to James Bay in Quebec. Other rail lines serve the region's southern margins, such as the mining districts on the Quebec-Ontario border. The Mackenzie Highway also reaches Great Slave Lake, and the Alaska Highway and some of its branches pass through the region's western margins.

Taken together, these conventional routeways provide access to only a small part of the Northlands. Small ships serve some coastal towns on an occasional basis. Elsewhere, however, the light airplane and its bush pilot is the only transportation link available. Nearly a score of carriers operate a relatively dense pattern of scheduled routes in the north. Although expensive, some studies suggest that air transport may be only twice as expensive as truck transport on the winter highways, and it cuts development costs tremendously. No other region in the world, with the possible exception of Russia's own northlands and the Australian outback, places greater reliance on the airplane for basic transportation.

Tourism

As North Americans have gained higher and higher incomes with an accompanying growth in leisure time, they have extended their need for recreation space. When areas close to where they live and work become overused, their search has reached farther and farther away from settled portions of the continent. No major national, provincial, or state park today is so isolated that it does not face the problem of overuse during peak tourist periods. The Northlands, the continent's largest empty space, is pictured as a region with almost unlimited tourist potential. This belief is only partly justified.

For many in the large urban centers just south of the Northlands' margins, a summer vacation in the North Woods has long been a standard part of life. Large sections of the southern boreal forest are heavily used by tourists. Northern Minnesota, Wisconsin, and Michigan are an easy day's drive from some of the largest U.S. cities. In Canada, Montrealers flock to the Laurentians, a short distance north. The jam of cars carrying weekend vacationers north of Toronto can extend for 300 kilometers (185 miles) on a fine Friday afternoon in summer.

Tourist volumes are much smaller in the central and northern boreal forests and in the tundra. While the big game and sport fish found there are justifiably famous and attract hunters and fishing enthusiasts willing to fly to the area's isolated lodges, the total number of visitors is small and will likely remain so. For the average vacationer, these Northlands areas offer nothing that cannot be found closer to home with less effort and expense.

From an environmental standpoint, low tourist numbers are good. The fragile environment cannot support anything approaching heavy use. This is an area where vehicle tire marks made on the tundra 65 years ago during World War II are still visible.

A Nodal Settlement Pattern

Although the total regional population is not large, the great majority of people in the Northlands live in villages, towns, and cities (Figure 17.6). Agriculture, a major support of dispersed settlement elsewhere, is only locally important. The principal agricultural settlement areas are the Lake St. John-Saguenay Lowland of Quebec, the Clay Belt along the Quebec-Ontario border, and the Peace River District on the British Columbia-Alberta border. (The last two areas were settled in the twentieth century and represent the last significant examples of something that was once a key part of North American life—frontier agricultural settlement.) Nearly all of the larger cities are dominated

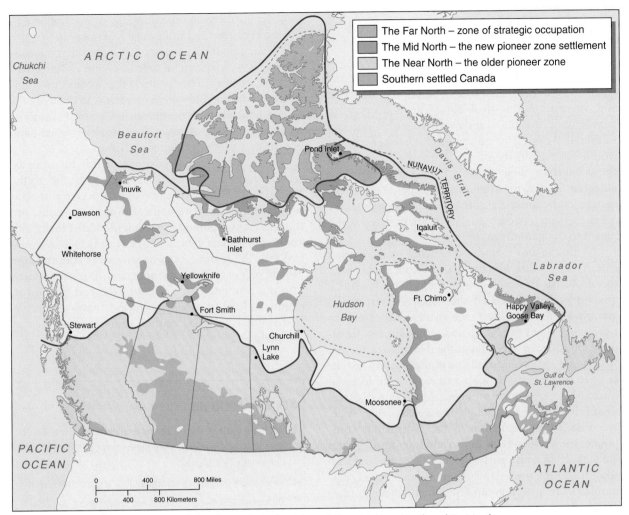

Figure 17.6   Settlement Zones of northern Canada.   Most of Canada's Near North and Far North were recently settled by migrants lured there by resource extraction activities. The Far North remains largely beyond the settlement frontier.

by a single major economic activity and are located in the south. Many, such as Sudbury or Flin Flon, are mining and smelting towns. Others, such as Duluth or Thunder Bay, are primarily transportation centers. Chicoutimi's economy focuses on aluminum smelting. Most smaller towns in the boreal forest are similarly unifunctional.

In the far north, European development has resulted in a few permanent settlements. Most of those of European descent in the area work, in some capacity or another, for the Canadian or U.S. governments or in resource exploitation. Far northern communities are extremely isolated, often with predominantly male populations, and the labor force frequently spends periods of weeks away from the communities for family visits and recreation. As with communities everywhere that are totally focused on minerals extraction, they are often short-lived and die when the resource is depleted.

The Inuit constitute the far north's majority population. Most live in small villages along the coastline, although more and more are drifting into towns in the area. Again, it is a nuclear settlement pattern; few people live away from these settlement nodes.

GOVERNMENT ATTITUDES AND NORTHLANDS DEVELOPMENT

The Northlands have presented both the U.S. and Canadian governments with a series of important and often perplexing problems. This is a vast area, largely unoccupied and only now is gradually being exploited. At an earlier time, perhaps neither country would have concerned itself with the integrity of these lands or their people. However, times have changed. Today, the region, still largely frontier, is controlled by industrial, urban economies that share an increasing concern for what their growth has meant to the natural environment.

With the Northlands so much a part of their country, Canadians in particular are concerned about the problems of development there. The Canadian government and many of the country's universities actively sponsor a wide range of research efforts to learn more about the Northlands environment. Canadians of all ranks are tempering their desire for the wealth unquestionably available in the region's resources with a concern for what extraction will do to an obviously fragile environment. The United States shares these concerns, though on a much smaller scale.

The native population in the Northlands compounds these issues. Both Indian and Inuit fertility rates are high. Even with a fairly high mortality rate and with some migration toward the margins of the region, their numbers are growing rapidly. Unemployment rates among these groups are already steep—often above 50 percent—and good jobs cannot be provided without substantial economic growth. Indians and Inuit are caught in a difficult situation; they have learned to desire many of the products of a developed economy, but they seldom have the means of attaining them.

The necessity of focusing development on mineral resources coupled with the high cost of transportation to widely scattered locations, suggests that even a decision to encourage extractive industries would have only a small and short-lived impact on the economy of most far northern communities. Economic opportunities are restricted and will probably remain so.

Several political geographers suggest that it is useful to look at what can be called a country's "effective" national territory. They contend that whatever a country's boundaries might seem to be, it is in control only of the land actually occupied and actively administered. Large areas of empty land are, by this measure, not firmly a part of a country's territory.

At least some repercussions of this idea can be seen in the American and Canadian approaches to the Northlands. The construction of improved transportation facilities, the development of military bases, and the study of development schemes for the Northlands are in many ways indicative of attempts to assert control over the region. As these endeavors continue, the Northlands will become more and more integrated into the economy and society of the rest of the two countries. It will not be set aside as a massive preserve. In the final analysis, neither country views that as acceptable.

Review Questions

1. What are some of the major environmental problems which have resulted from large-scale mining and petroleum operations in the Northlands?

2. What is the basis for the Northlands as a distinctive region?

3. Differentiate between "tundra" and "taiga" and discuss the pattern in the Northlands.

4. What is meant by "effective national territory," and how is the concept applicable to the Northlands?

5. What are some of the major changes that have occurred in the Inuit way of life over the past couple of decades?

6. Discuss the economic structure of the Northlands.

7. Is tourism a viable alternative for economic development in the Northlands? Explain your answer.

HAWAII

The Hawaiian archipelago is a string of islands and reefs 3300 kilometers (2050 miles) long that form a broad arc in the mid-Pacific. The archipelago begins in the east with the big island of Hawaii and ends almost at the International Date Line beyond Midway Island with a small speck in the ocean called Kure Atoll (Figure 18.1). Because most islands in the archipelago are small and widely spaced, only those in the easternmost 650 kilometers (400 miles) comprise the portion usually considered the actual Hawaii.

Hawaii's eight main islands contain more than 99 percent of the state's land area and all but a handful of its people. The big island of Hawaii, at 8150 square kilometers (4021 square miles), comprises nearly two-thirds of the state's landmass. Kahoolawe, the smallest of the eight, at 125 square kilometers (45 square miles), is uninhabited and used only as a military bombing and firing range.

LOCATION AND PHYSICAL SETTING

Hawaii is near the middle of the Pacific Ocean. Honolulu, the state capital, is 3850 kilometers (2400 miles) west of San Francisco, 6500 kilometers (4000 miles) east of Tokyo, and roughly 7300 kilometers (4500 miles) northeast of the Australian coast. Honolulu is far, too, from the swath of islands stretching across the southern and southwestern Pacific. With few small but largely inconsequential exceptions, the

South Pacific atolls are well over 3500 kilometers (2175 miles) away.

Until the last two centuries, Hawaii's location kept it relatively isolated. But as countries around the Pacific Basin began to communicate more with one another and to use the ocean's resources, these islands became an important focus of shipping, whaling, and political maneuvering. The strategic value of Hawaii's centralized oceanic location grew as the economic and political affairs of the central and north Pacific drew in European and Pacific Rim powers.

Geology

The Hawaiian island chain is the visible portion of a series of massive volcanoes. The ocean floor

The spectacular beauty and awesome fury of Hawaii's landscape has captured the imagination, and the tourist dollars, of millions of "Mainland" Americans. Here, lava emerges from Kilauea Volcano.

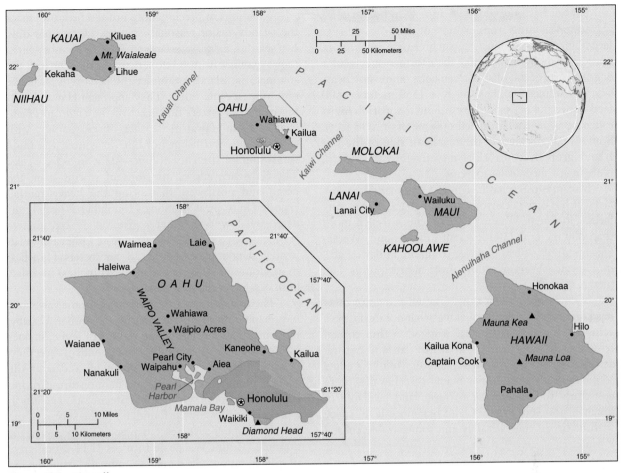

Figure 18.1   Hawaii.

supporting them is 4000 to 5000 meters (13,000 to 15,000 feet) below sea level. This means that for a volcano to break the water's surface, it already forms a mountain almost 5 kilometers (3 miles) in height. Indeed, if a mountain's height from its base is the sole criterion for size, Mauna Kea, on the "Big Island" of Hawaii, is one of the world's tallest. Standing 4528 meters (13,784 feet) above sea level, its base sits another 5400 meters (18,000 feet) below sea level. With a total height of nearly 10,000 meters (32,000 feet), Mauna Kea is almost 1000 meters (3000 feet) taller than Mount Everest.

The volcanic activity that created the islands and that continues there today has, for the most part, not been of the explosive variety, where large pieces of molten material and lethal gases violently burst from cones and are thrown great distances. Instead, the islands grew incrementally from repeated flows of lava from fissures. Volcanic cones resulting from explosive eruptions do exist on the islands. Diamond Head, the famous Honolulu landmark, is the highest at about 240 meters (810 feet). More common, however, are features formed from a gradual buildup of material as a sequence of lava flows piled up one layer on top of another. The usual shape of volcanic mountains formed in this way is dome-like, with the main feature being undulating slopes instead of steep cliffs. Such rolling highlands are found in many areas on the islands.

Several volcanoes on the Big Island remain active. Mauna Loa spouts lava about once every 4 years. In 1950, an eruption covered some 100 square kilometers (36 square miles) with lava. The threat of damage to Hilo, the island's largest town, is pervasive. Kilauea, a volcano on the island's southeast side, is usually active, but lava flows from it only about once every 7 years. Kilauea's 1960 flow covered 10 square kilometers (4 square miles), adding nearly a square mile to the Big Island's size.

Hawaii has rugged slopes and abrupt changes in elevation resulting from erosion of the volcanic surfaces by moving water. One such erosive feature are sea cliffs cut by waves, and they form a spectacular edge to parts of the islands. Such cliffs on the northeast side of Molokai stand as much as 1150 meters (3600 feet) above the ocean and are among the world's highest; others on Kauai exceed 600 meters (2000 feet). Some small streams on the northeast side of the Big Island drop over sea cliffs directly into the ocean, plunging in spectacular torrents.

Stream erosion has heavily cut into many lava surfaces. It has carved canyons that gouge the domes' surfaces. On Kauai, Waimea Canyon cuts a trench more than 800 meters (2500 feet) into the ground. Waterfalls hundreds of feet high are common. Hiilawe Falls, in Hawaii's Waipio valley, drop a precipitous 300 meters (950 feet).

Such intense erosion has limited the amount of level land on the islands, making what does exist very valuable. Kauai is particularly rugged. A thin coastal fringe on Kauai; a narrow swath of flat land cutting across Maui's center; a small plain on Molokai's edge; some limited coastal lava plains on Hawaii—these constitute the smattering of level land available in much of the state. Only Oahu has more level space with its broad central valley and some sizable coastal lowlands.

Climate

Hawaii's oceanic location impacts its climate. The ocean fills the winds with moisture that brushes the islands' mountains. The ocean moderates temperature extremes—Honolulu's record high of 31°C (88°F) faces off against a record low of only 13°C (57°F). The maritime influence tends to delay seasons on the islands. Whereas most places in the United States have their highest temperatures in July and August and their lowest in January and February, Hawaii's highs occur in September and October. The islands' lows may be delayed until early March, a lag reflecting slower changes in the ocean's water temperatures.

Hawaii has a tropical location. Honolulu's latitude of 20° N is the same as Calcutta's and Mexico City's. This position on the globe means Hawaii experiences little change in day length or sun angle from one season to another. It means, too, there is little seasonal variation in temperature. In January, Honolulu averages 22°C (72°F) and in July, 26°C (78°F).

Variations in precipitation, not temperature, mark the islands' major seasonal changes. Hawaii has only two seasons, a dry summer from May to October and a moist winter from October to April. During the summer, the islands are under the persistent influence of northeast trade winds. They approach Hawaii blowing over cool waters northeast of the chain to create characteristic Hawaiian weather—breezy, sunny with some clouds, warm but not hot. In winter, the trade winds stop for stretches of time—sometimes for weeks. Their absence allows invasions of storms from the north and northwest that bring the heavy rainfall expected in tropical locations. Honolulu has been inundated by downpours that registered 43 centimeters (17 inches) in a single 24-hour period. Other Hawaiian weather stations have recorded 28 centimeters (11 inches) in an hour and 100 centimeters (40 inches) in a day, both of which rank near world records.

The islands' topography creates extreme variations in precipitation from one location to another. Usually, most precipitation falls along the north and east sides of the mountains as a result of the dominant directions of wind flow. As the air is blown up and over the islands' mountains, it cools, and its moisture-carrying capacity is reduced. Orographic precipitation results in a process similar to that found along the shores of the North Pacific Coast (see Chapter 16). On Kauai, for example, Mount Waialeale receives 1234 centimeters of precipitation (486 inches) annually, thus laying claim to being the world's wettest spot. But the town of Waimea, also located on Kauai and only 25 kilometers (15 miles) from Mount Waialeale, receives just 50 centimeters (20 inches) annually. Oahu presents another example of rainfall contrast. Within the metropolitan area of Honolulu, people live near the beach in a semiarid climate that has less than 50 centimeters (20 inches) of annual rainfall while others live inland near Pali on the margins of a rainforest drenched by 300 centimeters (120 inches) of precipitation a year.

Unlike the Pacific Northwest, the greatest precipitation on Hawaii's mountains occurs at fairly low, not high, elevations. In Hawaii, air masses tend to push around the isolated mountains instead of up and over them. Hence, precipitation usually falls between 600 and 1200 meters (2000 to 4000 feet).

Biota

Volcanic soil tends to be permeable, so water percolates rapidly and drains beyond the reach of many plants. Thus, Hawaiian areas with moderate to low precipitation are arid. Expanses of desert and semi-desert are present on most of the islands.

Coupled with environmental variation and generally temperate climate, the Hawaiian Islands' isolation has fostered vastly diverse plant and bird communities, the great majority of which are unique to the islands. Several thousand plants are native there and found naturally nowhere else; 66 uniquely Hawaiian land birds have also been identified. Interestingly, no land mammals existed on the islands until humans arrived and introduced them.

Numerous studies have shown how fragile and jeopardized the island ecosystems are. Many island species developed in small, isolated locations in response to particular ecological frameworks and to the absence of competition. Many of these small niches are disappearing because of human intrusions; when the niches do disappear, resident plant and animal species disappear, too. Of Hawaii's 66 unique birds, more than one-third are extinct. Most of the rest are listed as rare and endangered species. Of the islands' plants, most indigenous cover below 450 meters (1500 feet) had been destroyed by 1900. New studies show that much of the flora at higher elevations—once protected by relative inaccessibility—is also succumbing to human encroachment.

Direct human destruction is not the only way Hawaii's native species are eliminated. Indirect causes are the animals and alien plants introduced to the islands when settlers arrived. The rats, dogs, pigs, and goats people brought with them have wreaked ecological havoc. Pigs and rats, the latter surreptitious stowaways, take a heavy toll on native birds by eating their eggs. Domestic-turned-feral pigs damage or destroy vegetation. Few Hawaiian plants have thorns or other protection from grazing animals. And many native plants could not compete with the hundreds of accidentally or intentionally imported plants. Government attempts to protect endangered species have resulted in a number of sanctuaries on the islands, but this is definitely a rearguard action.

POPULATING THE ISLANDS

The Polynesians

The Polynesians' settlement of Hawaii was one of humankind's most audacious ventures in ocean voyaging. From their isle of origin, these people set out in open canoes to cross broad oceanic expanses separating small island clusters. Historical clues suggest the first Polynesians arrived in Hawaii perhaps 1500 years ago, and scholars think they came from the Marquesas, 4000 kilometers (2500 miles) southwest of Hawaii. Scholars also believe a pre-Polynesian group lived on the islands when the Polynesian settlers arrived, but it was probably absorbed by the newcomers. A second wave of Polynesian migrants made the long voyage 400 or 500 years later.

Established in their new home, the Hawaiians solidified a complicated social organization in which hereditary rulers held absolute sway over their populations and owned all of the land. By the late eighteenth century when Europeans found the islands, the Hawaiians' numbers had grown to about 300,000.

Early European Impact

Captain James Cook was the first European to visit Hawaii. Casting anchor in 1778, he dubbed them the Sandwich Islands. Although Cook himself was killed on the shore of the Big Island, news of his discovery spread rapidly after reaching Europe and North America. Hawaii and its people were on the brink of change.

Entrepreneurs and government officials quickly recognized that the islands were ideally located to serve as a way station to exploit the trade developing between North America and Asia. Then, by the 1820s, the whaling industry moved into the North Pacific and for the next half-century, Hawaii became the principal rest and resupply center for Pacific whalers. About the same time, conservative Protestant missionaries came to the islands. Mostly from New England, they had a major influence on the islanders for decades.

The effect on the Polynesians and their culture was catastrophic. As Hawaiians adopted aspects of European culture, their own political and economic traditions suffered, sometimes disintegrating. Famine, heretofore unknown, became a part of life as traditional food gathering and distribution systems were upset. Infectious diseases—measles, leprosy, smallpox, syphilis, tuberculosis—endemic among Europeans but previously unknown to Hawaiians swept the islands. One outbreak of either cholera or bubonic plague reportedly halved the population in 1804.

Within three generations, the economic and social order of the Polynesian Hawaiians was destroyed. Their numbers fell from 150,000 in 1804 to perhaps 75,000 in 1850. Intermarriage led to still smaller

Some of the ethnic diversity of Hawaii's residents is captured in this scene from Oahu, which includes students of Filipino, Chinese, Japanese, European, and Hawaiian ancestry.

numbers. There are fewer than 10,000 Polynesian Hawaiians today, although State of Hawaii statistics indicate that an additional one-sixth of the state's population is part Polynesian.

The Asian Arrival and Mainland Migration

The first Hawaiian sugar plantation was established in 1837. Between then and the end of the century, Hawaii grew to be a major world sugar exporter. This development led to a great need for agricultural laborers. Native Hawaiians filled the ranks for a time. But they were unwilling workers, and their declining numbers provided nothing like the labor force needed. The plantation owners, who were Americans and Europeans, thus turned to Asia as a source of cheap, abundant labor. The first few hundred contract workers came from China in 1852. Japanese contract laborers began arriving in 1868, and Filipinos in 1906.

Between 1852 and 1930, sugar growers brought 400,000 agricultural laborers to Hawaii, most of them Asian men. Many chose to leave the plantations but remain on the islands when their contracts ended.

By repeatedly plying Asia for plantation laborers, sugar growers ushered in profound demographic change for Hawaii. Because many of the workers remained rather than returned home, they transformed Hawaii's ethnic face. In 1852, Hawaiians of Polynesian heritage represented more than 95 percent of the islands' population. By 1900, they were less than 15 percent of the total, and East Asians comprised nearly 75 percent. Japanese were the largest ethnic group at the time, comprising more than 50 percent of the islands' 150,000 residents.

This trend shifted again several decades into the twentieth century. The United States mainland was the main source of Hawaii's new residents after 1930 (Figure 18.2). In 1910, only about one Hawaiian resident in five was of European ancestry (referred to in Hawaii as Caucasian). Today, that share has grown to nearly 40 percent of the state's current population.

Hawaii's total population fell from its pre-European peak of about 300,000 to a low of 54,000 in 1876 before beginning to grow again. By the early 1920s, the state's population had reached pre-European levels. More than 400,000 service personnel, stationed on the islands during World War II, temporarily pushed the total to about 850,000. In 2006, the state had an estimated 1,285,498 residents. Because of immigration, Hawaii's annual rate of population growth is above the national average. Although the pre-European population was spread across the islands, with the Big Island occupied by the largest number of people, the islands' population has been concentrated on Oahu since the arrival of the Europeans (Table 18.1). Honolulu, with its fine harbor, became the island grouping's principal port city and the state's main growth center.

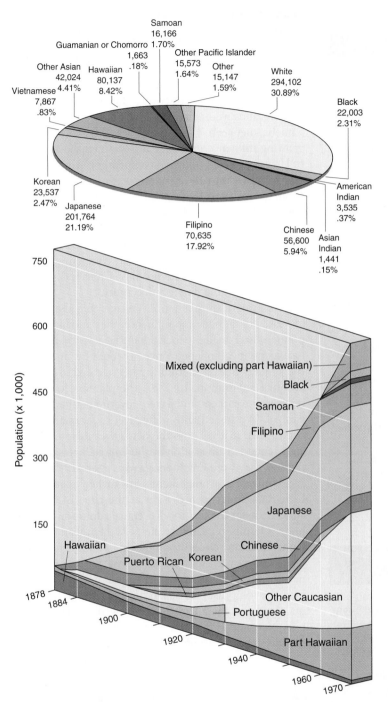

Figure 18.2 **Ethnic diversity in Hawaii.**
Present-day residents of Hawaii can trace their roots to many parts of the world.

Implications of Ethnicity

Although by no means a completely integrated society, Hawaiian residential integration easily surpasses that in other parts of the United States. No census tract anywhere in Hawaii has more than 40 percent of its people with Chinese ancestry, and none is more than 70 percent Japanese. Within Honolulu, where most of Hawaii's people live, a clear majority of tracts have a population mixture that includes at least 10 percent of each of the city's three major ethnic groups—Caucasians, Japanese, and Chinese. The poorer ethnic groups are less well integrated.

TABLE 18.1

Location	Date						
	1831	1878	1910	1940	1970	1990	2000
Oahu	23	35	43	61	82	79	72
Honolulu	10	24	27	42	42	33	31
Rest of Oahu	13	11	16	19	40	42	42
Other islands	77	65	57	39	18	21	28

Percentage Distribution of the Hawaiian Population

Source: R. Warwick Armstrong, ed., *Atlas of Hawaii* (Honolulu: University Press of Hawaii, 1973), p. 100; *1990 U.S. Census of Population; 2000 U.S. Census of Population.*

People of Japanese and particularly Chinese ancestry have done well in the state's economy. Educational standards among both groups are high, and average Chinese income levels have surpassed Caucasian averages in recent decades. Native Hawaiians and newer immigrant groups, like Filipinos, have comparatively low incomes and lower-status occupations.

In some respects, discussions of ethnicity and ethnic grouping have become moot issues in Hawaii. Individuals may raise these topics occasionally, but so many people with so many blends of mixed ethnic heritage live in the state and make up Hawaii's present cultural tapestry that identification of this strand or that becomes mainly of historic interest (Table 18.2).

TABLE 18.2

Subject	% of Total Population (1,285,498)
Total Population	100.0
One Race	79.9
White	26.8
Black or African American	2.3
American Indian & Alaska Native	0.3
Asian	41.5
Native Hawaiian & other Pacific Islander	9.0
Two or More Races	20.1

Population by Race for All Ages for the State of Hawaii: 2006

Source: U.S. Census Bureau, State & County Quick Facts.

POLITICAL INCORPORATION INTO THE UNITED STATES

The political history of Hawaii was turbulent during the 120 years after Cook's discovery. The various kingdoms of the islands were eliminated by a strong chief, Kamehameha, between 1785 and 1795. The missionaries' growing influence gradually undermined the authority of Hawaiian rulers, allowing competing European interests to move into the resulting vacuum in the 1800s. From the 1820s to 1850s, the French pushed their influence in the region. Much of the Hawaiian monarchy seemed to favor a relationship with Britain, however, and for a brief period in 1843 Britain annexed Hawaii. It was only the balance between the competing powers operating in Hawaii that kept the islands independent during much of the nineteenth century.

As American plantation owners on the islands increased in number and influence, their mounting economic power made it inevitable that if Hawaii was to lose its political independence, it would be annexed by the United States. Certainly, they cleared a path for eventual annexation. In 1887, the planters forced the monarchy to accept an elected government under their control. Weakened and ineffective, the monarchy was overthrown in 1893, and the new government immediately requested annexation by the United States. Initially refused, they were finally accepted as a territory in 1898.

Unlike its arrangements for earlier acquisitions, Congress made no provision at the time of Hawaii's annexation for the territory's eventual admission to statehood. Despite repeated requests by Hawaiian officials, Congress long refused to grant the islands state status. The reasons for this steadfast refusal were varied and never clear, but they likely centered around concerns over the long distance separating the islands from the mainland and the allegiance of Hawaii's Asian population. Not until 1959, after Alaska was admitted as a state and all economic arguments against Hawaiian statehood were exhausted, did it become the fiftieth state.

THE HAWAIIAN ECONOMY
Land Ownership

Roughly half of all land in Hawaii is government owned. This in itself is not unusual, because large proportions of all the continental United States' western states and Canada's western provinces are governmentally owned. What is unusual is that the state, not the federal government, controls 80 percent of that land. Most of Hawaii's state-owned portion is in the islands' agriculturally less desirable areas, and the bulk is in forest reserves or conservation districts. The remaining federal lands are primarily in national parks on the "Big Island" of Hawaii and Maui or in military holdings on Oahu and Kahoolawe.

Setting the state off from the rest of the country is the unusually large concentration of private lands in the hands of relatively few major landowners. About 95% of all of Hawaii's privately owned land belongs to only 72 owners. Six landowners each control more than 40,000 hectares (100,000 acres) out of a state total of about 1,040,000 hectares (2,570,000 acres). The largest, the Kamehameha Schools, holds almost 9 percent of all the land in Hawaii and 15 percent of all private lands.

Smaller unit ownership of private land is most extensive on Oahu, but even there the large holders control more than two-thirds of all privately owned land. Two of the islands, Lanai and Niihau, are each almost entirely controlled by a single owner, and on all other islands except Oahu, major landowners control about 90 percent of private property.

Most of Hawaii's large landholdings were created in the nineteenth century, when free-wheeling exploitation was handmaiden to Euro-American economics. Before then, only monarchies held land. When the monarchy and its political power declined, land passed into the hands of non-Hawaiian private owners. Then, when the early owners died, most estates were given over to trusts to administer rather than passing directly to heirs. This has made it difficult to break up ownership patterns.

The concentration of land ownership represents a major problem for Hawaii, especially in urban areas. The way the scenario now plays out, landowners lease land for urban use. When the leases expire, they are renewed, typically at a much higher rate because urban development generated by the lessees has greatly increased the land's value. Whether developer or small homeowner, the security of land ownership, so much a part of the American psyche, is effectively denied to many in Hawaii. Where smaller holdings are available and population growth is rapid, as on Oahu, a tremendous increase in land prices has occurred.

In 1967, Hawaii attempted to moderate the restrictive and financial impacts of large landholdings through the Land Reform Act. This act extended the state's power of eminent domain by allowing Hawaii's Housing Authority to condemn individual parcels of land when their leases expire. The Authority would then transfer title to tenants at fair market values determined by arbitration or negotiation. Soon after enactment, the lower courts declared the law unconstitutional. But in 1984 the U.S. Supreme Court overturned that ruling, thus paving the way for change.

Agriculture

Sugar, and later pineapple, fueled Hawaii's economy for decades after the 1860s. The economy remained primarily agricultural until World War II. In recent decades, agriculture has shown modest gains in income, but its relative importance has declined greatly. Sugar's share of income generated in the state fell from 20 percent in 1950 to 4 percent in 1985. During the same time span, the income from pineapples dropped from 16.5 to 2 percent of the total. Even diversified agriculture, such as livestock raising on the island of Hawaii, which has been the prime growth component of the agricultural industry, had its income share decline from 4.6 to about 2 percent.

Today, only one Hawaiian worker in 30 is employed in agriculture. Once the kingpin of Hawaii's economy, agriculture ranks no better than fourth or fifth.

In some ways, an emphasis on declining agricultural importance is misleading. Even with the move of some sugar or pineapple plantations out of production in the face of declining profits, Hawaii still provides a substantial share of the world's sugar harvest. But pineapple production has declined dramatically. Just a few years ago, Hawaii produced over 350,000 tons a year of the fruit, but production was just 89,000 in 2006, down from 106,000 tons in 2005. Del Monte, a leading world producer of pineapples, announced in 2006 that it would cease production in Hawaii. Acreage in pineapples dropped from 19,100 acres in 2002 to just 13,900 acres in 2006.

Current gross economic statistics overwhelmingly emphasize Oahu's position, where more than 80 percent of Hawaii's economy is concentrated. The role of agriculture remains great on the other islands. Lanai depends on pineapples for much of its employment and income. Livestock and sugar form the backbone of the Big Island's economy, as do sugar and pineapples on Maui and Kauai. Only on Oahu has agriculture

A substantial ranching economy exists on many of the drier portions of the islands, such as this location on Maui. Despite the state's small size, the pattern of concentrated private landholding has spawned a number of ranches covering thousands of acres each.

been replaced as the most important sector of the economy.

The Federal Government

As agriculture declined and lost dominance in the Hawaiian economy, other sectors gained. The federal government was first to step up. During the past several decades, governmental expenditures have increased in Hawaii at a rate comparable to the growth of Hawaii's total economy, maintaining about a one-third share of all state expenditures.

Most of this increase has come from the military. The strategic value of Hawaii's location in the mid-Pacific is obvious. It is headquarters for the U.S. Pacific Command, center of Pacific operations for all branches of the armed forces, and home of a number of major military bases. The military controls almost 25 percent of Oahu, including the land around Pearl

Harbor. Today, nearly one Hawaiian worker in four is a military employee. Military personnel and their dependents represent more than 10 percent of Hawaii's population. The armed forces are also the largest civilian employer in the state.

The armed forces have long been a substantial part of the Hawaiian community, and they are generally well received. Still, the heavy dependence on the military is often viewed as unfortunate, especially when periods of military cutbacks adversely affect the local economy. The military's relative importance is likely to decline somewhat in the future, although it will almost surely remain a major presence.

Tourism

Hawaii represents an exotic, distant tropical paradise to most Americans. Mainlanders dream of visiting its balmy shores. The first regularly scheduled

By the 1990s, decades of inexpensive air flights and effective advertising had spurred Honolulu's tremendous urban growth behind Waikiki and to the base of Diamond Head.

trans-Pacific flights began in 1936. As late as 1951, however, nearly half of that year's 52,000 visitors reached Hawaii by ship. Then came larger airplanes, lower fares, and a booming national economy. The number of visitors increased markedly in every decade since the 1950s. In 1965, the islands hosted 320,000 people, and in 1971, 1,819,000. Now, over 7.4 million people (most of them tourists) visit the state each year. Most visitors to Hawaii come from mainland United States, especially the West Coast. About a quarter of all tourists come from Asia, mostly Japan.

Hawaii's government and businesses encourage out-of-state visitors through the Hawaii Visitors Bureau, which widely advertises the islands' tropical and exotic image. Tourism has become the principal growth sector of the economy, increasing its share of total island income from 4 percent in 1950 to 18 percent in 1970 and over 22 percent today. Tourism produces over $11 billion per year for the Hawaiian economy and employs more than 171,000 people, nearly 22% of all jobs.

To many Hawaiians, the influx of tourists is a mixed blessing. In 1952, only 2800 hotel rooms existed on the islands. Now, there are more than 50,000. More than 60 percent of the existing units are on Oahu, crowding along Honolulu's famous beaches of the Waikiki-Kahala coast. On a typical day, 60,000 tourists flock to the high-rise hotels and rental condominiums lining the beach. Another 50,000 fan out across the rest of the state. The lower elevations of Diamond Head, long Honolulu's most famous landmark, are now pocked with hotels. The congestion, the decline in scenic quality, and the pollution resulting from crowded development are great concerns to many. Recent expansion is focusing on the outer islands, and the fear is that, without careful control, these problems will spread.

Another problematic issue islanders must deal with in a tourist economy is the industry's unstable

Honolulu's Waikiki Beach and Diamond Head were still relatively uncrowded in the 1950s.

nature. Downturns in the national economy in the early and mid-1970s and early 2000s, for example, had a heavy impact on Hawaiian tourism because many Americans chose to cut travel costs by vacationing closer to home. Seasonality also affects tourism. Although Hawaii has striven to develop a year-round tourist flow, the summer months remain the peak period; this results in higher unemployment each winter. Global terrorism has also become an issue for Hawaii's tourism industry. The attacks of September 11, 2001, reduced the number of people willing to travel by air, a major problem for isolated Hawaii. Even the 2003 SARS epidemic hurt Hawaii's tourist traffic for a period.

Hawaii's attractiveness to Japanese tourists has generated mixed feelings. Large Japanese tour operators, in an attempt to control the total tour package and increase profits, have bought hotels and other tourist facilities in Hawaii. Other kinds of Japanese firms, simply looking for good investments, have bought a variety of Hawaiian businesses. For example, to satisfy popular golfing tour demands, Japanese businesses developed many new golf courses in Hawaii. Growth was so sudden and brisk that Hawaiian officials decided to limit the number of new courses that could be constructed each year. With a certain historic irony, some Hawaiians have therefore urged control of land and business sales to foreigners, arguing that the state's economy could become controlled by outsiders. Others, however, downplay this concern, looking instead to the dollars Asian tourists bring to the islands. Japanese tourists, for example, tend to spend more than twice as much daily as their American counterparts, and they spend far more for gift and souvenir items. Although Japan's economic troubles in 1998 curtailed tourism and reduced Hawaii's income from this source for a time, more than 1.3 million Japanese still visit the islands annually.

Transportation

Hawaii's insular nature creates a transport situation that is far different from that of the mainland United

States. It is entirely dependent on air and sea movement; overland transport is of only local importance. This has several repercussions. Both everyday and luxury items are comparatively more expensive or slow to arrive. Hawaii imports much of its food, nearly all of its energy supplies and vehicles, and a host of other commodities. In spite of the transportation effects of its insularity, Hawaii's agricultural exports still amount to more than $96 million annually (2006).

HONOLULU

Honolulu dominates Hawaii. Over 72 percent of the state's residents and 80 percent of its economy are concentrated in the city and its suburbs. No other state or region in the United States or Canada is so totally dominated by a single urban center. Hawaii's second largest city is Hilo on the island of Hawaii. Although increasing its population at a more rapid rate than Honolulu, Hilo has only one-twentieth Honolulu's 2006 metropolitan area population of 909,853.

Mountains, ocean, and federally held land hug Honolulu, crowding it into a series of narrow valleys and a fringe along the coast. Recent construction of two highways and tunnels has increased the city's connections with Oahu's northeast coast and stimulated growth in the Kailau-Kaneohe urbanized area, now home to more than 100,000 residents. Construction is also increasing along Oahu's central valley.

Honolulu's problems are like those of other American cities, although somewhat magnified. Living costs are high, almost 20 percent above the national average. Among U.S. cities, only Anchorage, Alaska, matches the expense of living in Honolulu. Housing costs are also high. With so much land held by estates, marketable residential land is scarce. Only about 40 percent of Honolulu's residents live in housing they own. Housing for low-income groups is inadequate.

Honolulu is also plagued by congestion. Traffic flow is inhibited by the city's high density of settlement and its layout, wrapped amid barriers provided by physical features and federal land. Climatically, the city's leeward location, though providing a generally dry climate, leads to occasional periods of little wind. Atmospheric pollution is especially evident during these times.

All of these features certainly do not limit Honolulu as a city. Instead, they emphasize that it is a city with a set of problems and attractions much like that of any other. Honolulu may be in paradise, but it is an urban place in paradise.

INTERISLAND DIFFERENCES

The major Hawaiian islands share similar geologic histories and close proximity in a vast ocean, but all have their own individual character.

Oahu is densely populated and intensely used, and it offers a view of bustle and confusion common to urban America.

The island of Hawaii, or the Big Island, has by comparison a sense of relative space and distance. Its land area is dominated by five huge shield volcanoes. Mauna Loa alone covers half the island, and Mauna Kea almost another quarter. Interspersed are large ranches; high, barren volcanoes; and large stretches of almost treeless land. Sugar, cattle ranching, and tourism are the Big Island's major industries.

Kauai is heavily eroded into spectacular scenes of mountains, canyons, cliffs, and waterfalls. The northernmost island, Kauai is exposed to northern Pacific storms that have moved south as they travel east toward the Americas. Thus, it can be drenched by orographic precipitation from the trade winds as well as frontal precipitation. Kauai is sometimes called the garden isle because of its lush tropical vegetation. It is becoming increasingly popular with tourists because of its dramatic physical environment.

Neighboring Niihau sits in the lee of the larger Kauai and is shielded from the prevailing trade winds. The result is a much drier climate. Niihau is privately owned and is operated by the Niihau Ranch Company. Most of its few hundred residents are native Hawaiians.

Maui, the second largest of the islands, offers a contrast between its central lowlands dotted with plantations and the rugged mountains to either side. Intense tourist development has concentrated along Maui's western coastal strip, resulting in Maui having the most rapid rate of population increase of any of the islands in the 1970s and 1980s. Still, much of the rest of Maui remains little changed and sparsely populated.

Molokai is half ranchland and half rugged mountains. Its north coast is dominated by spectacular sea cliffs, while the south shore is a broad coastal plain. It is perhaps the least economically developed of the populated Hawaiian Islands. Molokai's economy was dominated by pineapple production until the Del Monte pineapple plantation closed in 1983.

Both Lanai and Kahoolawe are in the lee of much higher Maui. Both are climatically dry, and neither has permanent streams. Pineapple production is the only important economic activity on Lanai. The U.S. Navy administered Kahoolawe and used it for decades as a bombing range and for military exercises but returned it to the state in late 2003.

An Island Paradise?

It has become fashionable to talk of trouble in paradise when discussing Hawaii. Certainly, the state has its share of problems. High living costs and high unemployment rates will continue. The combined impact of tourism, economic development, and changes in the physical environment due to development pressure has Hawaiian leaders groping for solutions.

Still, Hawaii's positive attributes cannot be overshadowed. No part of America, and few parts of the world, can match its level of racial and ethnic assimilation. Much of Hawaii remains a place of great beauty, and its climate is often superlative. Concerns about the negative impacts of excessive tourist-oriented development on the outer islands have led to exertion of a strong counterbalance. Few other states in the country, for example, can match Hawaii's attempts at controlling growth to maintain the quality of their environment.

Review Questions

1. What are the eight largest islands of Hawaii, and how is the state's population distributed among them?

2. What are the geologic processes that produced the islands and what patterns are visible?

3. Explain the basic climate processes at work and describe the existing pattern.

4. What is unique about Hawaii's biota?

5. What is the dominant land ownership pattern in Hawaii?

6. Explain the pro's and con's of Honolulu's dominance.

7. How can tourism be considered a "mixed blessing" to the islanders?

Absolute Humidity   the mass of water vapor in the atmosphere per unit of volume of space.

Accessibility   a locational characteristic that permits a place to be reached by the efforts of those at other places.

Accessibility Resource   a naturally occurring landscape feature that facilitates interaction between places.

Acid Rain   rain that has become more acidic than normal (a pH below 5.0) as certain oxides present as airborne pollutants are absorbed by the water droplets. The term is often applied generically to all acidic precipitation.

Acculturation   the process by which members of two or more cultures living together gradually adopt each others' beliefs and practices. If one group is numerically dominant, the acculturation process may be asymmetrical.

Air Mass   a very large body of atmosphere defined by essentially similar horizontal air temperatures. Moisture conditions are also usually similar throughout the mass.

Alluvial Soils   soils deposited through the action of moving water. These soils lack horizons and are usually highly fertile.

Antebellum   before the war; in the U.S., belonging to the period immediately prior to the Civil War.

Arete   a sharp, narrow mountain ridge. It often results from the erosive activity of alpine glaciers flowing in adjacent valleys.

Arroyo   a deep gully cut by a stream that flows only part of the year; a dry gulch. A term normally used only in desert areas.

Badlands   very irregular topography resulting from wind and water erosion of sedimentary rock.

Base Level   the lowest level to which a stream can erode its bed. The ultimate base level of all streams is, of course, the sea.

Batholith   a very large body of igneous rock, usually granite, that has been exposed by erosion of the overlying rock.

Bedrock   the solid rock that underlies all soil or other loose material; the rock material that breaks down to eventually form soil.

Bilingual   the ability to use either one of two languages, especially when speaking.

Biota   the animal and plant life of a region considered as a total ecological entity.

Boll Weevil   a small, grayish beetle of Mexico and the southeastern United States with destructive larvae that hatch in and damage cotton bolls.

Break-in-Bulk Point   commonly, a transfer point on a transport route where the type of carrier changes and where large-volume shipments are reduced in size. For example, goods may be unloaded from a ship and transferred to trucks at an ocean port.

Caprock   a strata of erosion-resistant sedimentary rock (usually limestone) found in arid areas. Caprock forms the top layer of most mesas and buttes.

Carrying Capacity   the number of people that an area can support given the quality of the natural environment and the level of technology of the population.

Chinook   a warm, dry wind experienced along the eastern side of the Rocky Mountains in Canada and the United States. Most common in winter and spring, it can result in a rise in temperature of 35° to 40°F in a quarter of an hour.

Clear-Cutting   the logging practice whereby all trees within an area are cut, regardless of the market value of the tree.

Climax Vegetation   the vegetation that would exist in an area if growth had proceeded undisturbed for an extended period. This would be the "final" collection of plant types that presumably would remain forever, or until stable conditions were somehow disturbed.

Confluence   the place at which two streams flow together to form one larger stream.

Coniferous   bearing cones; from the conifer family.

Continental Climate   the type of climate found in the interior of the major continents in the middle, or temperate, latitudes. Characterized by a great seasonal variation in temperatures, four distinct seasons, and a relatively small annual precipitation.

Continental Divide   the line of high ground that separates the oceanic drainage basins of a continent; the river systems on opposite sides of a continental divide flow toward different oceans.

Core Area   the portion of a country that contains its economic, political, intellectual, and cultural focus. It is often the center of creativity and change (see **Hearth**).

Coulee   a dry canyon eroded by Pleistocene floods which cut into the lava beds of the Columbia Plateau.

Crop-lien System   a farm financing scheme whereby money is loaned at the beginning of a growing season to pay for farming operations with the subsequent harvest used as collateral for the loan.

Culture   the accumulated habits, attitudes, and beliefs of a group of people that define for them their general behavior and way of life; the total set of beliefs and learned activities of a people.

Culture Hearth   the area from which the culture of a group diffused (see **Hearth**).

Deciduous Forest forests where the trees lose their leaves each year.

De Facto Segregation the spatial and social separation of populations that occurs without legal sanction.

Degree Day deviation of one degree temperature for one day from an arbitrary standard, usually the long-term average temperature for a place.

De Jure Segregation the spatial and social separation of populations that occurs as a consequence of legal measures.

Demography the systematic analysis of population.

Dome an uplifted area of sedimentary rocks with a downward dip in all directions, typically caused by molten rock pushing upward from below. The sediments have often eroded away, exposing the rocks that resulted when the molten material cooled.

Double-Cropping the farming practice of harvesting two crops within a single growing season; generally, this is accomplished by selecting two crops that will have time to mature sequentially given local climatic conditions.

Dry Farming a type of farming practiced in semiarid or dry grassland areas without irrigation using such approaches as fallowing, maintaining a finely broken surface, and growing drought-tolerant crops.

Economies of Agglomeration the economic advantages that accrue to an activity by locating close to other activities; benefits that follow from complementarity or shared public services.

Economies of Scale savings achieved in the cost of production by larger enterprises because the cost of initial investment can be defrayed across a greater number of producing units.

Emergent Coastline a shoreline resulting from a rise in land surface elevation relative to sea level.

Erratic a boulder that has been carried from its source by a glacier and deposited as the glacier melted. Thus, the boulder is often of a different rock type from surrounding types.

Estuary the broad lower course of a river that is encroached on by the sea and affected by the tides.

Evapotranspiration the water lost from an area through the combined effects of evaporation from the ground surface and transpiration from the vegetation.

Exotic Stream a stream found in an area that is too dry to have spawned such a flow. The flow originates in some moister section.

Exurbia the area farthest from a large urban place in which people working in the urban place live; beyond suburbia. Exurban residents live in a rural or small town setting but do not draw their economic or cultural sustenance from the setting.

Fall Line the physiographic border between the piedmont and coastal plain regions. The name derives from the river rapids and falls that occur as the water flows from hard rocks of the higher piedmont onto the softer rocks of the coastal plain.

Fallow agricultural land that is plowed or tilled but left unseeded during a growing season. Fallowing is usually done to conserve moisture.

Fault Block Mountain a mountain mass created either by the uplift of land between faults or the subsidence of land outside the faults.

Fault Zone a fault is a fracture in the earth's crust along which movement has occurred. The movement may be in any direction and involve material on either or both sides of the fracture. A "fault zone" is an area of numerous fractures.

Federation a form of government in which powers and functions are divided between a central government and a number of political subdivisions that have a significant degree of political autonomy.

Feral Animal a wild or untamed animal, especially one having reverted to such a state from domestication.

Fish Ladder a series of shallow steps down which water is allowed to flow; designed to permit salmon to circumvent artificial barriers such as power dams as the salmon swim upstream to spawn.

Glacial Till the mass of rocks and finely ground material carried by a glacier, then deposited when the ice melts. This deposition creates an unstratified material of varying composition sometimes called glacial drift.

Great Circle Route the shortest distance between two places on the earth's surface. The route follows a line described by the intersection of the surface with an imaginary plane passing through the earth's center.

Hazardous Waste unwanted byproducts remaining in the environment and posing immediate or potential hazard to humans and wildlife.

Hearth the source area of any innovation. The source area from which an idea, crop, artifact, or good is diffused to other areas.

Heavy Industry manufacturing activities engaged in the conversion of large volumes of raw materials and partially processed materials into products of higher value; hallmarks of this form of industry are considerable capital investment in large machinery, heavy energy consumption, and final products of relatively low value per unit weight (see **Light Industry**).

Hinterland the area tributary to a place and linked to that place through lines of exchange or interaction.

Horizon a distinct layer of soil encountered in vertical section.

Humus partially decomposed organic soil material.

Hydrography the study of the surface waters of the earth.

Hydroponics the growing of plants, especially vegetables, in water containing essential mineral nutrients rather than in soil.

Igneous Rock rock formed when molten (melted) materials harden.

Industrial Linkages the network connecting industries that depend on each other for supplies or product sales.

Intervening Opportunity the existence of a closer, less expensive opportunity for obtaining a good or service, or for a migration destination. Such opportunities lessen the attractiveness of more distant places.

Intracoastal Waterway System a waterway channel, maintained through dredging and sheltered for the most part by a series of linear offshore islands, that extends from New York City to Florida's southern tip and from Brownsville, Texas, to the eastern end of Florida's panhandle.

Isohyet a line on a map connecting points that receive equal precipitation.

Jurisdiction the right and power to apply the law; the territorial range of legal authority or control.

Karst an area possessing surface topography resulting from the underground solution of subsurface limestone or dolomite.

Kudzu a vine native to China and Japan but imported into the United States; originally planted for decoration, for forage, or as a ground cover to control erosion, it now grows wild in many parts of the southeastern United States.

La Belle Province the former nickname of the province of Quebec, still commonly used.

Lacustrine Plain a nearly level land area that was formed as a lake bed.

Latitude a measure of distance north or south of the equator. One degree of latitude equals approximately 110 kilometers (69 miles).

Leaching a process of soil nutrient removal through the erosive movement and chemical action of water.

Legume a plant, such as the soybean, that bears nitrogen-fixing bacteria on its roots and thereby increases soil nitrogen content.

Light Industry manufacturing activities that use moderate amounts of partially processed materials to produce items of relatively high value per unit weight (see **Heavy Industry**).

Lignite a low-grade, brownish coal of relatively poor heat-generating capacity.

Lithosphere the rock portion of the earth's mantle, in contrast to the atmosphere and the hydrosphere.

Loess a soil made up of small particles that were transported by the wind to their present location.

Longitude a measure of distance east and west of a line drawn between the North and South Poles and passing through the Royal Observatory at Greenwich, England.

Magma molten rock, generally drawn from deep within the earth.

Maritime Climate a climate strongly influenced by an oceanic environment. Found on islands and the windward shores of continents, it is characterized by small daily and yearly temperature ranges and high relative humidity.

Mediterranean Climate a climate characterized by moist, mild winters and hot, dry summers.

Metamorphic Rock rock that has been physically altered by heat and/or pressure.

Metes and Bounds a system of land survey that defines land parcels according to visible natural landscape features and distance. The resultant field pattern is usually very irregular in shape.

Metropolitan Coalescence the merging of the urbanized areas of separate metropolitan regions; Megalopolis is an example of this process.

Moraine the rocks and soil carried and deposited by a glacier. An "end moraine," either a ridge or low hill running perpendicular to the direction of ice movement, forms at the end of a glacier when the ice is melting.

MSA Metropolitan Statistical Area. A statistical unit of one or more counties that focuses on one or more central cities larger than a specified size, or with a total population larger than a specified size. A reflection of urbanization.

Municipal Waste unwanted byproducts of modern life generated by people living in an urban area.

Open Range a cattle or sheep ranching area characterized by a general absence of fences.

Orographic Rainfall precipitation that results when moist air is lifted over a topographic barrier such as a mountain range.

Outwash rocky and sandy surface material deposited by meltwater that flowed from a glacier.

Overburden material covering a mineral seam or bed that must be removed before the mineral can be removed in strip mining.

Permafrost a permanently frozen layer of soil.

Physiographic Region   a portion of the earth's surface with a basically common topography and common morphology.

Plate Tectonics   geologic theory that the bending (folding) and breaking (faulting) of the solid surface of the earth results from the slow movement of large sections (plates) of that surface.

Pleistocene   a period in geologic history (basically the last 1 million years) when ice sheets covered large sections of the earth's land surface not now covered by glaciers.

Plural Society   a situation in which two or more culture groups occupy the same territory but maintain their separate cultural identities.

Postindustrial   an economy that gains its basic character from economic activities developed primarily after manufacturing grew to predominance. Most notable would be quaternary economic patterns.

Precambrian Rock   the oldest rocks, generally more than 600 million years old.

Presidio   a military post, especially garrisons built by the Spanish in the Southwest.

Primary Product   a product that is important as a raw material in developed economies; a product consumed in its primary (i.e., unprocessed) state (see **Staple Product**).

Primary Sector   that portion of a region's economy devoted to the extraction of basic materials (e.g., mining, lumbering, agriculture).

Pueblo   a type of Indian village constructed by some tribes in the Southwestern United States. Characteristically, it is a large community dwelling, divided into many rooms, up to five stories high, and usually made of adobe. Also, a Spanish word for town or village.

Pull Factors   pertaining to migration, those events, beliefs, or experiences in and about a place that attract an individual and may lead to a decision to choose that place as a new home.

Push Factors   pertaining to migration, those events, beliefs, or experiences in or about a place that support an individual's decision to leave that place for a different home.

Quaternary Sector   that portion of a region's economy devoted to informational and idea-generating activities (e.g., basic research, universities and colleges, and news media).

Rainshadow   an area of diminished precipitation on the lee (downwind) side of a mountain or mountain range.

Rangs   lines of long, narrow fields in French Canada, created to maximize access to rivers or roads for transportation.

Region   an area having some characteristic or characteristics that distinguish it from other areas. A territory of interest to people and for which one or more distinctive traits are used as the basis for its identity.

Resource   anything that is both naturally occurring and of use to humans.

Riparian Rights   the rights of water use possessed by a person owning land containing or bordering a water course or lake.

Riverine   located on or inhabiting the banks or the area near a river.

Scots-Irish   the North American descendants of Protestants from Scotland who migrated to northern Ireland in the 1600s.

Secondary Sector   that portion of a region's economy devoted to the processing of basic materials extracted by the primary sector.

Second Home   a seasonally occupied dwelling that is not the primary residence of the owner. Such residences are usually found in areas with substantial opportunities for recreation or tourist activity.

Sedimentary Rock   rock formed by the hardening of materialdeposited in some process such as by wind or settled out from ancient lakes; most common types are sandstone, shale, and limestone.

Seigneuries   large land grants in French Canada possessing feudal privileges that were awarded by the kings of France to noblemen and to the Church. These lands were then parceled out to individual settlers.

Sharecropping   a form of agricultural tenancy where the tenant pays for use of the land with a predetermined share of his crop rather than with a cash rent.

Shield   a broad area of very old rocks above sea level; usually characterized by thin, poor soils and low population densities.

Silage   fodder (livestock feed) prepared by storing and fermenting green forage plants in a silo.

Silo   usually a tall, cylindrical structure in which fodder (animal feed) is stored; may be a pit dug for the same purpose.

Sinkhole   crater formed when the roof of a cavern collapses; usually found in areas of limestone rock.

Site   features of a place related to the immediate environment in which the place is located (e.g., terrain, soil, subsurface geology, ground water).

Situation   features of a place related to its location relative to other places (e.g., accessibility, hinterland quality).

Smog   mixture of particulate matter and chemical pollutants in the lower atomsphere, usually over urban areas.

Soluble   capable of being dissolved; in this case, the characteristic of soil minerals that leads them to be carried away in solution by water (see **Leaching**).

Space Economy   the locational pattern of economic activities and their interconnecting linkages.

Spatial Complementarity   the occurrence of locational pairing such that items demanded by one place can be supplied by another.

Spatial Interaction   movement between locationally separate places.

Staple Product   a product that becomes a major component in trade because it is in steady demand; thus, a product that is basic to the economies of one or more major consuming populations (see **Primary Product**).

Subduction   the process by which part of the earth's mantle is caused to be drawn downward beneath an adjacent portion of the mantle.

Temperature Inversion   an increase in temperature with height above the earth's surface, a reversal of the normal pattern.

Tertiary Sector   that portion of a region's economy devoted to service activities (e.g., transportation, retail and wholesale operations, insurance).

Township and Range   the rectangular system of land subdivision of much of the agriculturally settled United States west of the Appalachians; established by the Land Ordinance of 1785.

Transferability   the extent to which a good or service can be moved from one location to another; the relative capacity for spatial interaction.

Transhumance   the seasonal movement of people and animals in search of pasture. Commonly, winters are spent in snow-free lowlands and summers in the cooler uplands.

Treeline   either the latitudinal or elevational limit of normal tree growth. Beyond this limit, closer to the poles or at higher or lower elevations, climatic conditions are too severe for trees to grow.

Tropics   technically, the area between the Tropic of Cancer (21½° N latitude) and the Tropic of Capricorn (21½° S latitude), characterized by the absence of a cold season. Often used to describe any area possessing what is considered to be a hot, humid climate.

Underemployment   a condition among a labor force such that a portion of the labor force could be eliminated without reducing the total output. Some individuals are working less than they are able or want to, or they are engaged in tasks that are not entirely productive.

Water Table   the level below the land surface at which the subsurface material is fully saturated with water. The depth of the water table reflects the minimum level to which wells must be drilled for water extraction.

Zoning   the public regulation of land and building use to control the character of a place.

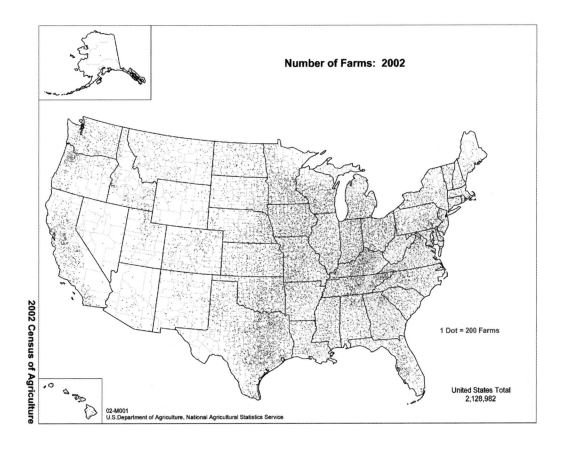

Number of Farms: 2002

1 Dot = 200 Farms

United States Total
2,128,982

2002 Census of Agriculture

02-M001
U.S.Department of Agriculture, National Agricultural Statistics Service

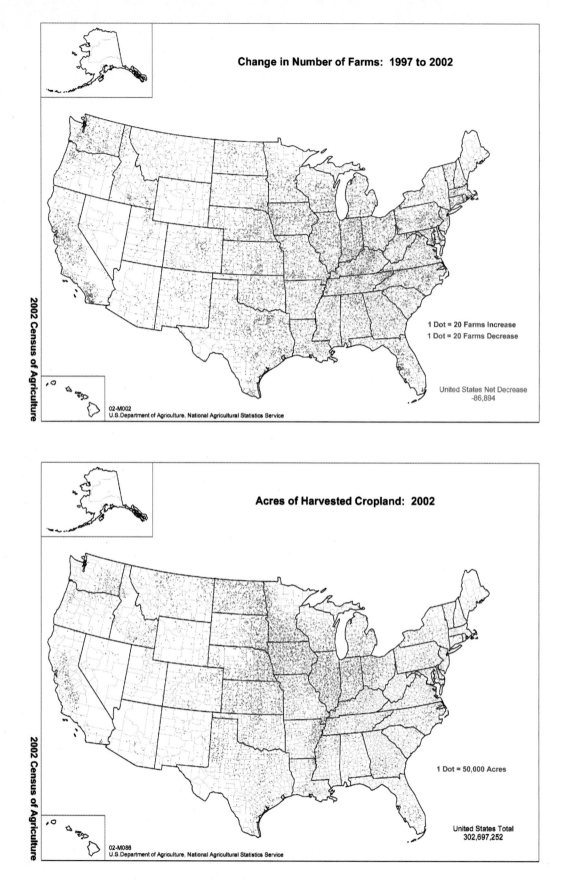

Change in Number of Farms: 1997 to 2002

1 Dot = 20 Farms Increase
1 Dot = 20 Farms Decrease

United States Net Decrease
-86,894

2002 Census of Agriculture

02-M002
U.S.Department of Agriculture, National Agricultural Statistics Service

Acres of Harvested Cropland: 2002

1 Dot = 50,000 Acres

United States Total
302,697,252

2002 Census of Agriculture

02-M086
U.S.Department of Agriculture, National Agricultural Statistics Service

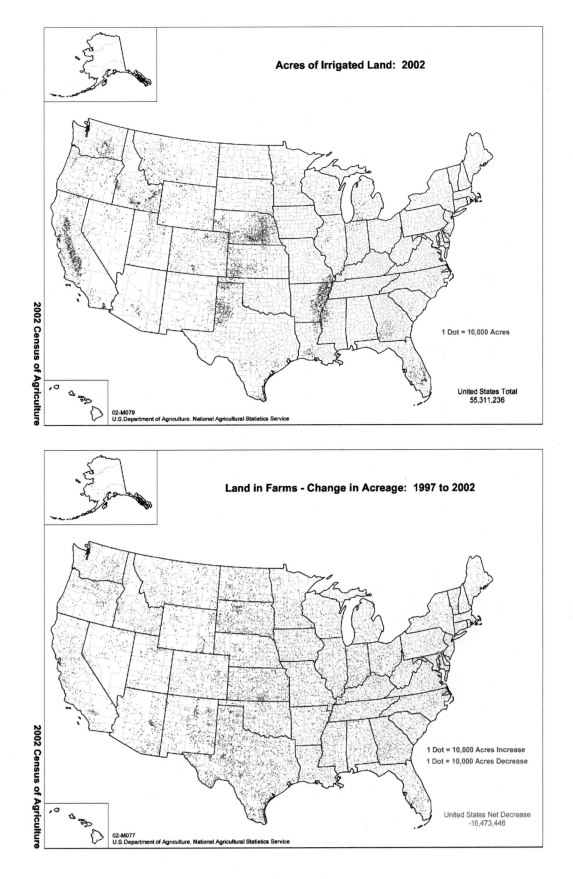

Acres of Irrigated Land: 2002

1 Dot = 10,000 Acres

United States Total
55,311,236

2002 Census of Agriculture

02-M079
U.S. Department of Agriculture, National Agricultural Statistics Service

Land in Farms - Change in Acreage: 1997 to 2002

1 Dot = 10,000 Acres Increase
1 Dot = 10,000 Acres Decrease

United States Net Decrease
-16,473,446

2002 Census of Agriculture

02-M077
U.S. Department of Agriculture, National Agricultural Statistics Service

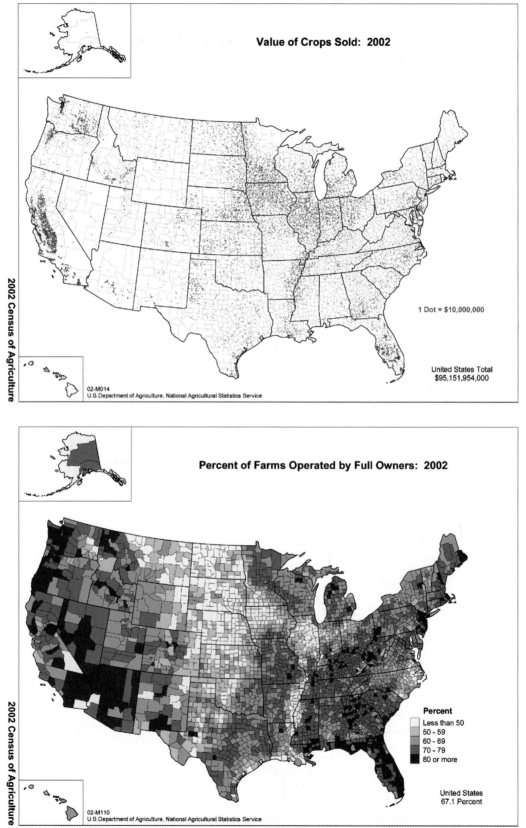

2002 Census of Agriculture

Value of Crops Sold: 2002

1 Dot = $10,000,000

United States Total
$95,151,954,000

02-M014
U.S.Department of Agriculture, National Agricultural Statistics Service

Percent of Farms Operated by Full Owners: 2002

Percent

Less than 50
50 - 59
60 - 69
70 - 79
80 or more

United States
67.1 Percent

02-M110
U.S.Department of Agriculture, National Agricultural Statistics Service

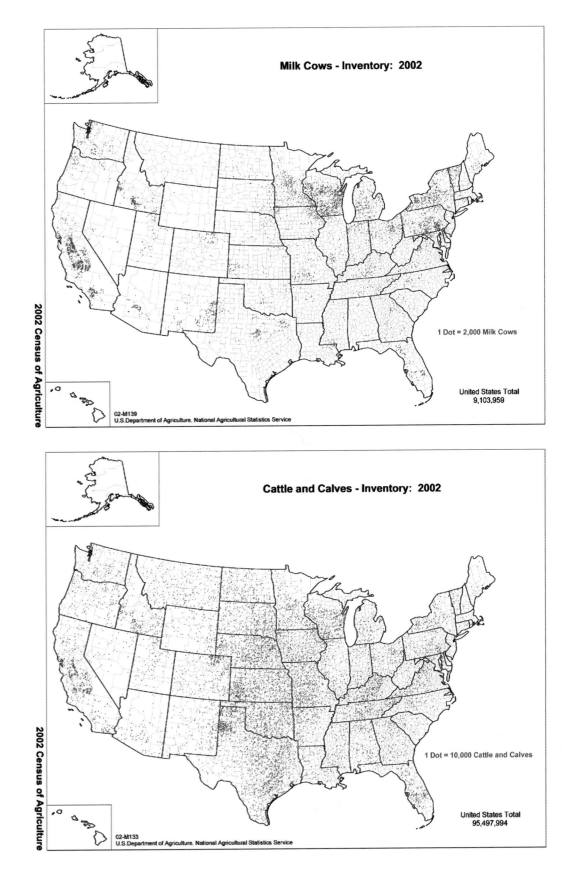

Milk Cows - Inventory: 2002

1 Dot = 2,000 Milk Cows

United States Total
9,103,959

02-M139
U.S.Department of Agriculture, National Agricultural Statistics Service

2002 Census of Agriculture

Cattle and Calves - Inventory: 2002

1 Dot = 10,000 Cattle and Calves

United States Total
95,497,994

02-M133
U.S.Department of Agriculture, National Agricultural Statistics Service

2002 Census of Agriculture

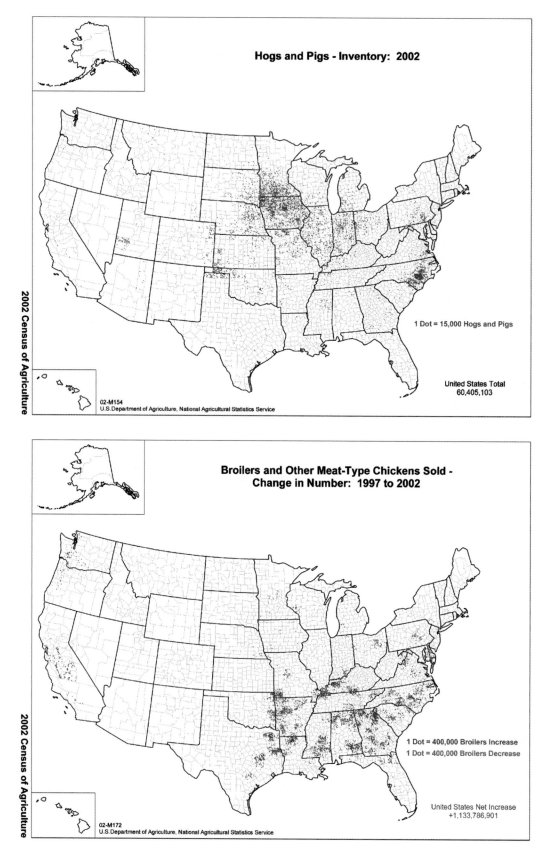

Hogs and Pigs - Inventory: 2002

1 Dot = 15,000 Hogs and Pigs

United States Total
60,405,103

2002 Census of Agriculture

02-M154
U.S.Department of Agriculture, National Agricultural Statistics Service

**Broilers and Other Meat-Type Chickens Sold -
Change in Number: 1997 to 2002**

1 Dot = 400,000 Broilers Increase
1 Dot = 400,000 Broilers Decrease

United States Net Increase
+1,133,786,901

2002 Census of Agriculture

02-M172
U.S.Department of Agriculture, National Agricultural Statistics Service

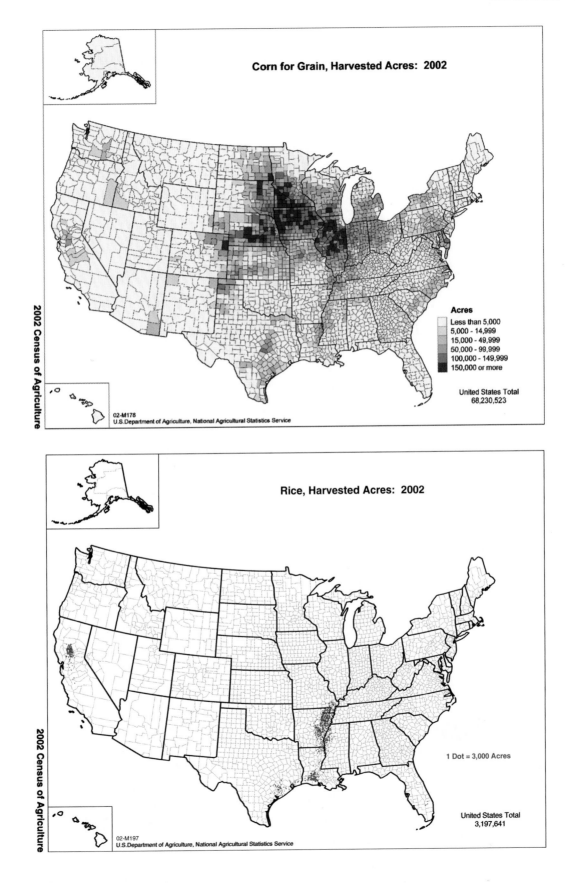

Corn for Grain, Harvested Acres: 2002

Acres
- Less than 5,000
- 5,000 - 14,999
- 15,000 - 49,999
- 50,000 - 99,999
- 100,000 - 149,999
- 150,000 or more

United States Total
68,230,523

02-M178
U.S.Department of Agriculture, National Agricultural Statistics Service

2002 Census of Agriculture

Rice, Harvested Acres: 2002

1 Dot = 3,000 Acres

United States Total
3,197,641

02-M197
U.S.Department of Agriculture, National Agricultural Statistics Service

2002 Census of Agriculture

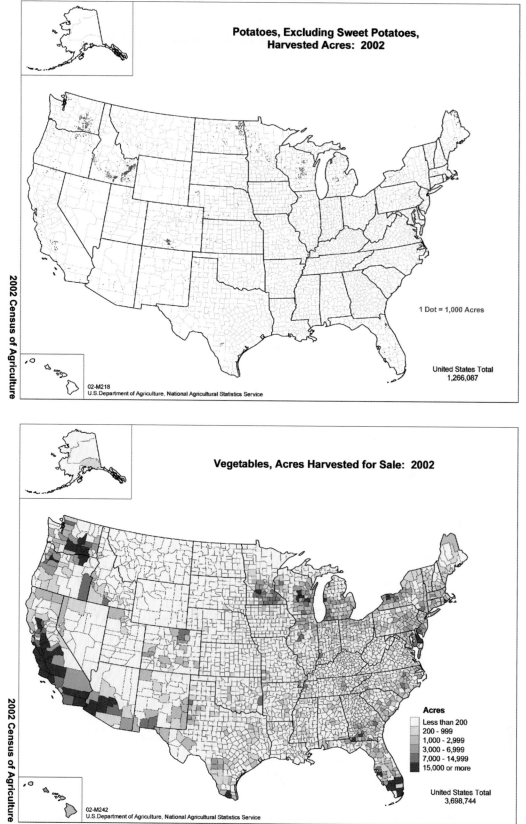

Potatoes, Excluding Sweet Potatoes, Harvested Acres: 2002

1 Dot = 1,000 Acres

United States Total
1,266,087

2002 Census of Agriculture

02-M218
U.S.Department of Agriculture, National Agricultural Statistics Service

Vegetables, Acres Harvested for Sale: 2002

Acres

	Less than 200
	200 - 999
	1,000 - 2,999
	3,000 - 6,999
	7,000 - 14,999
	15,000 or more

United States Total
3,698,744

2002 Census of Agriculture

02-M242
U.S.Department of Agriculture, National Agricultural Statistics Service

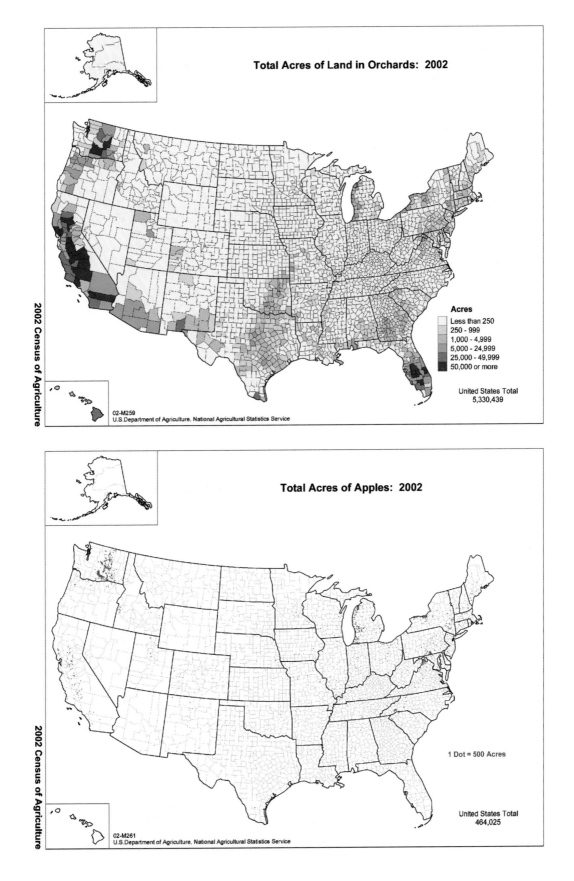

Total Acres of Land in Orchards: 2002

Acres

Less than 250
250 - 999
1,000 - 4,999
5,000 - 24,999
25,000 - 49,999
50,000 or more

United States Total
5,330,439

02-M259
U.S.Department of Agriculture, National Agricultural Statistics Service

2002 Census of Agriculture

Total Acres of Apples: 2002

1 Dot = 500 Acres

United States Total
464,025

02-M261
U.S.Department of Agriculture, National Agricultural Statistics Service

2002 Census of Agriculture

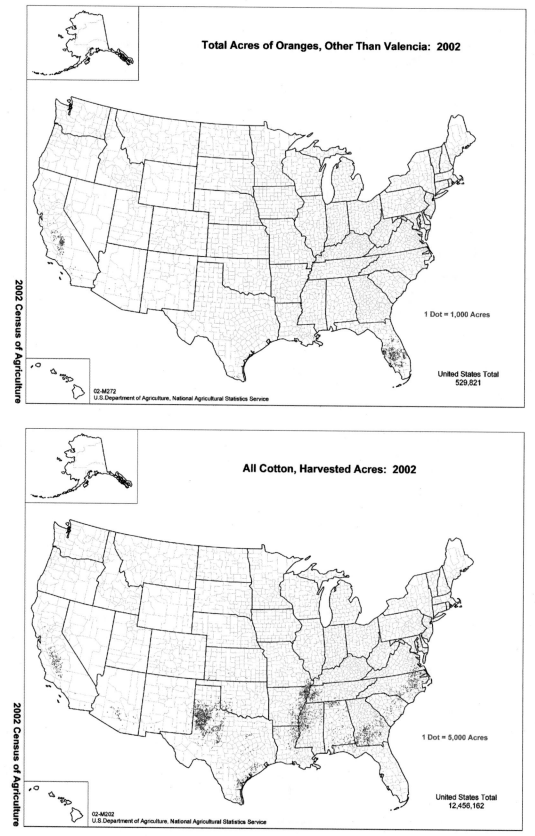

Total Acres of Oranges, Other Than Valencia: 2002

1 Dot = 1,000 Acres

United States Total
529,821

2002 Census of Agriculture

02-M272
U.S.Department of Agriculture, National Agricultural Statistics Service

All Cotton, Harvested Acres: 2002

1 Dot = 5,000 Acres

United States Total
12,456,162

2002 Census of Agriculture

02-M202
U.S.Department of Agriculture, National Agricultural Statistics Service

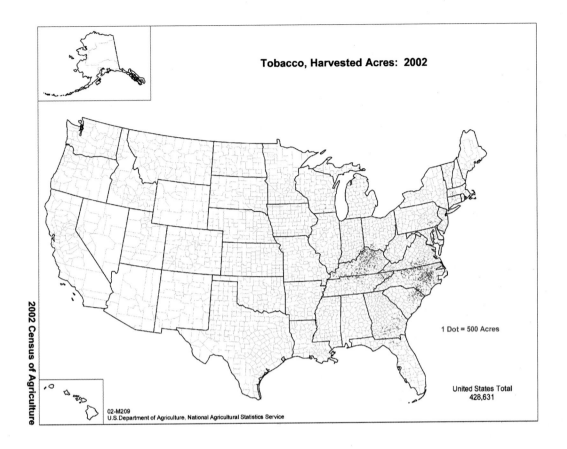

Tobacco, Harvested Acres: 2002

1 Dot = 500 Acres

United States Total
428,631

2002 Census of Agriculture

02-M209
U.S.Department of Agriculture, National Agricultural Statistics Service

Tobacco, Harvested Acres: 2012

PHOTO CREDITS

Richard Hamilton Smith/Corbis Images, Page 244: Courtesy USDA NRCS, Page 245: Digital Vision, Page 248: Courtesy of Jon Malinowski, Page 249: Courtesy of Steve Birdsall.

Chapter 13
Page 251 & Page 252: Courtesy of Steve Birdsall, Page 255: Phil Schermeister/Corbis Images, Page 261: Photographer's Choice/Media Bakery, Page 262: Gerald Hoberman/drr.net, Page 266: Calvin Larsen/Photo Researchers, Page 267: Courtesy of Jon Malinowski, Page 268: Jamie and Judy Wild/Danita Delimont, Page 269 (top): Geoffrey Clifford/ The Image Bank/Getty Images, Page 269 (bottom): Calvin Larsen/Photo Researchers, Page 270: Royalty-free/Corbis Images, Page 272: Ted Wood/Stone/Getty Images, Page 273: George Ranalli/Photo Researchers.

Chapter 14
Page 275 & Page 276: Earl S. Cryer, UPI/drr.net, Page 278: Royalty-free/Corbis Images, Page 281: Macduff Everton/ Corbis Images, Page 282: Shepard Sherball/Corbis Images, Page 285: Stephen Trimble/DRK Photo, Page 286: Neil Rabinowitz/Corbis Images, Page 287: Jack Hollingsworth/ PhotoDisc Green/Getty Images, Page 289: Sandy Felsenthal/Corbis Images.

Chapter 15
Page 293 & 294: Bilderlounge/Media Bakery, Page 297: Digital Vision, Page 299: Krista Kennell/drr.net, Page 301: Royalty-free/Corbis Images, Page 303: Courtesy of Gene

Palka, Page 304: Mark E. Gibson/Corbis Images, Page 307: Royalty-free/Corbis Images, Page 309: John K. Humble/ Photographer's Choice/Getty Images, Page 313 (top): Ringo Chiu/Zuma Press, Page 313 (bottom): Royalty-free/Corbis Images, Page 316: Gunter Marx Photography/Corbis Images, Page 317: Lloyd Cluff/Corbis Images.

Chapter 16
Page 321 & Page 322: Harald Sund/The Image Bank/Getty Images, Page 326 (left): Vince Streano/Corbis Images, Page 326 (right): Corbis-Bettmann, Page 331 (top): age fotostock/ SUPERSTOCK, Page 331 (bottom): Robert Glusic/ PhotoDisc, Inc./Getty Images, Page 334: Alan Kearney/ Taxi/Getty Images, Page 336: Royalty-free/Corbis Images, Page 337: Neil Rabinowitz/Corbis Images.

Chapter 17
Page 341 & Page 342: B & C Alexander/Photoshot, Page 346 (top): Joseph Van Os/The Image Bank/Getty Images, Page 346 (bottom): Brian A. Vikander/Corbis Images, Page 351: Paul A. Souders/Corbis Images, Page 352: Paul A. Souders/Corbis Images, Page 354: Harv Sawatzky Photography, Page 355: Corbis Images.

Chapter 18
Page 359 & Page 360: G. Brad Lewis/Stone/Getty Images, Page 364: Jeff Greenberg/Danita Delimont/drr.net, Page 368: Phil Schermeister/Corbis Images, Page 369: Heeb/laif/ Redux, Page 370: State of Hawaii Public Archives.